Maxwell Displacement Current and Optical Second-Harmonic Generation in Organic Materials

Analysis and Application for Organic Electronics

Maxwell Displacement Current and Optical Second-Harmonic Generation in Organic Materials

Analysis and Application for Organic Electronics

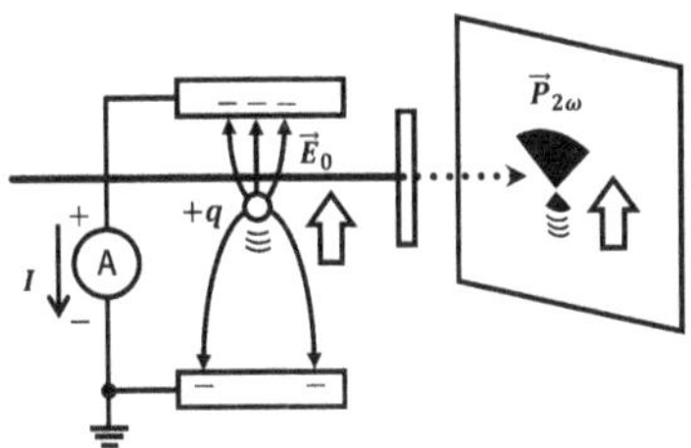

Mitsumasa Iwamoto
Dai Taguchi
Tokyo Institute of Technology, Japan

World Scientific

NEW JERSEY • LONDON • SINGAPORE • BEIJING • SHANGHAI • HONG KONG • TAIPEI • CHENNAI • TOKYO

Published by

World Scientific Publishing Co. Pte. Ltd.
5 Toh Tuck Link, Singapore 596224
USA office: 27 Warren Street, Suite 401-402, Hackensack, NJ 07601
UK office: 57 Shelton Street, Covent Garden, London WC2H 9HE

British Library Cataloguing-in-Publication Data
A catalogue record for this book is available from the British Library.

MAXWELL DISPLACEMENT CURRENT AND OPTICAL SECOND-HARMONIC GENERATION IN ORGANIC MATERIALS
Analysis and Application for Organic Electronics

ISBN 978-981-123-694-5 (hardcover)
ISBN 978-981-123-695-2 (ebook for institutions)
ISBN 978-981-123-696-9 (ebook for individuals)

For any available supplementary material, please visit
https://www.worldscientific.com/worldscibooks/10.1142/12278#t=suppl

Typeset by Stallion Press
Email: enquiries@stallionpress.com

Preface

The probing and modeling of carrier transport in materials is a fundamental research subject in electronics and materials science. According to Maxwell electromagnetic field theory, there are two kinds of currents, *i.e.*, conduction current and Maxwell displacement current (MDC). The conduction current flows when electronic charges, *e.g.*, electrons and holes, are conveyed in solids, whereas MDC is the transient current that is generated due to the change of electric flux density. The source of conductive current is charged particles, *i.e.*, electrons, holes, ions, *etc.*, and the source of MDC is also the charged particles. The most conventional way to study the carrier transport in materials is to explore the conduction current flowing across materials by applying external electric field. Therefore, lots of experimental works on the conduction current flowing through solids have been reported. As a result, there are many books that describe carrier conduction mechanism of solids, including metals, insulators, conductors, semiconductors and so forth, on the basis of established physics of carrier conduction mechanism of metals, conductors, and semiconductors. On the other hand, MDC is the transient current, and it ceases to flow when materials are charged. Therefore, MDC is being paid much attention in terms of dielectric polarization phenomena. Accordingly, there are many books that describe dielectric polarization of solids, including insulators, dielectrics, ferroelectric materials, ceramics, *etc.* However, we cannot find books that describe carrier transport on the basis of dielectric physics, that is, on the basis of physics of MDC. The authors named this approach as "Dielectric Physics Approach". As the source of conductive current and MDC is the same, it is anticipated that we can probe and model carrier transport in materials, in terms of "MDC". In other words, we can find a novel way for modeling and analyzing materials on the basis of Dielectric Physics Approach, on focusing dielectric polarization phenomena. "*Maxwell Displacement Current and Optical Second-Harmonic Generation*" are basically dielectric phenomena. The aim of this book is to show the Dielectric Physics Approach for the study of molecular materials and organic electronics devices related to carrier transport and dielectric polarization, on focusing "*Maxwell Displacement Current and Optical Second-Harmonic Generation in organic materials* from viewpoints of *Analysis and Application for organic electronics*".

As an introduction to the Dielectric Physics Approach, the authors show that the conceptual idea of using Maxwell displacement current (MDC) and Optical second-harmonic generation (SHG) for studying carrier transport mechanism of solids, besides for studying dielectric phenomena such as charging and discharging. Dielectric Physics Approach is particularly useful for studying new materials and organic electronic devices, which includes molecular films, *etc.* The details are described in this book, theoretically and experimentally. For the past several decades, many new materials have been synthesized and have been discovered. Among them are superconductive materials, organic semiconductors, functional materials and molecules, bio-materials and so on. These materials exhibit a variety of carrier transport behaviors. Nevertheless, Dielectric Physics Approach can be commonly adopted for analyzing them. This means the Dielectric Physics Approach can bridge many research fields, *e.g.*, electronics and chemistry, biology, *etc.* At the same time, using the methodology based on Dielectric Physics Approach, we can bridge electronics and optics. This is also a very important point of this book.

This book is intended to give a fundamental Dielectric Physics Approach, based on the probing of *Maxwell Displacement Current and Optical Second-Harmonic Generation in organic materials,* and to show *Analysis and Application for organic electronics.* Those who want to deeply understand carrier behaviors in materials and devices will be inspired from this book. The level of this book is suitable for graduate course students and experts in the field of condensed matter physics, materials science, electrical & electronic engineering. This book will be also useful for students who are studying academic courses such as Advanced Electronic Materials, Organic electronic devices, Condensed matters, *etc.*

This book could not have been completed without the recommendation and long-term substantial support by Prof. Z.-C. OuYang, Institute of Theoretical Physics (ITP), Academia Sinica, People's Republic of China. M. I. used much time at ITP for discussion with him and writing this book, while M.I. was working at Tsinghua University as a professor of Joint Graduate Course program between Tokyo Institute of Technology, Japan and Tsinghua University, China. M.I. is grateful to the Late Emeritus Prof. Taro Hino of Tokyo Institute of Technology, for guiding to the research field using *Thermally Stimulated Current.* Gratitude is also given to staffs, colleagues, visiting scientists, and graduate and undergraduate course students, who were once engaged in their researches and studies in Iwamoto's laboratory in Tokyo Institute of Technology, particularly, to Prof. Takaaki Manaka of Tokyo Institute of Technology. Some chapters are based on the work with him. D.T. thanks to Dr. Takashi Nakagiri for the introduction of many books and kind support as a mentor from the beginning of the Ph. D study of D.T. at Iwamoto Lab.

We will be very appreciated to hear comments, criticisms, suggestions from any readers.

Emeritus professor of Tokyo Institute of Technology
Mitsumasa Iwamoto

Assistant professor of Tokyo Institute of Technology
Dai Taguchi

September, 2020

Contents

List of Figures

Chapter 1

Introduction

The probing and modeling of carrier transport in materials is a fundamental research subject in electronics and materials science. This book describes *Maxwell Displacement Current and Optical Second Harmonic Generation in organic materials* from viewpoints of *Analysis and Application for organic electronics.* In this chapter, as an introduction to this book, we briefly discuss why *Maxwell Displacement Current* and *Optical Second Harmonic Generation* are important for studying carrier behaviors in solids.

1.1 Currents flowing solids

According to the Maxwell electromagnetic field theory [1, 2], the total current flowing across solids is the sum of the conduction current and Maxwell displacement current (MDC). The conduction current flows when electrons and holes are conveyed under the external electric field formed by applied potentials on electrodes, and it flows steadily. On the other hand, MDC is the transient current that is generated due to the change of electric flux density in solids with time. That is, the total current flows in solids by using two routes, independently. Figure 1.1 illustrates this situation, where resistor R and capacitor C represent the route of conduction current and the route of MDC, respectively. We usually measure current–voltage (I–V) characteristics in steady state to study the carrier injection and carrier transport mechanisms. This current I is conduction current that flows crossing the resistor R in steady. There are many

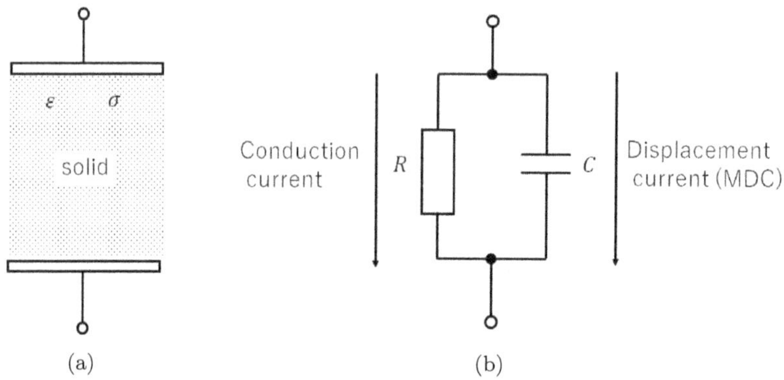

Figure 1.1. Current flowing solids is the sum of conduction current and Maxwell displacement current (MDC). The flowing routes of the two currents are represented by R (resistor) and C (capacitor), respectively. ε and σ represent dielectric constant and electrical conductivity of solids, respectively.

books and review articles that describe carrier conduction mechanism of solids, including metals, insulators, conductors, semiconductors and so forth, where we can find established physics of carrier conduction mechanism of metals, conductors, and semiconductors [3–5]. To measure I–V characteristics is one of important ways to study carrier transport in solids. In this book, we stress dielectric physics approach, and show that the MDC that flows through the capacitor C is also useful for studying carrier transport mechanism of solids, besides for studying dielectric phenomena such as charging and discharging. In Chap. 4 and Chap. 7, we briefly discuss carrier conduction mechanism of solids from viewpoint of insulators and dielectrics, on focusing the conduction current and MDC current.

The MDC is generated when the electric flux density in solids, which originates from electrons, holes, and dipoles, changes with time, and it is basically a transient current and flows whilst the electric flux density changes with time. Consequently, we can study migration behavior of electrons and holes, orientational dipolar motions in solids, *etc.*, by probing the MDC [6, 7]. The MDC current is the transient current that flows through the capacitor C. In other words, on paying attention to the dielectric phenomena, we can find a way to study dynamical carrier motion in solids. We measure transient currents that are generated by external stimuli, such as pulse voltage, light illumination, pressure, and heating. Main reason is that the current flowing through C appears only in the transient-state. In Chap. 2, we briefly discuss the conduction current and the MDC flowing across electrically insulating solids.

1.2 MDC and dipolar motion

Thermally stimulated current (TSC) measurement is employed for studying dipolar depolarization in solids which are sandwiched between two electrodes, where a short circuit current is measured during heating [8–12]. In some cases, TSC due to dipolar depolarization is called as TSDC (Thermally Stimulated Depolarization Current). This short circuit current is the MDC current and it flows whilst samples are being heated. What we measure by the TSC measurement is the change of the amount of induced charges on the electrodes connected to an electrometer due to the disordering of aligned dipoles in solids sandwiched between two electrodes. That means we measure the electric flux density distribution change related to dipolar depolarization in the solids with time. The recorded short circuit current, *i.e.*, MDC, traces the dynamical dipolar motion in the solids. Figure 1.2 illustrates this situation. This TSC technique is very simple, and we can extend the concept of TSC measurement of solids to the case for studying dipolar motion in thin films, including organic monolayers with nano-meters in thickness [13]. Technically, it will be

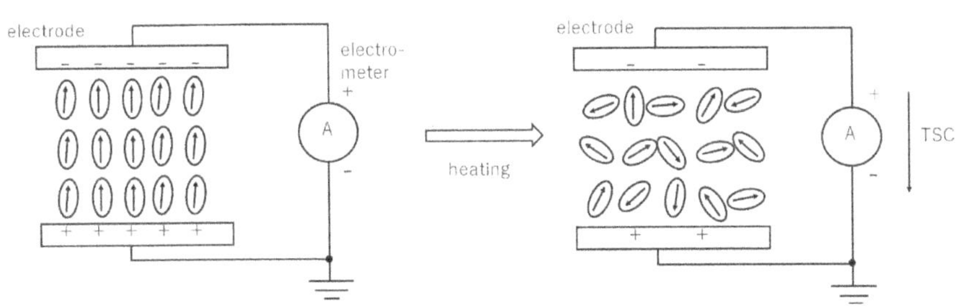

Figure 1.2. Thermally stimulated current (TSC) traces dipole rotation in a space sandwiched between two electrodes during heating. TSC measurement is basically MDC measurement, and it can also probe detrapping charges, ion migration, *etc.*

a very hard task to prepare metal–insulator–metal (MIM) samples with a structure of monolayers sandwiched between two electrodes. However, this technical problem does not matter in the MDC measurement and it is easy to overcome, because we do not need to contact the two electrodes with the monolayers for detecting the amount of polarization change of monolayers caused by the electric flux density distribution change. Actually, to probe dynamical dipolar motion in monolayers on the water surface, a suspended electrode connected to a short circuit is used to measure the MDC [6, 12, 13]. That is, the change of the amount of charge induced on the suspended electrode above the monolayer on the water surface is measurable during the course of monolayer compression and monolayer expansion. It is noteworthy that the change of the amount of induced charges on the suspended electrode will also appear due to the structural change of monolayer films such as phase transition, photo-isomerization, *etc* [6, 14]. A dipole is a pair of positive and negative charges separated in space, and it is represented in a vector form. The ensemble of dipoles determines the direction of polarization of solids. Therefore, the concept of orientational order parameters that represent ensemble system is very helpful. We will discuss the order parameters used for specifying molecular systems in Chap. 5. In Chap. 2 of this book, we will discuss the principle of the MDC measurement and show the detection of Maxwell's displacement current generated from monolayers and solids. Further we should note that the short-range and long-range movement of carriers in solids in transient state also contribute to the MDC, and this idea is well adopted for analyzing the behavior of trapped carriers in solids and in electronic devices, and also for analyzing carrier mobility of solids. We discuss these topics very shortly in the following section, and the details in Chap. 3.

1.3 MDC and carrier transit

In a way similar to the rotational motion of dipoles that results in depolarization in solids as illustrated in Fig. 1.2, short-range and long-range movements of carriers inside of solids can contribute to the generation of MDC. Actually, TSC measurement is used for the

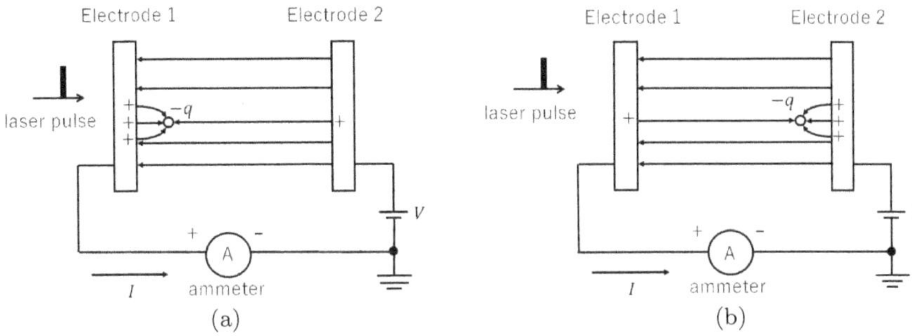

Figure 1.3. Time-of-flight (TOF) measurement traces transit of electronic charges crossing from one electrode (electrode 1) to the other counter electrode (electrode 2). TOF measurement is basically MDC measurement. Electric field distribution and electrode charges change in accordance with the location of a charged particle such as the electron ($-q$) meanwhile MDC flows in TOF measurement. (a) Electric field distribution when $-q$ is located nearby electrode 1. (b) Electric field distribution when $-q$ is located nearby electrode 2.

study of carriers released from traps in solids, for the study of carrier motion of impurity ions in solids, for the study of phase transition of solids, and so on. That is, TSC is useful for the study of dynamical behaviors of charged particles in solids. We will discuss the TSCs due to carrier detrapping and ion movements in Chap. 3.

By the way, in order to probe transits of electrons and holes across solids sandwiched between two electrodes, time-of-flight (TOF) technique is very often utilized [3, 15, 16]. Figure 1.3 illustrates the principle of this TOF measurement. What we measure by using this TOF technique is the change of the amount of induced charge on the electrode connected to an ammeter whilst electrons are traveling from one electrode to the counter electrode.

Hence, the principle of the TOF measurement is basically the same as that of the MDC measurement used for studying monolayers associated with dipolar motion, where the change of induced charge on the electrode is measured. In the conventional laser-TOF measurement [3, 15, 16], the long-range carrier motion in solids is recorded as a transient current, from which we determine the transit time of carriers and then analyze the carrier mobility. This technique is very popular, and well utilized for the determination of carrier mobility of solids. In experiment, we need to generate carriers at a location very close to electrode 1 in solids by laser irradiation, and must convey the generated carriers toward the counter electrode at a constant speed under a non-zero d.c. external electric field formed by a potential difference between electrodes. It is noteworthy that the key point of the TOF measurement is to utilize the generation of carriers in solids and the succeeding transfer of carriers under the external electric field. Consequently, other techniques that can catch these two processes, *i.e.*, generation and transport, are available for the determination of carrier mobility. The details will be discussed in Chap. 2.

1.4 SHG and carrier transit

Laser-TOF is useful, but there is a drawback in this measurement. For example, the actual location of carriers traveling across solid materials cannot be determined directly from the trace of transient current recorded by ammeter. This is the reason why we need to carry experiments under assumption that carriers are generated at a location very close to electrode 1 in solids by laser radiation, and that the generated carriers must be conveyed at a constant speed toward the counter electrode 2 under the constant external electric field formed by potential difference between electrodes 1 and 2. Noteworthy the polarity of the potential difference is important to convey electrons and holes, selectively. In other words, we must assume the location of carriers and the polarity of transported carriers when we employ the TOF measurement. However, actual carrier motions in solids will be always different from the assumed ones, owing to the presence of space charges formed by carriers, the penetration of laser beam into the solids, *etc.* Consequently, the conventional laser-TOF measurement forces an elaborate mathematical approach on us for analyzing the actual traveling carriers, and this analysis leads to a puzzling situation in which there are many possible mathematical solutions to reproduce the observed transient currents. This means that it is quite a task to trace the actual carrier behavior in solids by only using the laser-TOF method. Furthermore, it is difficult to distinguish the carrier species, *i.e.* electrons or holes, when we employ the e-TOF measurement [17, 18] instead of laser-TOF method, because there is a possibility that electrons and holes inject into the solids from electrodes 1 and 2 (double injection), respectively, and travel in opposite directions each other along the electric field formed by the external electric field to generate the transient currents. The best way to overcome this problem is to develop a technique that can directly probe the traveling carriers themselves. The direct probing of the carrier motion itself is indeed helpful to discuss the detailed carrier transport mechanism in devices as well as in solids. One may wonder the state-of-art technology, *e.g.*, nanoscope and its family instruments, is available for this purpose, because nano-meter size particles and molecules are visualized with help of related advanced nano-technology. However, this is not a case for observing electrons and holes, because they are too small to observe. The size of electrons , holes, dipoles is much smaller, *i.e.*, on the femto-meter order scale, according to the estimation using the theory of relativity [19]. In this situation, the Gauss law in the Maxwell's electromagnetic field theory [1, 2] suggests us an insightful idea for probing the carrier motion in materials. By probing the electric field associated with moving electrons and holes, the carrier motion can be explored if the polarization induced in the solid material surrounding the moving electrons and holes is directly probed. That is, we can probe carrier motion by probing dielectric polarization locally induced by the electric field arising from moving carriers.

Figure 1.4 illustrates this situation [13, 20]. Time-resolved microscopic optical second harmonic generation (TRM-SHG) is a technique that enables us to probe the propagation of the dielectric polarization induced by electrons and holes which are moving across the organic materials, where nonlinear polarization induced in solids by coupling with incident electromagnetic waves, *e.g.*, laser beam, and the d.c. electric field generated from moving carriers is probed in real time [13, 21]. Furthermore, using a charge-coupled device (CCD) imaging system directly visualizes their motion. In Chap. 6, we will discuss the technique to probe carrier motion in solids by using electric-field-induced optical second-harmonic generation,

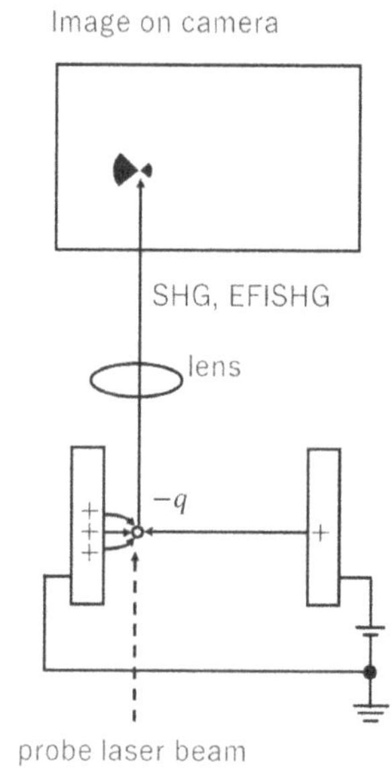

Figure 1.4. Concept of electric-field-induced optical second-harmonic generation (EFISHG) measurement. EFISHG visualizes charge motions by probing the polarization caused by the coupling of electric field originating from electrons with electric field of incident laser beam. This incident laser beam does not generate electron–hole pairs in solids during the irradiation.

and in Chap. 7 and Chap. 8, we will show novel findings obtained by the EFISHG observation, *e.g.*, carrier motions in organic field-effect transistors (OFETs), carrier behaviors in organic light-emitting diodes (OLEDs), organic memory devices using ferroelectric layers, and organic solar cells. It is noteworthy that the idea of using EFISHG for the study of carrier transport is based on dielectric physics where *polarization* is a most important factor, and the EFISHG measurement for probing carrier motions has been originally developed by the authors' group.

Further it will be instructive here to note that the general solution of Poisson's equation of charges suggests us that the general electric field surrounding charges are given as sum of the electric field by point charges, dipoles, quadrupoles, *etc.* This means if we can find a way to probe the field by these charges individually, we can study dynamical motion of these charges. We should note here that the

TOF described in Sec. 1.3 is a technique to probe the electric field by point charges, and the MDC measurement described in Sec. 1.2 is a technique to probe the field by dipoles.

1.5 Maxwell–Wagner effect and carrier behaviors

According to the Gauss law in the Maxwell's electromagnetic field theory [1, 2], charged particles such as electrons, holes, ions, dipoles, *etc.*, are source of the electric field. This electric field has an electric effect on the surrounding materials, and interesting electrical phenomena will be induced. Dielectric polarization is one of important phenomena. The surrounding materials will be subjected to a high electric field by the electric field diverging from charges if charges are accumulated locally in some places of solids, and this charge accumulation will be important to study electro-static phenomena and electrical breakdown. Therefore, probing electric field in solids is important. Surface potential measurement based on the Kelvin probe method is widely used, but other method such as optical EFISHG method will be available. In Chap. 6, we will focus on the probing of the electric field in solids and discuss the EFISHG measurement [22, 23]. By the way, there is one basic question; where charges are accumulated in solids? If solids are homogeneous, it will be difficult to intentionally accumulate charges in arbitrary positions in the solids, and charges will spread along the electric field generated from the charges. Consequently, "interfaces" and "traps" play important roles for charge accumulation in solids and in devices. We often utilize "interface" in device applications. For example, for metal–insulator–semiconductor (MIS) devices, we purposefully regulate the amount of accumulated charges at the insulator–semiconductor interface region for the operation of the device. The theory of Maxwell–Wagner effect is helpful for understanding the charge accumulation at the interface. As shown in Fig. 1.1, the macroscopic electrical property of solid is basically represented by using a parallel electrical circuit, comprised of two elements, *i.e.*, R and C. The product of R and C, given by $RC(=\varepsilon/\sigma)$, represents a response time, and this value specifies the spreading time of charges in solids, where ε and σ are dielectric constant of solids and conductivity of solids, respectively.

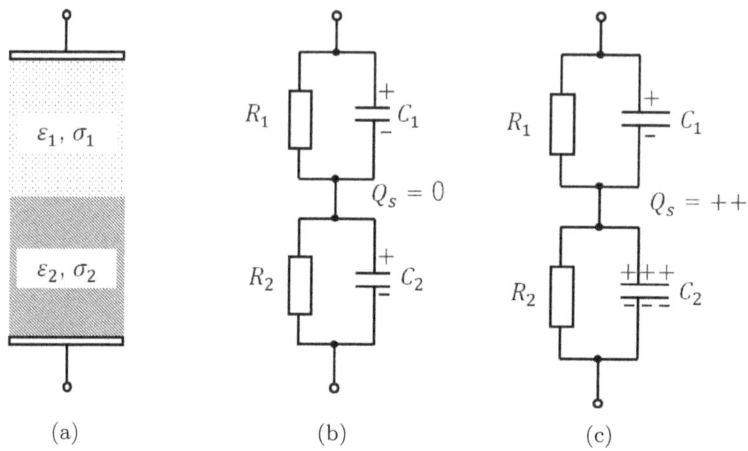

Figure 1.5. Charge accumulation at the interface between two adjacent materials with different time constant $RC = \varepsilon/\sigma$, where $\sigma(= 1/\rho)$ is conductivity (ρ: resistivity) and ε is dielectric constant. (a) A two-layer device. Equivalent circuit model for two-layer device where (b) dielectric relaxation time of the adjacent two layers is identical ($R_1C_1 = R_2C_2$), and (c) dielectric relaxation time of the adjacent two layers is different ($R_1C_1 \neq R_2C_2$).

According to the theory of Maxwell–Wagner effect [24, 25], charge accumulation happens at the interface between two solids only when the two solids have different relaxation time $\tau = \varepsilon/\sigma$. Figure 1.5 illustrates the interface charging due to the Maxwell–Wagner effect. The charge accumulation at the interface between insulator and semiconductor in MIS devices is due to this effect. The accumulation of charges on metal electrodes of metal–insulator–metal (MIM) elements is also due to the Maxwell–Wagner effect, where response time of metal (M) is extremely short in comparison with that of insulator (I). Once charges are accumulated at the interface, the electric field is formed due to the accumulated charge at interfaces in solids and in devices, according to the Gauss Law, and we can study the contribution of the accumulated charge by means of EFISHG, Surface potential measurement, and other methods. Noteworthy that "traps" are local positions where carriers are captured. The location of "traps" is physically different from other places in solids; that is, the "trap" is the place, very likely interface from viewpoint of the Maxwell–Wagner effect. In Chap. 7 and Chap. 8, we study the device operation, on focusing the electric field generated from accumulated charges at the interface. We also discuss the electrostatic phenomena and electrical breakdown, on focusing the electric field probed by EFISHG in Chap. 9.

1.6 Dielectric physics approach for analyzing carrier transport

As described in this chapter, according to the Maxwell electromagnetic field theory [1, 2], there are two kinds of currents. One is conduction current and the other is Maxwell displacement current (MDC), and they flow independently using two routes. The conduction current flows when electrons and holes are conveyed in solids steadily. MDC is the current that is generated due to the change of electric flux density and this is the transient current. Nevertheless, the source of conductive current is of course charged particles, *i.e.*, electrons, holes, ions, *etc.*, and the source of MDC is also the charged particles. We therefore can probe and model carrier transport in materials from viewpoint of "MDC" in the same way as from viewpoint of "conductive current". "*Maxwell Displacement Current and Optical Second Harmonic Generation*" are basically dielectric phenomena, but can apply to study carrier transport in solids. This is the dielectric physics approach which the authors call, and it can find a novel way for modeling and visualizing carrier motions and related phenomena in solids and devices.

References

1. Maxwell, J. C. (1954). *A Treatise on Electricity and Magnetism*, 3rd Ed. Volume 1 (Dover, New York).
2. Heaviside, O. (1893). *Electromagnetic Theory*, Vol I (The Electrician publishing, London).
3. Lampert, M. A. and Mark, P. (1970). *Current Injection in Solids* (Academic, New York).
4. Sze, S. M. (1969). *Physics of Semiconductor Devices* (Interscience (Wiley), New York).
5. Mott, N. F. and Gurney, R. W. (1964). *Electronic Process in Ionic Crystals*, 2nd Ed. (Dover, New York) Chap. V.
6. Iwamoto, M. and Wu, C.-X. (2001). *The Physical Properties of Organic Monolayers* (World Scientific, Singapore), https://doi.org/10.1142/4575.

7. Iwamoto, M., Yamamoto, T. and OuYang, Z.-C. (in preparation) *Electrical and Geometrical Properties of Organic Monolayers* (World Scientific, Singapore).
8. Sessler, G. M. (1979). *Electrets*, Topics in Applied Physics 33 (Springer Verlag, Berlin Heidelberg New York).
9. Chen, R. and Kirsh, Y. (1979). *Analysis of Thermally Stimulated Processes* (Pergamon Press, Oxford).
10. Turnhout, J. V. (1975). *Thermally Stimulated Discharge of Polymer Electrets* (Elsevier, New York).
11. Hino, T. (1986). *IEEE Trans. Electr. Insul.*, EI-21, 1007, https://doi.org/10.1109/TEI.1986.349015.
12. Iwamoto, M. and Taguchi, D. (2014). *Thermally Stimulated Current in Electrical and Electronic Materials — Analysis and Application —* (Corona publishing, Tokyo) (in Japanese).
13. Iwamoto, M., Manaka, T. and Taguchi, D. (2015). *J. Phys. D*, 48, 373001 (Topical review), https://doi.org/10.1088/0022-3727/48/37/373001.
14. Iwamoto, M., Majima, Y., Naruse, H., Noguchi, T. and Fuwa, H. (1991). *Nature*, 353, 645, https://doi.org/10.1038/353645a0.
15. Kepler, R. G. (1960). *Phys. Rev.*, 119, 1226, https://doi.org/10.1103/PhysRev.119.1226.
16. Meier, H. (1974). *Organic Semiconductors — dark and photoconductivity of organic solids —* (Verlag Chemie, Weinheim) Chap. 4.
17. Basu, D. and Dodabalapur, A. (2009). *Drift Velocity and Drift Mobility Measurement in Organic Semiconductors Using Pulse Voltage*, in Adv. Polym. Sci. no. 223, 73 (2010) (Springer–Verlag, Berlin Heidelberg), https://doi.org/10.1007/12_2009_4.
18. Weis, M., Lin, J., Taguchi, D., Manaka, T. and Iwamoto, M. (2009). *J. Phys. Chem. C*, 113, 18459, https://doi.org/10.1021/jp908381b.
19. Einstein, von A. (1905). Zur Elektrodynamik bewegter Körper, Annalen der Physik, Band 17, pp. 891–921, https://doi.org/10.1002/andp.19053221004; Japanese translation by Uchiyama, R. (1988). *Einstein Soutaisei Riron*, Iwanami Bunko No. blue 934–1 (Iwanami Shoten, Tokyo) (in Japanese).
20. Iwamoto, M., Manaka, T. and Taguchi, D. (2016). *IEEJ Trans. Fundam. Mater.*, 136, 678 (in Japanese, Review), https://doi.org/10.1541/ieejfms.136.678.
21. Manaka, T., Lim, E., Tamura, R. and Iwamoto, M. (2007). *Nature Photon.*, 1, 581, https://doi.org/10.1038/nphoton.2007.172.
22. Boyd, R. W. (2008). *Nonlinear Optics*, 3rd Ed. (Academic Press, Massachusetts California London) Sec.1.1.
23. Shen, Y. R. (1984). *The Principles of Nonlinear Optics* (John Wiley & Sons, New York).
24. Fröhlich, H. (1949). *Theory of Dielectrics* (Oxford at the Clarendon Press, London).
25. Oka, S. and Nakata, O. (1960). *Kotai Yudentai–Ron* (*Theory of Dielectrics on Solids*) (Iwanami Shoten, Tokyo) (in Japanese), https://doi.org/10.11501/1378002.

Chapter 2

Maxwell Displacement Current

As described in Chap. 1, *Maxwell displacement current* is the transient current and it flows due to the change of electric flux density distribution in solids with time. By measuring Maxwell displacement current flowing across solids, we can study short-range and long-range movements of charges in solids, orientational motion of dipoles, and so forth. As the electric field distribution changes in accordance with these carrier motions, it is important to know the relationship between the electric field distribution and the location of charges in solids. In this chapter, we discuss the principle of the generation of Maxwell displacement current, due to the transfer of single carrier crossing the film, dipolar rotational motion, *etc.* After that we discuss Time-of-Flight (TOF) measurement that is used for the evaluation of carrier mobility and MDC measurement used for determination of dipole moment and analysis of dipolar rotational motion in monolayers.

2.1 Induced charge and Maxwell displacement current due to single charge movement — long-range charge motion

As displayed in Fig. 2.1, let us assume here that a point charge $+q$ is displaced from electrode 1 (Left) and then positioned at $x = +x$ between a set of two parallel electrodes 1 and 2, which are separated by insulating layer with a thickness of d and electrically shorted through an ammeter A. In this situation, the electric field generated

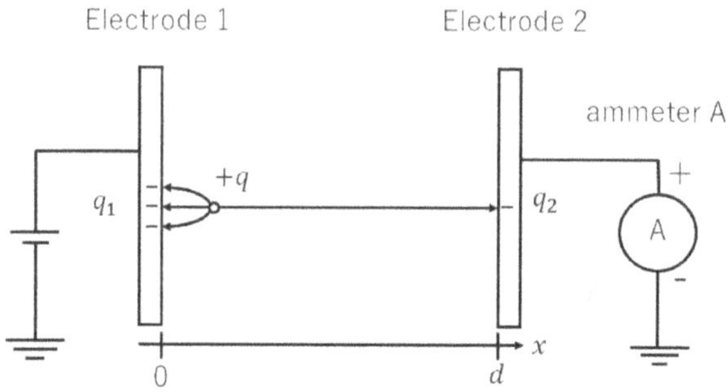

Figure 2.1. TOF measurement. Charge $+q$ is the source of electric field and the electric field diverging from $+q$ terminates on electrodes 1 and 2, where counter negative charge q_1 and q_2 are induced, on satisfying the charge neutrality condition; $q_1 + q + q_2 = 0$. Maxwell displacement current flows through an ammeter A due to the change of charge q_2 on electrode 2.

from the point charge $+q$ (>0) falls on electrodes 1 and 2, on satisfying the following physical condition called "neutrality of charges" of this electrode system, which is given by:

$$q_1 + q + q_2 = 0 \tag{2.1}$$

q_1 and q_2 are negative charges induced on electrodes 1 and 2 [1–3]. The amount of charges q_1 and q_2 changes in accordance with the location of the charge $+q$. The amount of charge q_1 and q_2, respectively, gradually decreases and increases along with the displacement of the charge $+q$ in the direction from electrode 1 ($x = 0$) toward electrode 2 ($x = d$). We call this charge transfer process "long-range" charge motion . Charge transfer in electric conduction, successive ion hopping described in Ion Hopping Model, *etc*, are examples of long-range carrier motion. In the following, we discuss how the amount of induced charge q_1 and q_2 changes in accordance with the displacement of charge $+q$ while the charge $+q$ is moving at a constant speed u, in the direction from electrode 1 to electrode 2.

Firstly, we consider a case of $t = 0^+$, when the charge $+q$ just leaves from electrode 1, as illustrated in Fig. 2.2(a). At this moment, the charge $+q$ is located at a position very close to electrode 1, *i.e.*, at $x = +\delta$ ($\delta \to 0$) from electrode 1. Therefore, we can assume the charge distribution on electrodes 1 and 2 are given as follows:

$$q_1 \approx -q \text{ and } q_2 \approx 0 \text{ (charge } +q \text{ at } x = +\delta\, (\delta \to 0)). \tag{2.2}$$

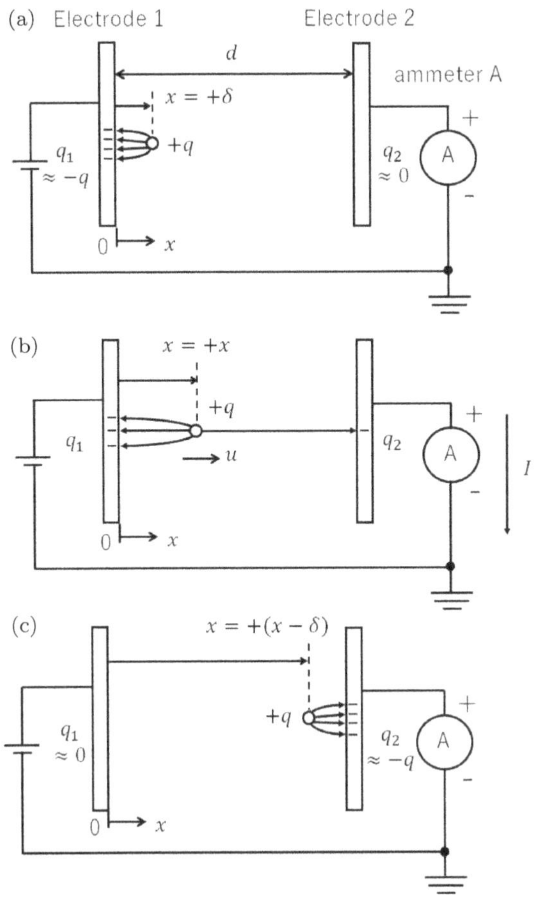

Figure 2.2. The electrode charge q_1 and q_2 in the TOF measurement. Transient current flows through an ammeter A according to the amount of induced charge q_2 on electrode 2. The charge q_2 is (a) $q_2 \approx 0$ at $t = 0^+$, (b) $q_2 = -(x/d) \cdot q$ at $t = t$ $(0 < t < t_{tr})$ (see Eq. (2.6)), and (c) $q_2 \approx -q$ at $t = t_{tr}$.

Then we consider the situation while charge $+q$ is moving from the electrode 1 to electrode 2 at a constant speed of u, and arriving at electrode 2 at $t = t_{tr}$ $(= d/u)$ (see Fig. 2.2(b) and 2.2(c)). At the time just before the charge $+q$ is reaching electrode 2 (Fig. 2.2(c)), *i.e.*, at the moment when charge $+q$ is located at $x = +(d - \delta)$, charges q_1 and q_2 induced on electrodes 1 and 2, respectively, become as follows:

$$q_1 \approx 0 \text{ and } q_2 \approx -q \text{ (charge } +q \text{ at } x = +(d - \delta)). \tag{2.3}$$

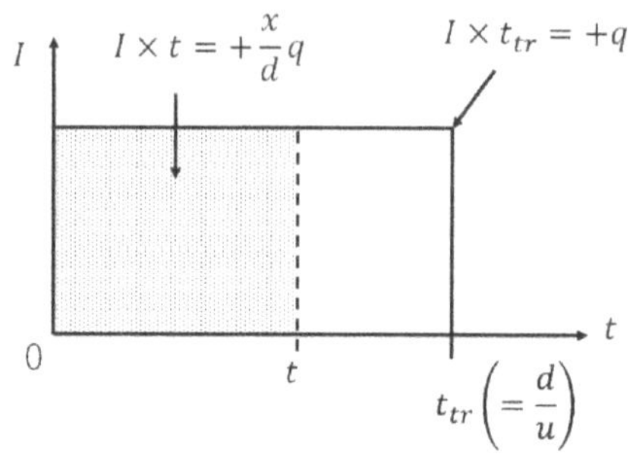

Figure 2.3. Transient current in TOF measurement. Maxwell displacement current I flows through an ammeter during charge transfer $+q$ from electrode 1 to electrode 2 in a film.

Therefore, in accordance with the transfer of charge $+q$ inside of the film in the direction from electrode 1 to electrode 2, charge $-q$ is transferred from electrode 1 to electrode 2 through the external short circuit with an electrometer A (see Fig. 2.2). This charge transfer process is recorded as a transient current by the ammeter A. Figure 2.3 shows this transient current. Noteworthy that in this electrode arrangement, the transient current $I \neq 0$ flows until the time $t = t_{tr}(= d/u)$ due to the charge transfer with a constant speed of u, and this current should be constant until $t = t_{tr}(= d/u)$, whereas $I = 0$ at $t > t_{tr}$.

Looking at Fig. 2.3, we can get the following relation

$$I \times t_{tr} = +q. \tag{2.4}$$

On the one hand, at $t = t$ $(0 < t < t_{tr})$, when charge $+q$ is displaced to the position $x = +x$ from electrode 1 $(x = 0)$ (see Fig. 2.2(b)), we find from Fig. 2.3 that the following relation must be satisfied:

$$I \times t = \frac{x}{d}(+q). \tag{2.5}$$

As the amount of the charge $-(x/d)q$ is transferred from electrode 1 to electrode 2 through external short-circuit with an ammeter by the time $t = t$, charge q_1 and q_2 induced on electrodes 1 and 2 are given by

$$q_1 = -\frac{d-x}{d}q \text{ and } q_2 = -\frac{x}{d}q \text{ (charge } +q \text{ at } x = x). \tag{2.6}$$

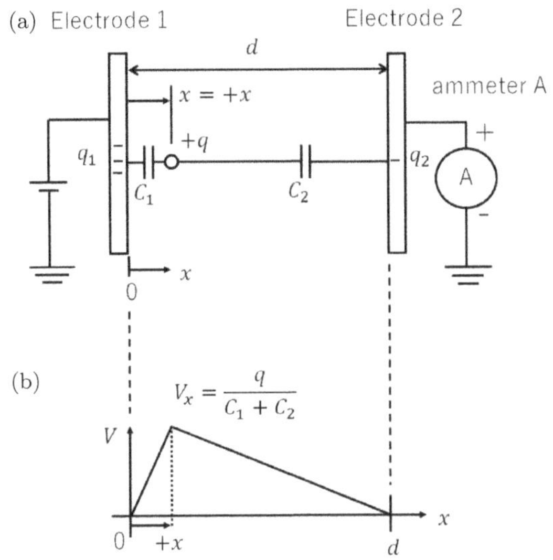

Figure 2.4. Electrode charge q_1 and q_2, expressed by using capacitor between the charge $+q$ and two electrodes. (a) Capacitance C_1 and C_2, respectively, represent the electric field distribution formed between the charge $+q$ and electrode 1 and between the charge $+q$ and electrode 2. (b) A potential V_x is built at the position $x = +x$ where charge $+q$ is located. $q_1 = -C_1V_x$ and $q_2 = -C_2V_x$ are electrostatically induced on electrodes 1 and 2.

Noteworthy that the situation of charge distribution described by Eq. (2.6) is illustrated using the equivalent capacitors C_1 and C_2 (see Fig. 2.4(a)). Here, C_1 (and C_2) represents the capacitor formed between the positive charge $+q$ at $x = x$ and the negative charge q_1 on electrode 1 ($x = 0$) (and C_2 the capacitor formed between the positive charge $+q$ at $x = +x$ and the negative charge q_2 on electrode 2 ($x = d$)). Owing to the presence of the charge $+q$ at $x = +x$ under the short-circuit condition with charges q_1 and q_2 on electrodes 1 and 2, respectively, the potential $V_x = +q/(C_1 + C_2)$ is built at this position $x = +x$ (Fig. 2.4(b)), and the charges $q_1 = -C_1V_x$ and $q_2 = -C_2V_x$ are electrostatically induced on electrodes 1 and 2, respectively. Consequently, we can find the relation related to the charges $q_1 = -qC_1/(C_1 + C_2) = -q(d - x)/d$ and $q_2 = -qC_2/(C_1 + C_2) = -qx/d$, induced on electrodes 1 and 2, respectively (see Eq. (2.6)). In other word, the term $(d - x)/d$ and x/d in Eq. (2.6) are given by the ratio of capacitance C_1 and C_2.

From the discussion above, we find that transient current flows through an ammeter in accordance with transfer of charge $+q$ in the film from electrode 1 to electrode 2. That is, the capacitance C_1 and C_2 change in accordance with the transfer of charge $+q$. Along with this charge transfer, the amount of charges q_1 and q_2, induced on electrodes 1 and 2, respectively, changes accordingly with time. Obviously we find from Eq. (2.6) that the transient current $I(t)$ is given by

$$I = \frac{d}{dt}q_1 = -\frac{d}{dt}q_2 = \frac{u}{d}q \, (0 < t < t_{tr}). \tag{2.7}$$

As illustrated in Fig. 2.5(a), electric flux falls on both electrodes 1 and 2 from charge $+q$ and then induces charges q_1 and q_2. From

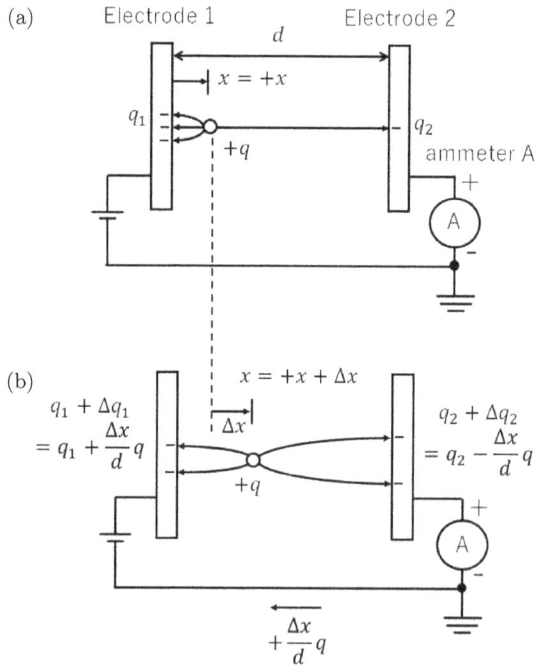

Figure 2.5. Charge q_1 and q_2 induced on electrode 1 and electrode 2, and its change Δq_1 and Δq_2 induced by a movement Δx of charge $+q$. (a) Charge $+q$ is located at $x = +x$ from electrode 1. Charge q_1 and q_2 are induced on electrodes 1 and 2, respectively. (b) The movement Δx of charge $+q$ causes charge Δq_1 and Δq_2 induced on electrodes 1 and 2, respectively, and charge $\Delta q_1 = -\Delta q_2$ flows through the ammeter.

the discussion above it is clear that the current I flows due to change of the electric field distribution formed in the insulator sandwiched between electrodes 1 and 2. The transient current given by Eq. (2.7) is the Maxwell displacement current (MDC), and we can study the carrier transit across the insulator by using the trace of recorded MDC. This idea is well utilized for the determination of carrier mobility μ of materials sandwiched between two electrodes, by using the transit time t_{tr} recorded, under the assumption that carrier speed u is given by $u = \mu(V/d)$, *i.e.*, $\mu = d^2/(Vt_{tr})$. This technique is known as Time-of-Flight (TOF) method [4–6]. The details on the TOF will be discussed in Sec. 2.3.

It is instructive here to note that using the relation Eq. (2.6), the changes of induced charge on electrodes 1 and 2, Δq_1 and Δq_2, respectively, caused by the positional shift of a point charge $+q$ located at $x = +x$ due to a displacement of $+\Delta x$ are given by (see Fig. 2.5(b))

$$\Delta q_1 = +\frac{\Delta x}{d}q, \quad \Delta q_2 = -\frac{\Delta x}{d}q. \tag{2.8}$$

That is, the displacement of point charge $+q$ with a distance of $+\Delta x$ gives rise to the change of induced charge $+(\Delta x/d)q$ on electrode 1 and the change of induced charge $-(\Delta x/d)q$ on electrode 2, and leads to flow of charge $-(\Delta x/d)q$ from electrode 1 to electrode 2 (or flows of $+(\Delta x/d)q$ from electrode 2 to electrode 1) through the closed circuit with an ammeter. That is, the generation of MDC is also allowed due to short-range movement Δx of charges in films.

2.2 Maxwell displacement current due to dipolar rotational motion — short-range charge motion

In this section, by extending the idea developed in Sec. 2.1, we discuss the Maxwell displacement current due to dipolar rotational motion. A dipole is defined as a pair of positive and negative charges as illustrated in Fig. 2.6. It is described in a vector form $\vec{\mu}$. The direction of dipole is defined to be parallel to the direction from the negative charge to the positive charge, and the magnitude $|\vec{\mu}| = \mu$, *i.e.*, dipole moment, is defined as the product of charge q and the spacing Δx

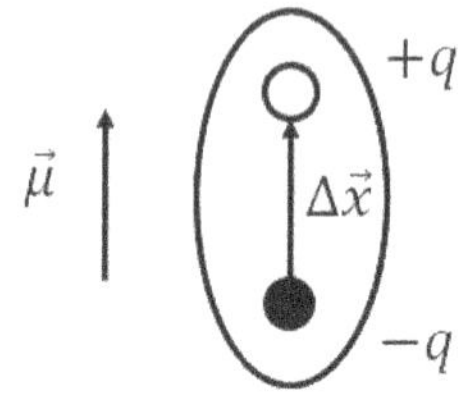

Figure 2.6. Electric dipole is a pair of positive charge $+q$ and negative charge $-q$, separated with a spacing of Δx. Dipole vector $\vec{\mu}$ is pointing in the direction $\Delta\vec{x}$ from negative charge $-q$ to positive charge $+q$, $\vec{\mu} = q\Delta\vec{x}$. Dipole moment of $\vec{\mu}$ is defined as $\mu = |\vec{\mu}| = q\Delta x$, where $\Delta x = |\Delta\vec{x}|$.

between the pair of positive and negative charges in a manner as $\mu = q\Delta x$ [7–9].

Let us assume here that a dipole is located at $x = +x$ in insulator, to be parallel to parallel electrodes 1 and 2, as shown in Fig. 2.7(a). This situation is the same as that a pair of positive and negative charges separated with a spacing of Δx are displaced to the position $x = +x$ in the direction from electrode 1 to electrode 2, in a manner like a single point charge $+q$ is displaced from electrode 1, as illustrated in Fig. 2.5(a). In this situation, as the distance between positive charge $+q$ and electrode 1 is the same as that the distance between negative charge $-q$ and electrode 1, no charge is induced on electrodes 1 and 2, because from Eq. (2.6) charge flowing through the ammeter should be sum of the negative charge $x/d(-q)$ (Fig. 2.7(b)) and the positive charge $x/d(+q)$ (Fig. 2.7(c)), resulting in no-flow of charge through the closed circuit with an ammeter.

On the other hand, when the dipole changes its orientation toward the direction pointing from electrode 1 to electrode 2, as illustrated in Fig. 2.8, we can assume that positive charge shifts its position with a distance of $+\Delta x/2$ and the negative charge shifts its position with a distance of $-\Delta x/2$, respectively. This is the "short-range" charge motion we call in this book. Dipolar rotation, microscopic displacement of a point charge (see Eq. (2.8)), *etc.*, are example of short-range charge motion. In this situation, we find from Eq. (2.8) that the change of induced charge Δq_1 on electrode 1 is given by

$$\frac{\Delta x/2}{d}(+q) + \frac{-\Delta x/2}{d}(-q) = +\frac{\mu}{d},$$

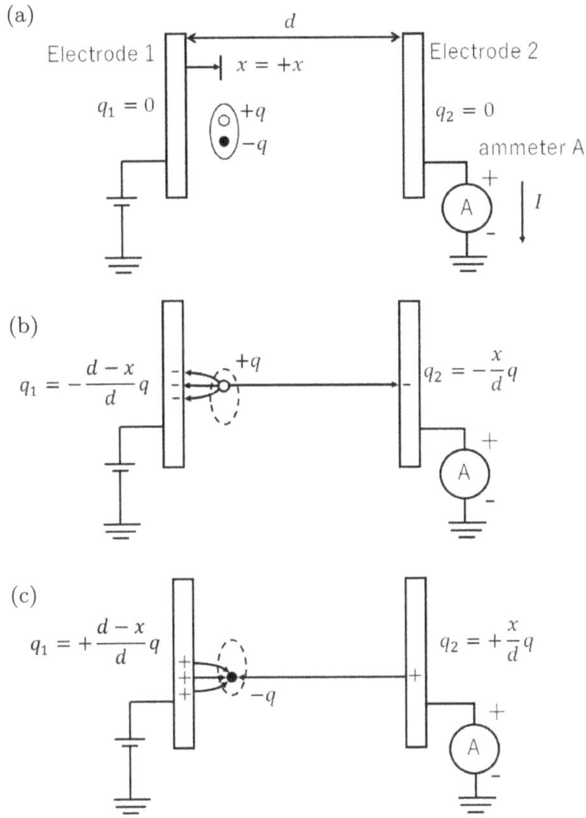

Figure 2.7. (a) Charge and electric field distribution when a dipole is located at $x = +x$ between electrode 1 and electrode 2, to be parallel to the surface of electrode 1. This situation is the superposition of two situations (b) and (c); (b) charge $+q$ is located at $x = +x$, and (c) charge $-q$ is located at $x = +x$.

and the change of induced charge Δq_2 on electrode 2 is given by $-\mu/d$. That is, the rotation of the dipole gives the change of induced charge on electrodes 1 and 2. Consequently, the charge μ/d flows due to the rotational motion through the external closed circuit. From the discussion above, it is clear that the MDC is generated due to the rotational motion of dipoles. This idea is available for exploring dipolar relaxation process in dielectric materials comprised of polar molecules with permanent dipoles. It is noteworthy that thc clcctric field surrounding a dipole changes according to the rotational motion

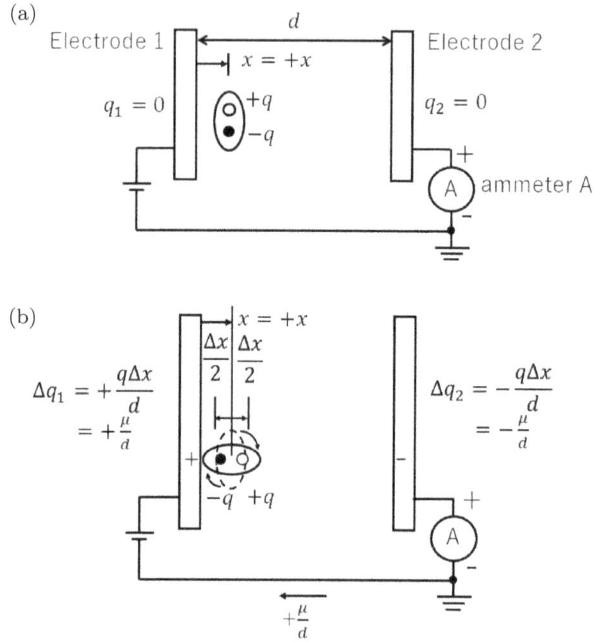

Figure 2.8. Generation of Maxwell displacement current by dipole rotation. (a) Dipole is orienting parallel to electrode 1. (b) Dipole rotates and points in the direction from electrode 1 to electrode 2. During the rotation process, Maxwell displacement current flows through an ammeter in closed short-circuit.

of dipoles, because the electric field is dependent on the orientational direction of dipole $\vec{\mu}$ (see Appendix A).

Thermally stimulated current (TSC) method is well-known as a technique to study dielectric relaxation process in dielectric materials, where MDC current that flows on accompanying dipolar rotational motion during heating process is recorded [10, 11]. This is the TSC. The details will be discussed in Chap. 3. In Sec. 2.4, we will show MDC current generated from monolayers on the water surface by monolayer compression [12].

2.3 TOF for carrier mobility measurement

Time-of-Flight (TOF) measurement is widely used to characterize carrier mobility of solids, including organic and inorganic films. Laser-TOF [4–6] is the most well-known method. e-TOF [4, 13] is also known as a method that can be utilized without using laser, meanwhile by using the applied voltage. As mentioned in Sec. 2.1, transient current I flows in accordance with flow of point charge in film sandwiched between two electrodes, as illustrated in Fig. 2.3. After recording such transient currents, we study carrier dynamics and determine carrier mobility, *etc.* For this purpose, we use the experimental set up illustrated in Figs. 2.9 and 2.10 for the laser-TOF measurement. In the conventional laser-TOF measurement, we firstly give a d.c. voltage U on electrode 2 with respect to the grounded electrode 1 to form a constant electric field $E = U/d$ between the two electrodes during the measurement (Fig. 2.9(a)). Then at $t = 0$, laser irradiation is employed to generate electron–hole pairs, only inside the organic layer near the electrode 1 (Figs. 2.9(b) to (d)). In order to make this situation, a thick film on the order of μm to mm thickness is chosen, much greater than laser beam spot size and also greater than laser penetration depth. After the dissociation of the generated electron–hole pairs in film, only electrons can move from electrode 1 to electrode 2 at a velocity of $u = \mu_e E$ for $U > 0$, whereas only holes can move from electrode 1 to electrode 2 at a constant velocity of $u = \mu_h E$ for $U < 0$ (electron movement is illustrated in Figs. 2.9(d)–(f)). Here, μ_e and μ_h are the electron mobility and hole mobility, respectively. We define the transit time by the relations

$$t_r = d/u = d^2/\mu_e U \; (U > 0),$$

or

$$t_r = d/|u| = d^2/\mu_h|U| \; (U < 0),$$

and we measure the carrier mobility of holes and electrons by changing the polarity of the applied voltage U. This is the basic principle of the TOF measurement, and many experiments have been carried

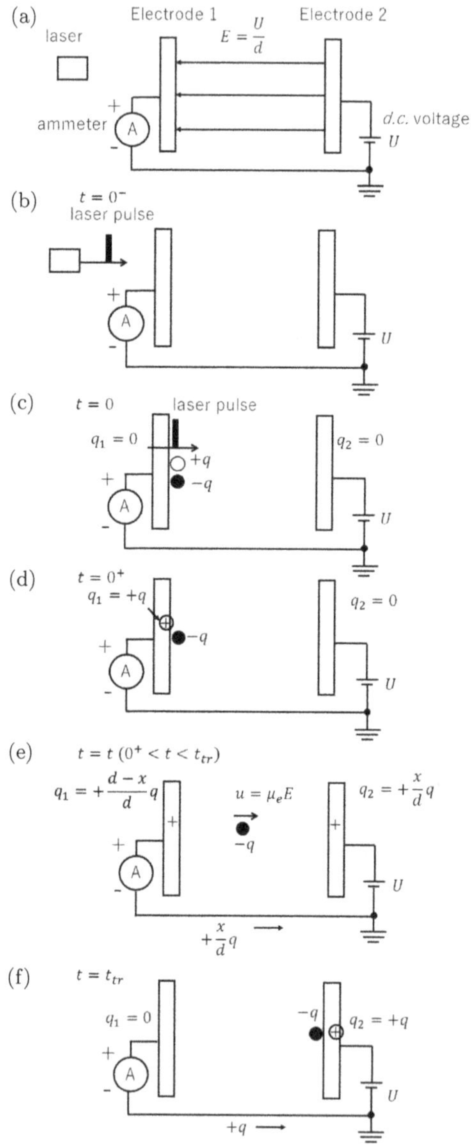

Figure 2.9. TOF measurement; (a) setup, (b) laser pulse is impinging on the sample, (c) laser pulse generates electron–hole pairs in the sample near electrode 1, (d) the electron–hole pair is separated to an electron and hole, (e) only electrons (holes) travel from electrode 1 to electrode 2 for $U > 0$ ($U < 0$). Meanwhile Maxwell displacement current flows though the ammeter in closed short-circuit, (f) electrons (holes) arrive at electrode 2 and current flow stops.

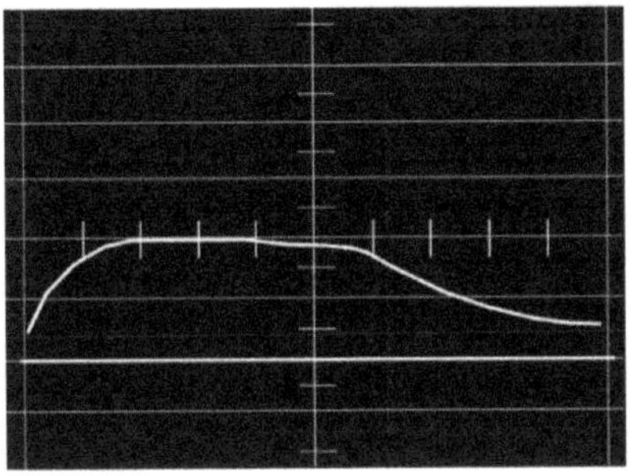

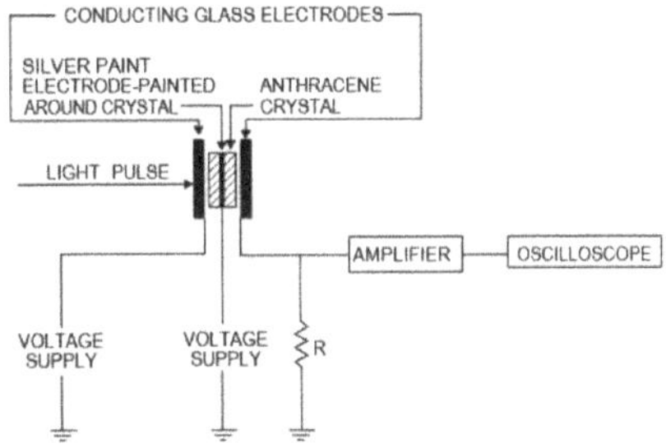

(a) (b)

Figure 2.10. TOF measurement of anthracene crystal (thickness 1–2 mm) by Kepler. (a) Trace of transient current in TOF measurement. Each division on the horizontal axis represents 50 μsec. (b) Experimental arrangement of the TOF measurement. Reprinted figure with permission from R. G. Kepler, Physical Review, Vol. 119, 1226, 1960. Copyright (1960) by the American Physical Society [5].

out for the determination of carrier mobility of polymers [14], insulators [15–17], organic semiconductors [18–20], liquid crystals [21, 22], *etc.*, since the pioneering work of the carrier mobility measurement on anthracene crystals by Kepler [5]. Figure 2.10 shows an example of TOF measurement that was employed for the determination of carrier mobility in anthracene crystals, where a 1–2 mm thickness sample was used, and electrons or holes were conveyed from electrode 1 to the counter electrode 2, depending on the polarity of applied voltage. In actual experiments, however, there are many technical and scientific problems. For example, it is very difficult to irradiate a laser beam only around electrode 1, in particular, in the case of very thin films. Also, the space charge field caused by generated carriers distorts the electric field distribution in the system and the carrier transport will be thus governed by this electric field. Noteworthy this situation would be serious when the number of generated carriers is much greater than CU/q. Here, C is the capacitance between the two electrodes. Furthermore, there is an upper limit on the transit time that can be measured in the TOF measurement, namely the ohmic relaxation time of the film under test. That is, we cannot observe carrier transits in the region $t_r > t_{di}$ by using TOF if carriers decay very fast with the ohmic relaxation time t_{di}.

In the e-TOF measurement, the basic principle is the same as that used for the detection of MDCs. Instead of laser irradiation to generate electrons and holes, we apply a step voltage U to the two-electrode system in order to inject carriers from electrodes and then transport them between electrodes 1 and 2. As we probe the change of induced charges on the electrodes along with the transport of injected carriers, this idea is easily extended to a three-electrode case such as field-effect transistors [13, 23], asymmetric needle–plane electrode systems [24] and others, although the amount of induced charge on the electrodes changes in a very complex manner, depending on the location and distribution of the charges. Figure 2.11 shows an example of e-TOF measurement that is carried out for the determination of hole mobility in pentacene by using pentacene field-effect transistors [25, 26]. As we can see in the figure, the transient current abruptly flows at around the time when the holes arrive at the counter (drain)

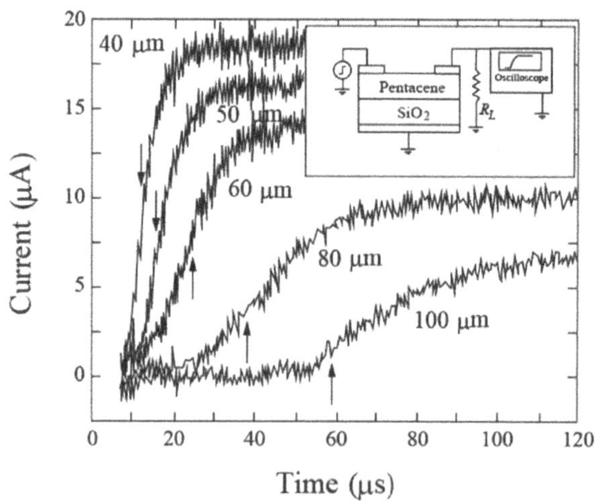

Figure 2.11. e-TOF measurement of pentacene field-effect transistor (FET) (channel length 40–100 μm, channel width 3 mm). Hole injected from source electrode travels FET channel. Upon arrival of holes at drain electrode of top-contact pentacene-FET, transient current abruptly increases with an inflection point as a transit time t_{tr}. The arrows indicate the inflection points. Inset illustrates an arrangement of e-TOF measurement. Reprinted with permission from M. Weis, J. Lin, D. Taguchi, T. Manaka, M. Iwamoto, J. Phys. Chem. C, Vol. 113, 18459, 2009. Copyright (2009) American Chemical Society [25].

electrode by crossing the channel. As shown in Fig. 2.11, e-TOF measurement is helpful for the determination of the carrier mobility of the materials, but it has a fundamental problem for usage. We cannot determine the polarity of the carriers injected from electrodes, which move across the film, merely from the trace of I–t curves of MIM elements, because positive and negative charges can move along the electric field in opposite directions each other. That is, the current is recorded in the same way as MDCs. In other words, we cannot distinguish whether electron injection happens or hole injection happens. Therefore, we need to determine the carrier polarity using other methods, before we carry out the e-TOF measurement. For example, in the three-electrode case such as field-effect transistors (FETs), we determine the polarity of carriers from the FET transfer curves. That is, by choosing the applied voltage on the gate electrode, we assign the polarity of main carriers, *i.e.*, electrons or holes. In the experiment illustrated in Fig. 2.11, holes are chosen by applying the positive potential on the source electrode in reference to gate and drain electrode grounded. As mentioned above, the TOF measurement is helpful for the determination of carrier mobility, but this is available under many physical boundary conditions.

2.4 MDC measurement for the determination of dipole moment of monolayer

Dielectric polarization is given as sum of the spontaneous polarization, linear polarization, and nonlinear polarization [27]. Among them, spontaneous polarization $\vec{P}_0$ is generated when permanent dipoles align in the same direction, as illustrated in Fig. 2.12(a). For example, the constituent rod-like polar molecules of Langmuir monolayers align on a water surface, pointing their long-axis direction towards air. Owing to this non-symmetric alignment of polar molecules on the water surface (symmetry breaking), spontaneous polarization is generated in the monolayer [28, 29]. According to the Gauss law, the electric flux density arises from the constituent polar molecules with permanent dipoles. As a result, when we placc a suspended electrode above the monolayer on a water surface, charges

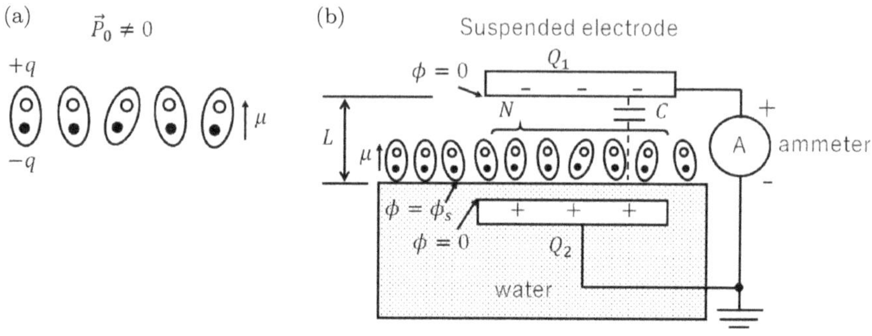

Figure 2.12. Maxwell displacement current (MDC) due to change of spontaneous polarization. (a) Alignment of dipoles $\vec{\mu}$ generates non-zero spontaneous polarization $\vec{P}_0$. (b) Spontaneous polarization is non-zero for monolayers on the water surface, and MDC flows due to change of $\vec{P}_0$ during a course of monolayer compression and expansion.

are induced on the electrode as shown in Fig. 2.12(b). The amount of induced charge is dependent on the orientational order of the constituent rod-like polar molecules, and can be described as [30, 31]

$$Q_1 = -\frac{N\mu S_1}{L} - C\phi_s \quad \text{with } S_1 = \langle \cos\theta \rangle. \tag{2.9}$$

Here, N is the number of molecules under the suspended electrode, μ is dipole moment, L is the spacing between the suspended electrode and the water surface, and S_1 is the orientational order parameter which stands for the average of tilt angle θ of the constituent polar molecules in space (θ is tilt angle from the normal direction to the water surface). ϕ_s is the surface potential of the water subphase in reference to the electrodes, C is the capacitance between the water surface and the suspended electrode. Electro-static dielectric constant of water $\varepsilon_w \approx 78$ is sufficiently large compared with that of air $\varepsilon_{air} \approx 1$, and induced charge due to water can be omitted. Noteworthy, the formalism of order parameters will be discussed in Chap. 5. The MDC flows when charge Q_1 changes with time. Consequently, MDC I is given in a manner as

$$I = -\frac{d}{dt}Q_1 = +\frac{B}{L}\frac{d}{dt}P_z + C\frac{d}{dt}\phi_s \tag{2.10}$$

where B is the electrode area of the suspended electrode, and P_z is the polarization of the monolayer given by $P_z = n\mu S_1$, with the surface density of molecules $n = N/B$. The change of surface potential ϕ_s is negligibly small by external stimuli, such as photoirradiation and monolayer compression. We therefore can pay more attention to the change of $P_z = n\mu S_1$ by external stimuli. P_z changes with n, μ, and S_1. For example, n and S_1 change due to the condensation and orientational ordering of constituent molecules, respectively, in accordance with the occupied area per molecule $A\,(n = 1/A)$ by monolayer compression. The magnitude of μ itself changes due to the phase transition of monolayer by compression. Further, for particular functional molecules, the change of μ will be induced due to photo-isomerization *etc.* As mentioned above, P_z changes by external stimuli, but basically the change is the result of the dynamical motion of dipoles such as rotational motion and positional shift. Figure 2.13 shows an example of MDCs generated from fatty acid monolayers by monolayer compression, where the MDC is generated in accordance with the change of surface pressure [32, 33]. Interestingly, the generation of MDC is very sensitive, because the MDC is a transient current and it is generated only when the electric flux density between the suspended electrode and the monolayer changes with time. Actually, the MDC is detectable even in the region of molecular area where the surface pressure is immeasurably very low. See region I in Figs. 2.13(b) and (c), the MDC is generated before the initial rise of surface pressure during the monolayer compression process. The details are as follows: the MDC current is described as

$$I = -\frac{B}{L}\left(\frac{d}{dA}P_z\right)\alpha = -\alpha\frac{B}{L}\left(-\frac{\mu S_1}{A^2} + \frac{\mu}{A}\frac{d}{dA}S_1 + \frac{S_1}{A}\frac{d}{dA}\mu\right) \quad (2.11)$$

$$\text{with } \alpha = -\frac{d}{dt}A$$

assuming that the change of ϕ_s is negligibly small during monolayer compression. In the MDC measurement, experiments are carried out by monolayer compression with a constant compression speed α. From Eq. (2.11), it is clear that the MDC gradually changes in region I of Fig. 2.13(b), in accordance with the condensation of molecules

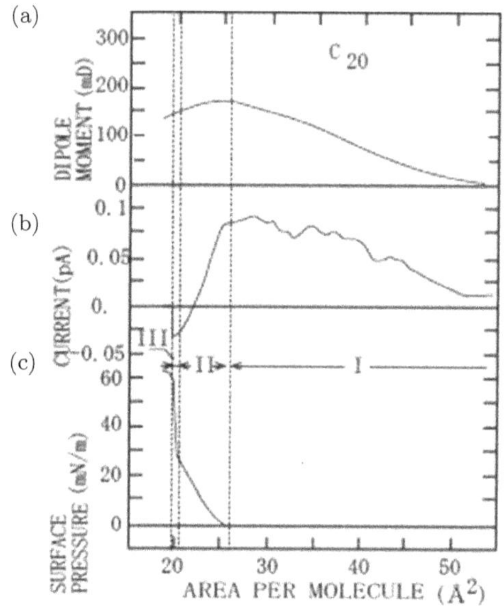

Figure 2.13. Example of MDC. Fatty acid is spread on a water surface and MDC is recorded during a course of monolayer compression. (a) Dipole moment component along water-surface normal direction calculated from MDC. (b) MDC. (c) Surface pressure. Reprinted from M. Iwamoto and Y. Majima, J. Chem. Phys., Vol. 94, 5135, 1991, with permission of AIP Publishing [33].

and gradual change of orientational order $S_1 = \langle \cos\theta \rangle$ by compression (1st and 2nd terms of Eq. (2.11), respectively). On the other hand, the abrupt change of MDCs appears in regions II and III due to the change of polarization caused by phase transition (3rd term). Using Eq. (2.11), we can determine the vertical component of the permanent dipole moment of fatty acid molecules, μS_1, as shown in Fig. 2.13(a).

Similarly, we can sensitively probe the change of molecular forms such as cis–trans photo-isomerization induced in azo-benzene monolayers by alternating photo-irradiation of ultra-violet and visible light [34, 35]. The trans- and cis-form isomers have the permanent dipole μ_a and $\mu_b\,(\mu_a \neq \mu_b)$, respectively. Accordingly, the MDC is generated during the photo-isomerization process. The detection of photo-switching due to cis–trans photo-isomerization in monolayers on solids as MDC signals will be available as a basic principle of organic memory elements as shown in Fig. 2.14, where alternating

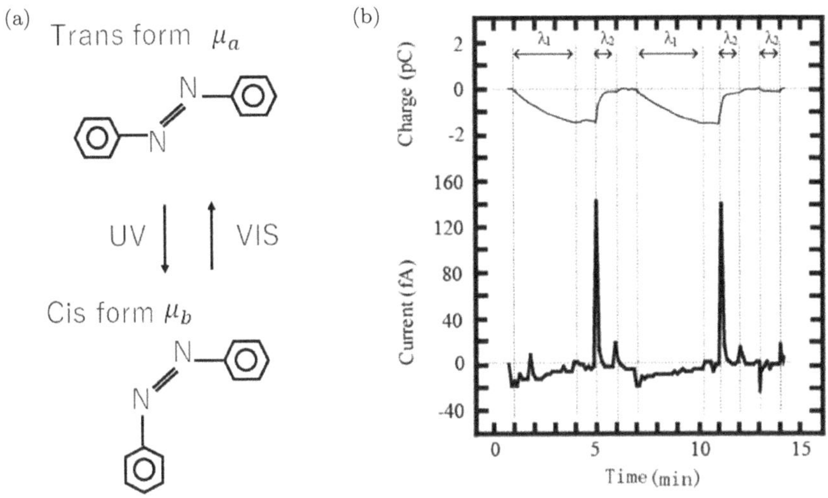

Figure 2.14. Organic memory. (a) Memory effect of azo-benzene molecule where trans form and cis form isomers are switched by ultraviolet (UV) and visible (VIS) light irradiation. (b) MDC detected by an ammeter. A sequential UV (λ_1) and VIS (λ_2) light irradiation switches molecular isomers and MDC flows. A successive VIS radiation (two successive regions λ_2 from Time 11 min and 13 min) never shows generation of MDC pulse, indicating a memory effect that the azo-benzene stores its state in trans-conformation. Reprinted by permission from Springer Nature: Nature (Generation of Maxwell displacement current across an azobenzene monolayer by photoisomerization, M. Iwamoto, Y. Majima, H. Naruse, T. Noguchi, H. Fuwa), COPYRIGHT (1991) [34].

ultra-violet (UV) and visible (VIS) light irradiation results in the generation of alternating MDCs, whereas successive irradiation never generates an MDC pulse [36]. It is instructive here to note that the surface potential measurement based on the Kelvin-probe method has been widely used for the study of monolayers at the air–water interface [37, 38]. What we record by using this method is a compensation voltage V_{comp} that is needed to make a zero-electric field between the monolayer and the suspended electrode. That is, a compensation voltage V_{comp} is applied to be $Q_1 = 0$ in the surface potential measurement. By adding the contribution of V_{comp} to Eq. (2.9), we obtain the following equation:

$$Q_1 = -\frac{N\mu S_1}{L} - C\phi_s + CV_{comp} = 0. \quad (2.12)$$

That is

$$V_{comp} \approx +\frac{P_z}{\varepsilon_0} + \phi_s.$$

Here ε_0 is the dielectric constant of the air gap between the suspended electrode and monolayer, that is, nearly equal to the dielectric constant of vacuum. Equation (2.12) tells us that the conventional surface potential measurement probes polarization P_z in accordance with the change of surface pressure by monolayer compression. In comparing Eq. (2.12) with Eq. (2.10), it is clear that the MDC measurement is very sensitive to the change of polarization P_z with molecular area A, while the surface potential measurement is not so sensitive. Actually, it is very difficult to probe the change of polarization in the region before the initial rise of surface pressure by means of surface potential measurement. On the other hand, we can clearly see the generation of the MDC in this region, as shown in region I of Fig. 2.13. Therefore, the MDC measurement has an advantage in probing dynamical dipolar motions in monolayers.

References

1. Purcell, E. M. (1963) *Electricity and magnetism*, berkeley physics course vol. 2, (McGRAW-HILL book company, New York) pp. 4–5, Sec. 1.2.
2. Maxwell, J. C. (1954) *Treatise on electricity & magnetism*, 3rd Ed. volume 1, (Dover, New York) Art. 30. Experiment V.
3. Faraday, M. (1839) *Experimental Researches on Electricity*, volume 1 (Bernard Quaritchii, London) on Internet Archive, Art. 1169–1177; Japanese translation by Yajima, S. and Inanuma M. (1945) *Faraday Denkigaku Jikken Kenkyuu*, vol. 2, series 11, ii, pp. 185–190, Art. 1169–1177, Iwanami Bunko No. 2910–2914 (Iwanami Shoten, Tokyo) (in Japanese).
4. Lampert, M. A. and Mark, P. (1970) *Current injection in Solids* (Academic, New York London) Chap. 6. pp. 113–139.
5. Kepler, R. G. (1960). *Phys. Rev.*, 119, pp. 1226–1229, https://doi.org/10.1103/PhysRev.119.1226.
6. Meier, H. (1974) *Organic Semiconductors — dark and photoconductivity of organic solids* (Verlag Chemie, Weinheim) pp. 98–104.
7. Fröhlich, H. (1958) *Theory of dielectrics, Dielectric constant and dielectric loss*, 2nd Ed. (Clarendon, Oxford); Japanese translation by Nagamiya, T. and Nakai, Y. (1960) *Fröhlich: Yuu den tai ron* (Yoshioka Shoten, Kyoto) (in Japanese); Nagamiya, T. and Nakai, Y. (2000) *Yuudentai Ron* [POD edition] (Yoshioka Shoten, Kyoto) (in Japanese).
8. Kinbara, T. (1972) *Electromagnetism (I) "Denjiki Gaku (I)"* (Shokabo, Tokyo) Sec. 6.2–6.4. pp. 132–136 (in Japanese).

9. McConnel, J. (1980) *Rotational Brownian Motion and Dielectric theory* (Academic, London New York) Chapter 1. pp. 1–11.
10. Hino, T. (1986). IEEE Trans. Electr. Insul., EI–21, 1007, https://doi.org/10.1109/TEI.1986.349015.
11. Iwamoto, M. and Taguchi, D. (2014) *Thermally Stimulated Current in Electrical and Electronic Materials — Analysis and Application* (Corona publishing, Tokyo) (in Japanese).
12. Iwamoto, M. and Wu, C.-X. (2001) *The Physical Properties of Organic Monolayers* (World Scientific, Singapore), https://doi.org/10.1142/4575.
13. Basu, D. and Dodabalapur, A. (2009) *Drift Velocity and Drift Mobility Measurement in Organic Semiconductors Using Pulse Voltage, Adv. Polym. Sci.*, 223, pp. 73–112, 2010. (Springer–Verlag, Berlin Heidelberg).
14. Abkowitz, M., Facci, J. S. and Stolka, M. (1993). *Appl. Phys. Lett.*, 63, 1892, https://doi.org/10.1063/1.110639.
15. Spear, W. E. (1957). *Proc. Phys. Soc. B*, 70, 669; Spear, W. E. (1969). *J. Non-crystal. Solids*, 1, 197, https://doi.org/10.1016/0022-3093(69)90001-5.
16. Petravichyus, A. D., Yushka, G. B. and Baubinas, R. V. (1976). *Sov. Phys. Semicond.*, 9, 1530. (translation from Fiz. Tekh. Polyprovodn., 9, 2372, 1975).
17. Iwamoto, M. (1996). *J. Appl. Phys.*, 79, 7936, https://doi.org/10.1063/1.360953.
18. Köhler, A. and Bässler, H. (2015) *Electronic processes in organic semiconductors: an introduction* (Wiley Online Library), https://doi.org/10.1002/9783527685172 Chap. 3.3, pp. 208-246.
19. Kajiwara, T., Inokuchi, H. and Minomura, S. (1967). *Bull. Chem. Soc. Japan*, 40, 1055, https://doi.org/10.1246/bcsj.40.1055.
20. Naka, S., Okada, H., Onnagawa, H., Yamaguchi, Y. and Tsutsui, T. (2000). *Synth. Met.*, 111–112, 331, https://doi.org/10.1016/S0379-6779(99)00358-6.
21. Funahashi, M. and Hanna, J. (1998). *Appl. Phys. Lett.*, 73, pp. 3733–3735, https://doi.org/10.1063/1.122896.
22. Iino, H., Ohno, A. and Hanna, J. (2005). *J. Chem. Phys.*, 123, 244701, https://doi.org/10.1063/1.2141507.
23. Burns, J. R. (1969). *RCA Review*, 30, pp. 15–35.
24. Iwamoto, M., Fukuzawa, M. and Hino, T. (1987) *IEEE Trans. Electr. Insul.*, EI–22, 419, https://doi.org/10.1109/TEI.1987.298902.
25. Weis, M., Lin, J., Taguchi, D., Manaka, T. and Iwamoto, M. (2009). *J. Phys. Chem. C*, 113, 18459, https://doi.org/10.1021/jp908381b.
26. Lin, J., Weis, M., Taguchi, D., Manaka, T. and Iwamoto, M. (2011). *Jpn. J. Appl. Phys.*, 50, 04DK01, https://doi.org/10.7567/JJAP.50.04DK01.
27. Iwamoto, M., Manaka, T. and Taguchi, D. (2016). *IEEJ Trans. Fundam. Mater.*, 136, pp. 678–684 (in Japanese, Review), https://doi.org/10.1541/ieejfms.136.678.
28. Roberts, G. G. and Pitt C. W. Eds. (1983) *Langmuir-Blodgett Films, 1982* (Elsevier, Amsterdam Oxford New York). Thin films science and technology Vol. 3 (Thin Solid Films, 99, nos. 1–3, special issues), https://doi.org/10.1016/0040-6090(83)90350-4.
29. Fukuda, K. and Sugi M. Eds. (1989) *Langmuir-Blodgett Films 4*, vol. 1-3, (Elsevier Applied Science, London New York). (Thin Solid Films, 178, nos.

1–2, 179, nos. 1–2, 180 nos. 1–2, special issues), https://doi.org/10.1016/0040-6090(89)90278-2.

30. Iwamoto, M., Manaka, T. and Taguchi, D. (2015). *J. Phys. D*, 48, 373001, https://doi.org/10.1088/0022-3727/48/37/373001.
31. Iwamoto, M. amd Wu, C.-X. (2001) *The Physical Properties of Organic Monolayers* (World Scientific, Singapore), https://doi.org/10.1142/4575.
32. Majima, Y. and Iwamoto, M. (1990). *Jpn. J. Appl. Phys.*, 29, 564, https://doi.org/10.1143/JJAP.29.564.
33. Iwamoto, M. and Majima, Y. (1991). Investigations of the dynamic behavior of fatty acid monolayers at the air–water interface using a displacement current-measuring technique coupled with the Langmuir-film technique, *J. Chem. Phys.*, 94, pp. 5135–5141, https://doi.org/10.1063/1.460551.
34. Iwamoto, M., Majima, Y., Naruse, H., Noguchi, T. and Fuwa, H. (1991). *Nature*, 353, pp. 645–647, https://doi.org/10.1038/353645a0.
35. Xu, X. and Iwamoto, M. (1997). *Jpn. J. Appl. Phys.*, 36, 7348, https://doi.org/10.1143/JJAP.36.7348.
36. Iwamoto, M., Kanai, Y. and Naruse, H. (1993). *J. Appl. Phys.*, 74, 1131, https://doi.org/10.1063/1.354938.
37. Evans, S. D. (1995) *Surface potential*, Chap. 6. in A. Ulman Ed., *Characterization of organic thin films* (Butterworth-Heinemann, Massachusetts Manning Connecticut).
38. Ulman, A. (1991) *An introduction to ultrathin films from Langmuir–Blodgett to self assembly* (Academic Press, San Diego New York London) pp. 21–28, Part one: Analytical tools, 1.1 Surface potential, https://doi.org/10.1016/C2009-0-22306-3.

Chapter 3

Thermally Stimulated Current due to Carrier Motion in Solids

Maxwell displacement current (MDC) is the transient current and it flows due to the change of electric flux density distribution in solids with time. By measuring Maxwell displacement current flowing across solids, we can study dipolar rotational motion and displacement of electronic charges in solids. Among the techniques using MDC, thermally stimulated method is one of the most successful methods, which have been developed for characterizing electrical properties of solids and device performance. In the measurement, thermal stimulation is applied to solids and devices, which are placed in closed circuits with an ammeter, to measure MDC current due to carrier motion during heating. Therefore, it is helpful to study thermally stimulated current and its related topics for understanding the MDC. In this chapter, we discuss the principle of the generation of thermally stimulated current due to dipolar depolarization, detrapping of electrons, and long-range ionic carrier motion in solids. After that, we give a brief review on recent research trend using TSC.

3.1 Introduction

Thermally stimulated current (TSC) measurement is widely used in a variety of research fields, *e.g.*, physics, electronics, electrical engineering, chemistry, ceramics, and biology. TSC is short-circuit current that flows during heating, due to displacement of positive and negative charges (*e.g.*, electrons, holes, ions, *etc.*), rotational motion

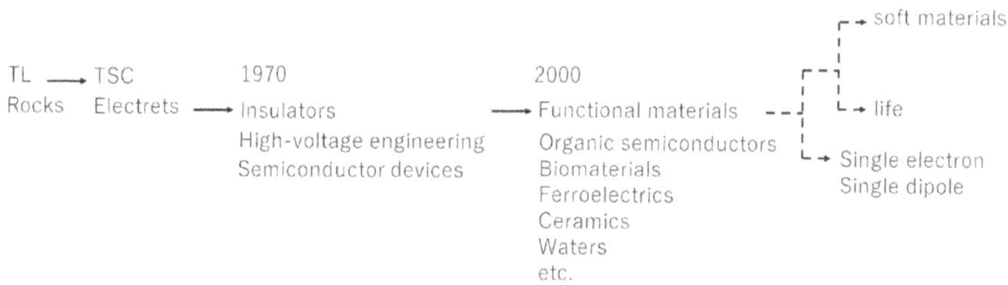

Figure 3.1. TSC research trend.

of permanent dipoles in samples, phase transition of materials, long-range ionic carrier motion in solids, and so forth. The electrical circuit used for TSC measurement is very simple, where only one electro-ammeter is used. Nevertheless, recorded TSC curves give important information on charge behaviors in solids. This motivates researchers to employ TSC measurement as an important method for characterizing materials and devices.

In the 1970s, the development of high-voltage engineering, integration circuit technology in semiconductor device electronics, and so on, were very important R&D subjects in industries, research laboratories, and universities in advanced countries (Fig. 3.1). In such a situation, the characterization of the electronic and electrical properties of materials and electronic devices being subjected to an external electric field under operational conditions was an urgent task. Thus, the TSC method was chosen as one of useful methods in research laboratories and industries, and it had contributed greatly to the development of electrical insulation engineering [1, 2]. Also, this method had been widely applied to the characterization of the bulk and interface properties of electronic devices such as metal–insulator–Si-semiconductor (MIS) diodes and metal–nitride–oxide–Si-semiconductor (MNOS) devices [3, 4]. At the same time, the experimental TSC technique and analytical TSC method for studying carrier behaviors in solids and devices had greatly advanced, and many important books and review articles on the TSC method had been published [5–7].

On the other hand, over the past decades, many new molecules have been discovered, and many advanced functional materials such as organic semiconductor materials, ferroelectric materials, biomaterials, ceramics *etc.*, have been developed [8–10]. The discovery of new molecules and the emergence of novel materials have highly motivated researchers to utilize them in the research fields of electronics, electrical engineering, chemical engineering, medical engineering, and so forth. In this situation, TSC measurement has attracted much attention in industries and academia as a tool for characterizing new materials and devices using novel materials. It is instructive here to note that the recently developed materials are never replacements of classical ones such as polyethylene and Si-semiconductor. They have the potential to open new research fields such as organic materials electronics, molecular electronics, nano-electronics, and bioelectronics. This means that there are many new technical and scientific issues that we should explore in order to utilize the TSC method to characterize new materials and to evaluate the performance of novel devices. In the last part of this chapter, the latest research trend in the TSC method for the development of materials and devices is also briefly reviewed, on focusing the progress of TSC studies in Japan [11–13].

3.2 Origin of TSC

Here, we briefly discuss the principles of TSC and TSC measurement. As mentioned in Sec. 3.1, TSC is short-circuit current that flows through samples during heating owing to the displacement of positive and negative charges, the rotational motion of permanent dipoles, and the long-range movement of ions in solids, and so forth. By heating, trapped carriers, dipoles, ions, and so on are thermally activated in samples, depending on material properties. This material-dependent thermal process is characterized by the physical quantity $+H$, called the activation energy, and it proceeds with the characteristic time τ, called relaxation time, given by

$$\tau = \tau_0 \exp\left(+\frac{H}{kT}\right). \tag{3.1}$$

Here, τ_0 is the pre-exponential factor, k is the Boltzmann constant, and T is the absolute temperature. The activation energy

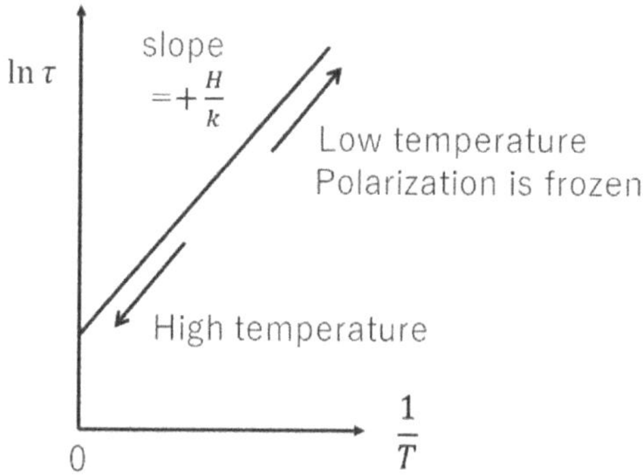

Figure 3.2. Relationship between relaxation time τ and temperature T, which is utilized in TSC measurement. The relationship is shown in $\ln\tau$–$1/T$ plot.

$+H$ represents the depth of carrier traps E_t for TSC due to carrier detrapping, whereas it represents barrier height for the rotational motion of dipoles for TSCs due to dipolar depolarization. Also, it represents height of potential hill for the long-range movement of ions. Therefore, it is very important to determine the activation energy $+H$ to study material properties. Figure 3.2 plots the relationship between $\ln\tau$ and $1/T$. This linear relationship suggests that the activation process proceeds very rapidly at high temperatures, whereas it proceeds very slowly at low temperatures. TSC measurement makes good use of this $\ln\tau$–$1/T$ relationship, and many methods have been developed to determine the activation energy $+H$ and the relaxation time τ from TSC curves over the past decades. The typical method is described in Sec. 3.3.2.

Figure 3.3 illustrates a typical experimental arrangement for the TSC measurement. V_b is a d.c. source battery used for biasing samples, whereas V_c is a collecting d.c. voltage source used during heating the samples. In electrical circuit, one electrometer is installed. For typical TSC measurement, firstly, samples are biased using the V_b at high temperatures to induce polarization in them within a very short time by applying d.c. voltage to the samples, where this polarization includes carrier trapping in the samples by injected electrons and holes, dipolar polarization, interfacial polarization, and ionic polarization, and others. Subsequently, the samples are very rapidly

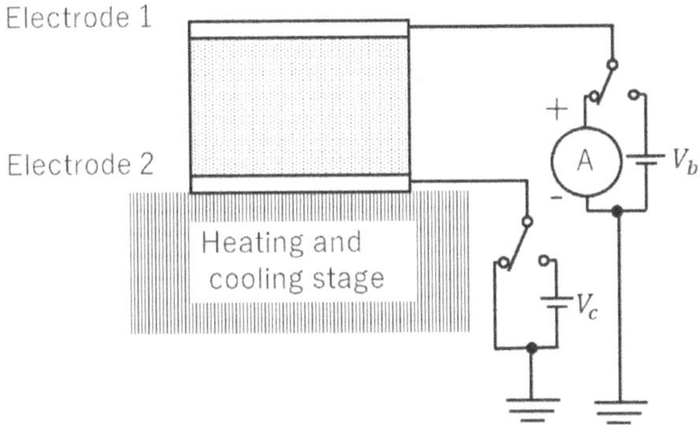

Figure 3.3. Typical experimental arrangement for TSC measurement. V_b is the external d.c. voltage used for biasing, and V_c is collecting voltage used in TSC measurement.

cooled to low temperatures to freeze the polarization formed at high temperatures, while taking account of the $\ln \tau$–$1/T$ relationship (see Fig. 3.2). After that, they are electrically shorted.

Subsequently, they are heated at a constant heating rate β, on applying collecting voltage V_c. Here $V_c = 0$ is most often chosen, but non-zero V_c is also adopted, depending on the TSC depolarization mechanisms. Note that the heating rate β does not need to keep constant in TSC measurement, but the use of a constant heating rate makes the TSC analysis easy. Consequently, constant heating is most often used in experiment. During thermal heating, short-circuit current flowing through the sample is measured. This short-circuit current is TSC, and it is obviously the MDC that flows due to the depolarization. It is noteworthy that photo-radiation, X-ray radiation, and so on are available in the biasing process, instead of d.c. biasing, to explore degradations, defects, and so forth for studying the electrical properties of materials.

Figure 3.4 illustrates an example of TSCs from a polyethylene (PE) film polarized by d.c. biasing, where a couple of TSC peaks are observed with the increase in temperature [14]. Interestingly, some TSC peaks (Peaks A and B) are positively generated in the depolarization direction, *i.e.*, in the direction of discharging, but there is a TSC peak (Peak C) that is negatively generated in the polarization

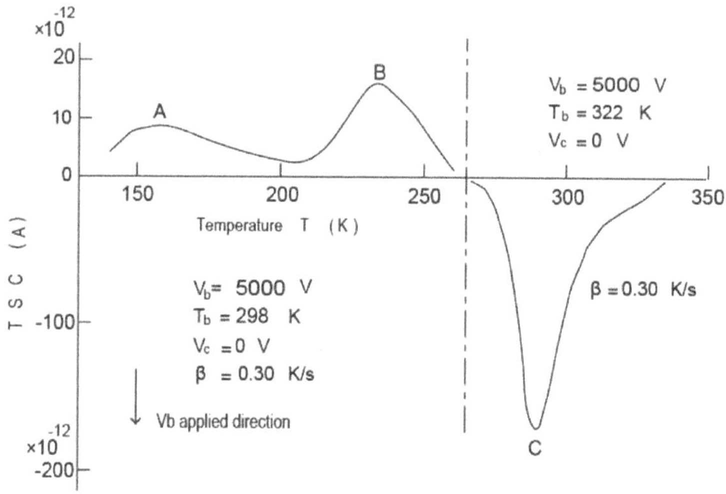

Figure 3.4. TSC of polyethylene film (thickness: 40 μm, electrode area: 2 cm^2) polarized by d.c. biasing. Reprinted by permission from the Institute of Electrical Engineers of Japan (T. Hino, K. Yamashita, IEEJ Trans. Fundam. Mater., Vol. 95, 79, 1975). [14]

direction, *i.e.*, in the direction of charging. One may wonder why TSC flows as illustrated in Fig. 3.4. The results should be discussed by assuming that TSC flows due to discharging during heating, that is, a film stressed by d.c. biasing will return to the initial state by heating, and the charge distribution in the film formed by d.c. biasing will return to the initial state that has been established before stress biasing. However, it will be difficult to understand the physical reason for the appearance of TSC peaks generated negatively in the manner as peak C, *i.e.*, in the direction of charging, because our general accepted idea derived from the theory on charging and discharging mechanism of an ideal capacitor is as follows: the discharging current flows in the direction opposite to the charging current, and the amount of charge flowing during the charging process is the same as that flowing during the discharging process. In other words, the charging and discharging currents flow reversibly, in opposite directions. Positive TSC peaks A and B that are generated in discharging process follow this general accepted idea, but why negative TSC peak C appears? To answer this question, we need to discuss the direction

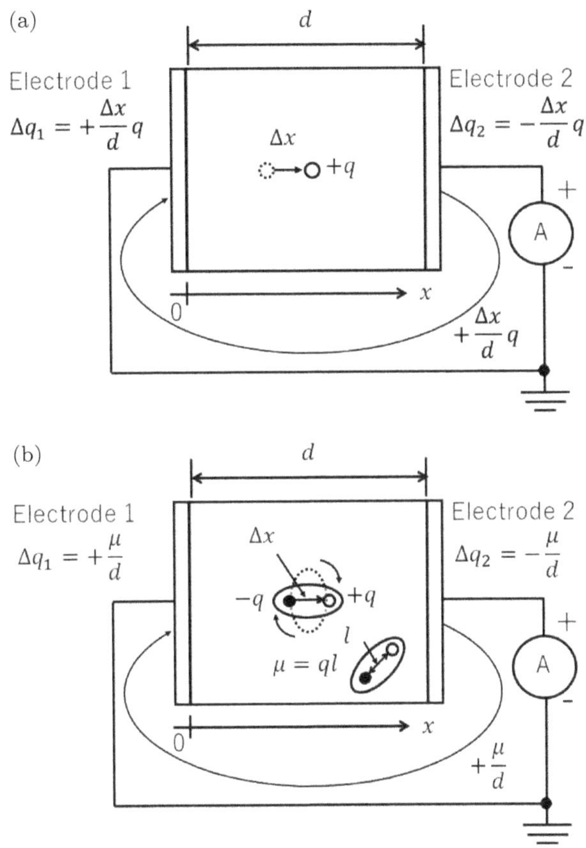

Figure 3.5. Principle of TSC measurement. (a) TSC due to charge displacement. (b) TSC due to dipolar rotation.

of electron movement in a short-circuit sample in terms of the electric field distribution in the sample, because electrons will be conveyed by the electric field formed in the samples. In other words, we need to find a mechanism that accounts for non-reversible carrier process. We can deliver similar discussion for hole movement in a short-circuit sample. The physical reason that accounts for the negative TSC is given in Sec. 3.4(i–1).

Figure 3.5(a) illustrates the principle of TSC, where a film with a thickness of d is sandwiched between two parallel electrodes 1 and 2, and the film is placed in a closed short-circuit with an ammeter.

According to the electromagnetism theory, if the charge $+q$ located inside the film is displaced at a distance of Δx in the film thickness direction as illustrated in Fig. 3.5(a), the amount of charge $+q(\Delta x/d)$ flows through the ammeter (see Chap. 2).

That is, the displacement Δx of the charge $+q$ is surely the origin of TSC, and this displacement happens in the film by heating in the TSC measurement. The displacement Δx of the charge $+q$ is possible to happen in the region $-d \le \Delta x \le +d$. Here, $\Delta x = +d$ represents the situation when the charge $+q$ moves from electrode 1 to electrode 2, whereas $\Delta x = -d$ represents the situation when the charge $+q$ moves from electrode 2 to electrode 1. This kind of displacement happens for TSCs due to detrapping of electrons (and holes), and also for TSCs due to the long-range movement of ions. For dipolar rotational motion, the original position of dipoles is fixed inside the film, but the rotational motion of dipoles is equivalent to the displacement of a pair of the positive charge $+q$ and negative charge $-q$ separated with a distance of $\Delta x = +l/2$ or $-l/2$ in the film thickness direction, respectively, where $\mu = ql$ is permanent dipole moment (see Fig. 3.5(b) and Chap. 2). The displacement Δx of the charge $+q$ happens in the region $-l \le \Delta x \le +l$. That is, owing to the rotational motion of one dipole, the amount of charge given by $+\mu/d$ flows through the ammeter if $\Delta x = l$ as illustrated in Fig. 3.5(b), whereas the charge $-\mu/d$ flows if the dipole rotates in opposite and $\Delta x = -l$.

3.3 TSC curves due to dipolar depolarization

3.3.1 *TSC curve*

For TSCs due to dipolar depolarization in films sandwiched between two identical electrodes, an external electric field caused by external voltage is formed in the films and this field is only acting to rotate dipoles. If only dipolar rotation happens in the absence of carrier injection from electrodes, no space charge field is formed in the films. Consequently, under the short-circuit condition with zero-external voltage, only dipolar rotation due to dipolar depolarization happens during heating. In this case, dipoles rotate and go back to their initial state, and finally the dipolar polarization P_0, which is initially formed in films, disappears. This means that the displacement of charge occurs in the direction opposite to the polarization process. Consequently, TSC is generated in the direction opposite to the charging direction. The two TSC peaks (Peaks A and B) observed in the low-temperature region in Fig. 3.4 correspond to this process.

For films polarized by d.c. biasing, dipolar depolarization proceeds on satisfying the following rate equation:

$$\frac{dP}{dt} = -\frac{P}{\tau}, \tag{3.2}$$

where P represents the polarization in film. Under short-circuit condition, charges are induced on electrodes as illustrated in Fig. 3.6(a). Noteworthy the amount of charges Q_1 and Q_2 induced on electrodes 1 and 2 per unit area are $Q_1 = +P$ and $Q_2 = -P$, respectively, and these are the same with the amount of polarization P formed in the film. After the completion of depolarization, $P = 0$, and $Q_1 = 0$ and $Q_2 = 0$ (Fig. 3.6(b)). As MDC flows through the short-circuit due to the change of induced charge on electrodes, MDC flows when the polarization P in the film changes with time and it is given by

$$MDC = -\frac{dP}{dt} \tag{3.3}$$

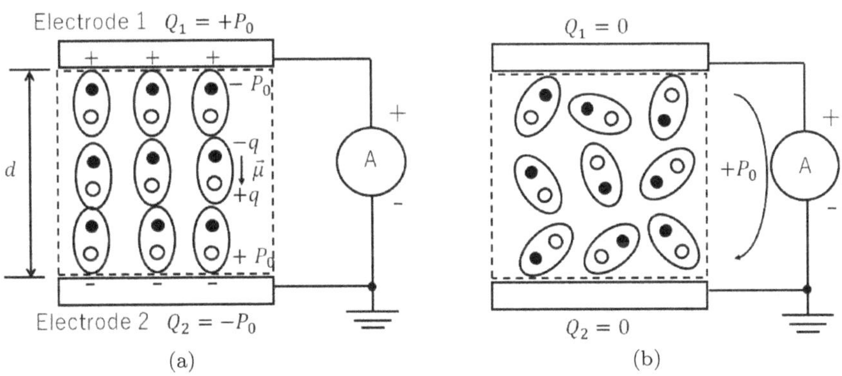

Figure 3.6. TSC due to dipolar depolarization. (a) Initial state after shortening the biased film, where dipolar polarization P_0 is formed and frozen at low temperature T_0. (b) During a course of heating, P_0 depolarizes and TSC flows through an ammeter.

TSC is basically the MDC current given by Eq. (3.3), while it flows during heating. In the TSC measurement, temperature T of film increases at a constant heating rate β, and given by $T = \beta t + T_0$ (t: time, T_0: initial temperature). Solving Eq. (3.2) with Eq. (3.1), and substituting the result of $P(T)$ into Eq. (3.3), we get a typical curve of TSC due to dipolar depolarization, described as [5–7, 11]

$$TSC(T) = \frac{P_0}{\tau_0} \exp\left(-\frac{H}{kT}\right) \exp\left[-\frac{1}{\tau_0 \beta} \int_{T_0}^{T} \exp\left(-\frac{H}{kT}\right) dT\right],$$

$$\text{with } P(T) = P_0 \exp\left[-\frac{1}{\tau_0 \beta} \int_{T_0}^{T} \exp\left(-\frac{H}{kT}\right) dT\right]. \tag{3.4}$$

Here P_0 is the initial polarization formed in samples by d.c. biasing, and k is the Boltzmann constant. TSC appears as illustrated in Fig. 3.7 with one peak at the temperature $T = T_m$ (maximum temperature). Here, the following relation

$$\tau(T_m) = \tau_0 \exp\left(+\frac{H}{kT_m}\right) = \frac{kT_m^2}{\beta H} \tag{3.5}$$

is satisfied at $T = T_m$. In TSC measurement, TSC flows owing to the depolarization of the initial polarization P_0. Consequently, the following relation is satisfied when TSC is measured using an electrically

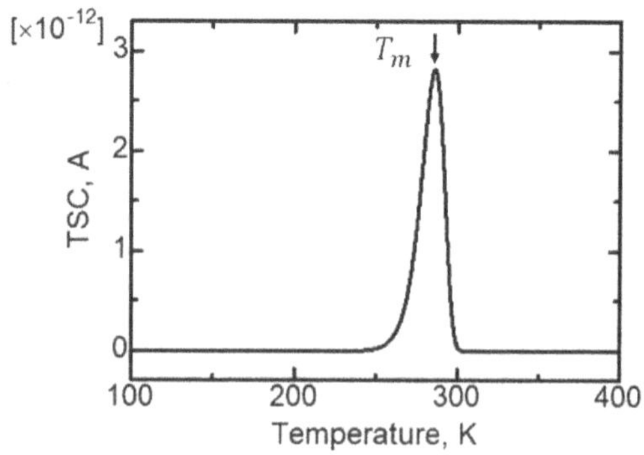

Figure 3.7. TSC of dipolar depolarization, calculated based on Eq. (3.4) at $P_0 = 1.0 \times 10^{-7}\,\text{C/cm}^2$, $\tau_0 = 1.0 \times 10^{-15}$ s, $H = 1\,\text{eV}$, $T_0 = 100\,\text{K}$, $\beta = 1\,\text{K/min}$. Electrode area $3.1\,\text{mm}^2$ was assumed. TSC curve gives a peak at $T_m = 286\,\text{K}$.

shorted circuit:

$$P_0 = \frac{1}{\beta}\int_{T_0}^{T_\infty} TSC(T)dT. \tag{3.6}$$

On the other hand, P_0 is given as (for example, see Sec. 2.2.1 of [11])

$$P_0 = \frac{N\mu^2}{3kT_b}E_b, \tag{3.7}$$

where N is the density of dipoles per volume, μ is permanent dipole moment, T_b is the biasing temperature, and E_b is the external electric field used for biasing. Noteworthy that the polarization should be $P_0 = N\mu$, if all dipoles of the film orient in the direction along the external electric field. This situation is expected to occur when the external electric field E_b is very high. However, in ordinary dielectric materials, the polarization P_0 is expressed using the Langevin function $L(x)\,(= \coth x - 1/x)$ in a manner as $P_0 = N\mu L(x)$ with $x \equiv \mu E_b/kT_b$ (see Fig. 3.8). Here, x represents the ratio of electric energy required for dipolar rotation μE_b and the thermal energy kT_b. In ordinary dielectric materials, all dipoles are randomly pointing and distributed with no interaction among dipoles owing to the thermal

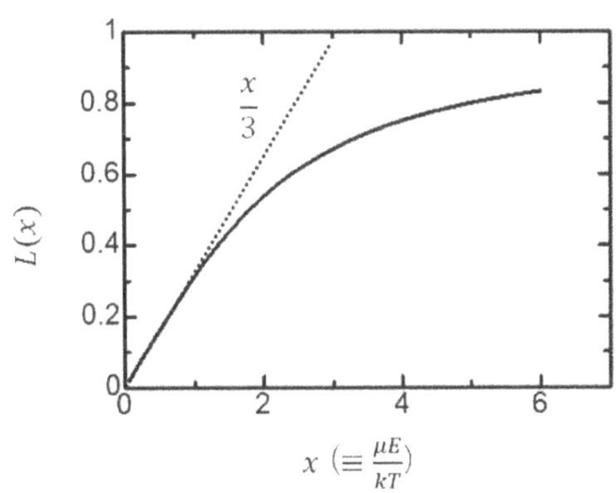

Figure 3.8. The solid curve is Langevin function $L(x)\,(= \coth x - 1/x)$, which represents characteristics of dipolar polarization in electric field E, under thermal equilibrium at temperature T. The dotted line is Eq. (3.7) derived for low electric field where $x \ll 1$.

energy kT_b, but they can rotate toward the electric field direction with help of the electric energy μE_b. Equation (3.7) is derived under the condition $x \ll 1$, where $L(x) \approx x/3$ (see Fig. 3.8). Equation (3.7) indicates that there is a linear relationship between the polarization P_0 and the external electric field E_b. This linear relationship is one of important characteristic properties of dipolar polarization of ordinary dielectric materials. Further it should be noted that if we apply the external electric field E_c during heating in TSC measurement, E_b in Eq. (3.7) should be replaced by $E_b - E_c$. That is, for TSC due to dipolar depolarization, an external applied electric field acts to change the polarization formed in samples.

3.3.2 *Determination of physical quantities*

Using TSC, we can evaluate physical quantities related to dielectric polarization as below.

(i) Activation energy H

The activation energy H is an important parameter for describing the rotational motion of dipoles. There are many ways to determine this parameter H, by making use well of Eq. (3.4), Eq. (3.5), and others. Typical methods are as follows:

(1) Initial rise method

The curve of TSC derived in Eq. (3.4) gives an initial rise followed by one peak in a manner as illustrated in Fig. 3.9. In the initial rise method, we pay attention to the depolarization that happens in the initial rise region, where most of the polarization remains as frozen, almost as in the initial state. That is, $P(T) \simeq P_0$, and Eq. (3.4) is approximately expressed as follows:

$$TSC(T) \approx \frac{P_0}{\tau_0} \exp\left(-\frac{H}{kT}\right). \tag{3.8}$$

From Eq. (3.8), we get the relation $\ln TSC(T) \approx const - H/kT$, with $const = \ln(P_0/\tau_0)$. Therefore, we can determine H from a slope of $\ln TSC(T) - 1/T$ plots, which are obtained from the experimental TSC curve in the initial rise region (Fig. 3.9(b)). Figure 3.10 shows

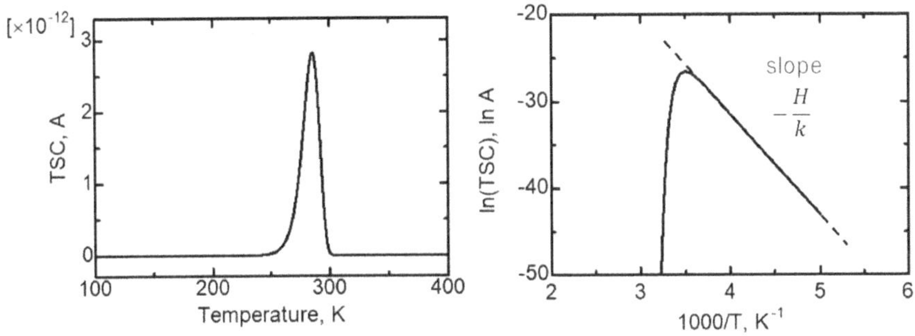

(a) (b)

Figure 3.9. Initial rise method to analyze TSC. Activation energy H is determined. (a) TSC curve is calculated with parameters $P_0 = 1.0 \times 10^{-7}\,\mathrm{C/cm^2}$, $\tau_0 = 1.0 \times 10^{-15}\,\mathrm{s}$, $H = 1\,\mathrm{eV}$, $T_0 = 100\,\mathrm{K}$, $\beta = 1\,\mathrm{K/min}$. Electrode area is $3.1\,\mathrm{mm}^2$. (b) $\ln TSC(T) - 1/T$ curve used in initial rise method. In this plot, initial current rise region $200 < T < 250\,\mathrm{K}$ $(0.004 < 1/T < 0.005)$ follows straight line and its slope is used to calculate activation energy H by using relation Eq. (3.8).

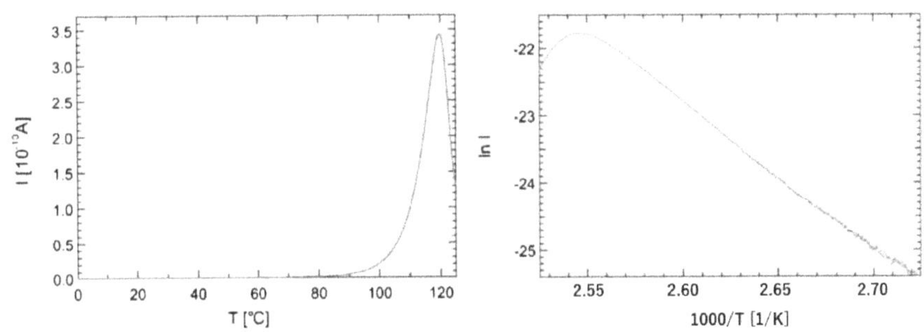

(a) (b)

Figure 3.10. Example of initial rise method applied for analyzing activation energy H from TSC curve. Dipole relaxation of electro-optic (EO) polymer, *i.e.*, poly(methyl methacrylate)-*co*-disperse red 1 (DR1) was evaluated. The sample thickness and electrode area are $3\,\mu\mathrm{m}$ and $20\,\mathrm{mm}^2$, respectively. TSC (a) and the activation energy $H = 1.91\,\mathrm{eV}$ of DR1 in copolymer is obtained from the slope of $\ln TSC(T) - 1/T$ plot (b). Copyright (2014) The Japan Society of Applied Physics [15].

an example of the determination of activation energy H due to dipolar polarization of dye molecules (dispersion red 1: DR1) in electro-optic (EO) polymer by using TSC, where the initial rise method was employed [15].

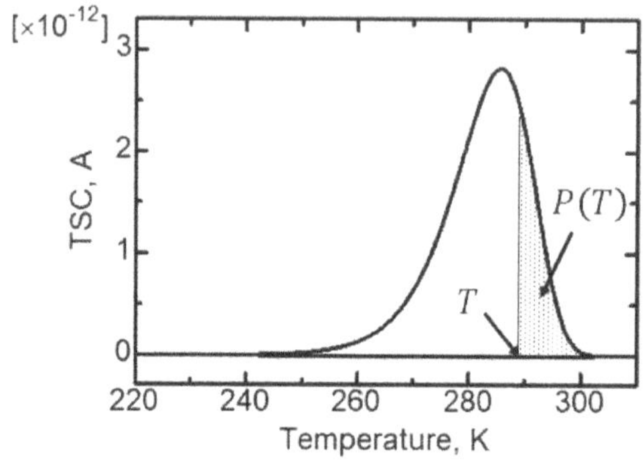

Figure 3.11. Polarization $P(T)$ remaining at temperature T during TSC measurement.

(2) Whole-curve method

In this method, we pay attention to the whole curve of TSC. TSC is the MDC current given by Eq. (3.3) with $T = \beta t + T_0$. Using Eq. (3.3) with Eq. (3.2), the following relation is derived

$$TSC(T) = -\beta \frac{dP}{dT} = \frac{P(T)}{\tau(T)} \tag{3.9}$$

$P(T)$ is the polarization at temperature T, and it represents the polarization remained in TSC samples at temperature T. With taking account of the relation given by Eq. (3.6), $P(T)$ can be described as follows (see Fig. 3.11)

$$P(T) = \frac{1}{\beta} \int_{T}^{T_\infty} TSC(T)dT. \tag{3.10}$$

Using Eqs. (3.1), (3.9) and (3.10), the following relation is obtained.

$$\tau(T) = \tau_0 \exp\left(+\frac{H}{kT}\right) = \frac{\frac{1}{\beta}\int_{T}^{T_\infty} TSC(T)dT}{TSC(T)} \tag{3.11}$$

From Eq. (3.11), we get the following relation

$$\ln \tau(T) = \ln \tau_0 + \frac{H}{kT} = \ln\left(\frac{\frac{1}{\beta}\int_{T}^{T_\infty} TSC(T)dT}{TSC(T)}\right). \tag{3.11$'$}$$

In the whole curve method, we plot the relationship given by

$$\ln\left(\frac{\frac{1}{\beta}\int_T^{T_\infty} TSC(T)dT}{TSC(T)}\right) \sim \frac{1}{T},$$

and evaluate the activation energy H from its slope (Fig. 3.12(c)). The plot of Fig. 3.12(c) is sometimes called Bucci plots after the name of Bucci who proposed this method [16]. Figures 3.12(a)–3.12(c) are computer simulations of TSC curve and its analysis using the whole curve method. Figure 3.12(d) is an example of the determination of the activation energy from the experimental TSC, by using the whole curve method [17].

(3) $\ln T_m^2/\beta \sim 1/T_m$ plot
A TSC curve gives one peak at a temperature of T_m as displayed in Fig. 3.7, and we can easily find the TSC peak temperature T_m in actual TSC measurements. As there is a relation given by Eq. (3.5) at this TSC peak temperature, we pay attention to the shift of peak temperature T_m caused by the change of the heating rate β. From Eq. (3.5), the following relation is derived

$$\frac{H}{kT_m} = \ln\frac{T_m^2}{\beta} + \ln\frac{k}{\tau_0 H}. \tag{3.12}$$

Therefore, by plotting the experimental relationship between $\ln T_m^2/\beta$ and $1/T_m$, the activation energy H is estimated from the slope $+H/k$ of this relationship. Figures 3.13(a) and (b) is an example of computer calculation for analyzing TSC by $\ln T_m^2/\beta \sim 1/T_m$ plot, where heating rate β was varied from 0.1 K/min to 1.2 K/min. Figure 3.13(c) shows an example of the experimental determination of the activation energy $+H$ on the basis of $\ln T_m^2/\beta \sim 1/T_m$ plots [18]. The activation energy $H = 0.19\,\text{eV}$ was obtained for internal polarization of Al–SiO2–Al structure device.

(ii) Polarization P and dipole moment μ
Equation (3.6) suggests that TSC curves are available for the evaluation of the amount of initial polarization P_0 formed in films by biasing. In the same way, TSC curves are available for the estimation of polarization $P(T)$ remained in the film at temperature T,

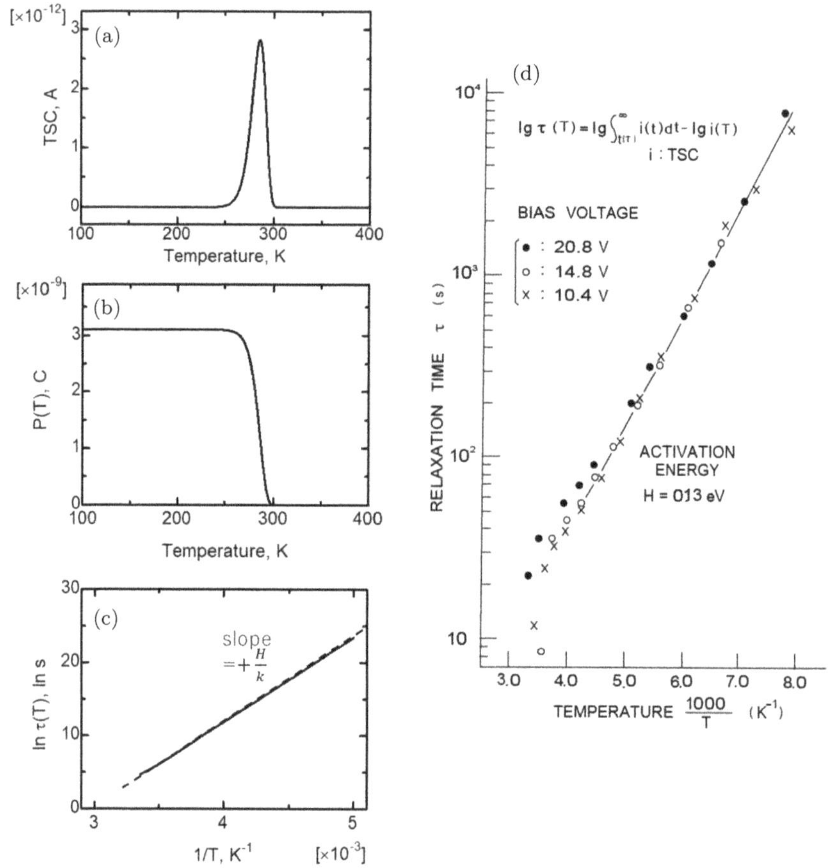

Figure 3.12. Whole curve method (Bucci plot) of TSC for analysis of activation energy H from TSC curve. (a) TSC calculated with parameters $P_0 = 1.0 \times 10^{-7}\ \mathrm{C/cm^2}$, $\tau_0 = 1.0 \times 10^{-15}$ s, $H = 1\,\mathrm{eV}$, $T_0 = 100\,\mathrm{K}$, $\beta = 1\,\mathrm{K/min}$. Electrode area $3.1\,\mathrm{mm}^2$ is assumed. (b) $P(T)$ from TSC. (c) $\tau(T)$–$1/T$ curve used for determining activation energy H by using whole curve method. The $\tau(T)$ – $1/T$ curve follows straight line with a slope $+H/k$. (d) Example of whole curve method. Copyright (1973) The Physical Society of Japan and The Japan Society of Applied Physics [17]. RF sputtered SiO_2 layer in Al–SiO_2–Al structure device was analyzed. The thickness of SiO_2 layer is 7500 Å.

during the TSC measurement (Eq. (3.10)). Further, using these relation Eq. (3.6) and Eq. (3.7), the dipole moment μ of the film material can be estimated if we have the information on the density of the dipole moment N.

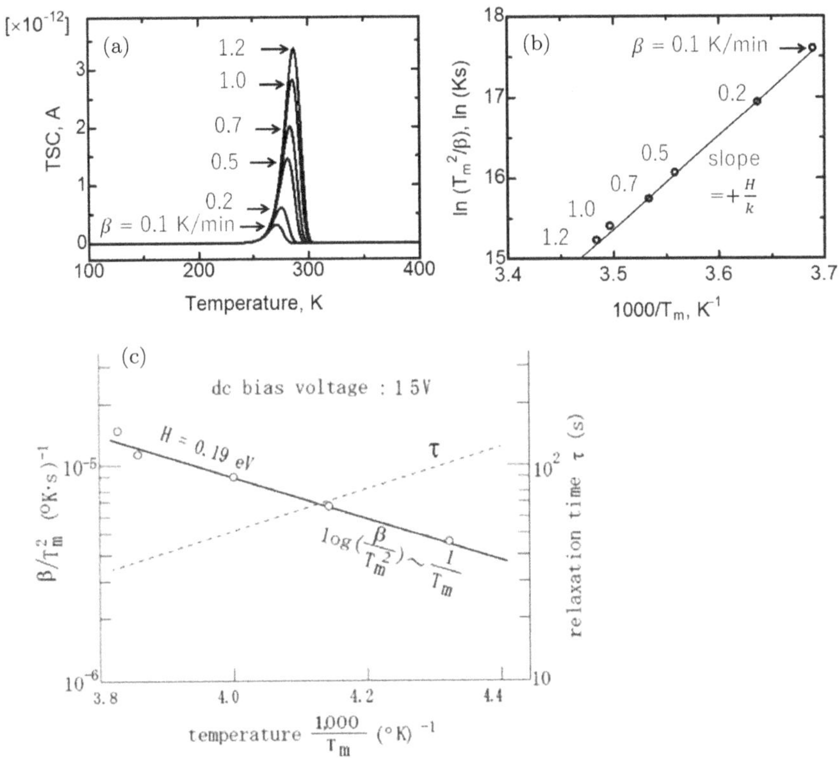

Figure 3.13. $\ln T_m^2/\beta \sim 1/T_m$ relation used for analyzing activation energy H. (a) TSC calculated with $P_0 = 1.0 \times 10^{-7}\,\mathrm{C/cm^2}$, $\tau_0 = 1.0 \times 10^{-15}$ s, $H = 1\,\mathrm{eV}$, $T_0 = 100\,\mathrm{K}$. Electrode area $3.1\,\mathrm{mm^2}$ is assumed. Heating rate β was 0.1–1.2 K/min. (b) $\ln T_m^2/\beta \sim 1/T_m$ relation. (c) Example of analysis of activation energy H of internal polarization in Al–SiO_2–Al structure device. Reprinted by permission from the Institute of Electrical Engineers of Japan (T. Hino, IEEJ Trans. Fundam. Mater., Vol. 93–A, 99, 1973). [18]

(iii) Relaxation time τ

Relaxation time τ is given by Eq. (3.1) with two parameters, τ_0 and H. Once we determine the activation energy H by using methods described in Sec. 3.3.2(i), the pre-exponential factor τ_0 is derived using Eq. (3.5). Consequently, we can determine the relaxation time at arbitrary temperature T as follows

$$\tau(T) = \tau_0 \exp\left(+\frac{H}{kT}\right) = \frac{kT_m^2}{\beta H} \exp\left(-\frac{H}{kT_m} + \frac{H}{kT}\right). \tag{3.13}$$

It is instructive to note here that it takes much longer time to estimate the relaxation time at low temperature T, including room temperature, by employing the isothermal measurement. At the low temperature T, relaxation time τ is very long owing to the linear relation between $\ln \tau$ and $1/T$ illustrated in Fig. 3.2. Nevertheless, once we employ TSC measurement and determine the activation energy H and the TSC peak temperature T_m at heating rate β, we can estimate the relaxation time $\tau(T)$ by using the relation Eq. (3.13). This idea is well utilized to evaluate the stability of dielectric materials and the lifetime of devices using the dielectric materials [15].

In this way, TSC measurement is effectively used for characterizing dipolar polarization in dielectric materials. It should be noted here that in actual TSC measurements, we often observe many TSC peaks due to many different sources of dipolar depolarization, whilst heating materials, *e.g.*, polymers. In some cases, overlapped TSC curves appear, but it does not matter if there is no interaction between origins of TSC peaks. We can wisely separate TSC peaks from overlapped TSC curves one by one, with making use well of the relationship between $\ln \tau$ and $1/T$ (see Fig. 3.2), that is, the relaxation proceeds very fast at high temperatures, whereas it proceeds very slow at low temperatures. For the determination of the activation energy distribution of such TSCs with many peaks, partial heating method is well-adopted, where heating and cooling cycles between $T = T_0$ and $T = T_l$ are employed successively, on increasing the temperature T_l ($T_0 < T_l < T_\infty$) step by step. That is, the dipolar depolarization with many TSC peaks is released one-by-one by employing the heating and cooling cycles in TSC measurement, on increasing the temperature T_l step by step. Figure 3.14(a) shows an example of calculated TSC, where TSC1 and TSC2 with activation energy $H_1 = 0.97\,\text{eV}$ and $H_2 = 1.00\,\text{eV}$, are superposed. By increasing T_l from 240 K to 305 K step by step, a series of partial TSCs is obtained (Fig. 3.14(b)). These partial TSCs are evaluated by using, for example, initial rise method where several lines with different slopes are plotted as Fig. 3.14(c). The activation energies are evaluated in Fig. 3.14(d), showing that activation energies are separated into H_1 and H_2. Experimental example for

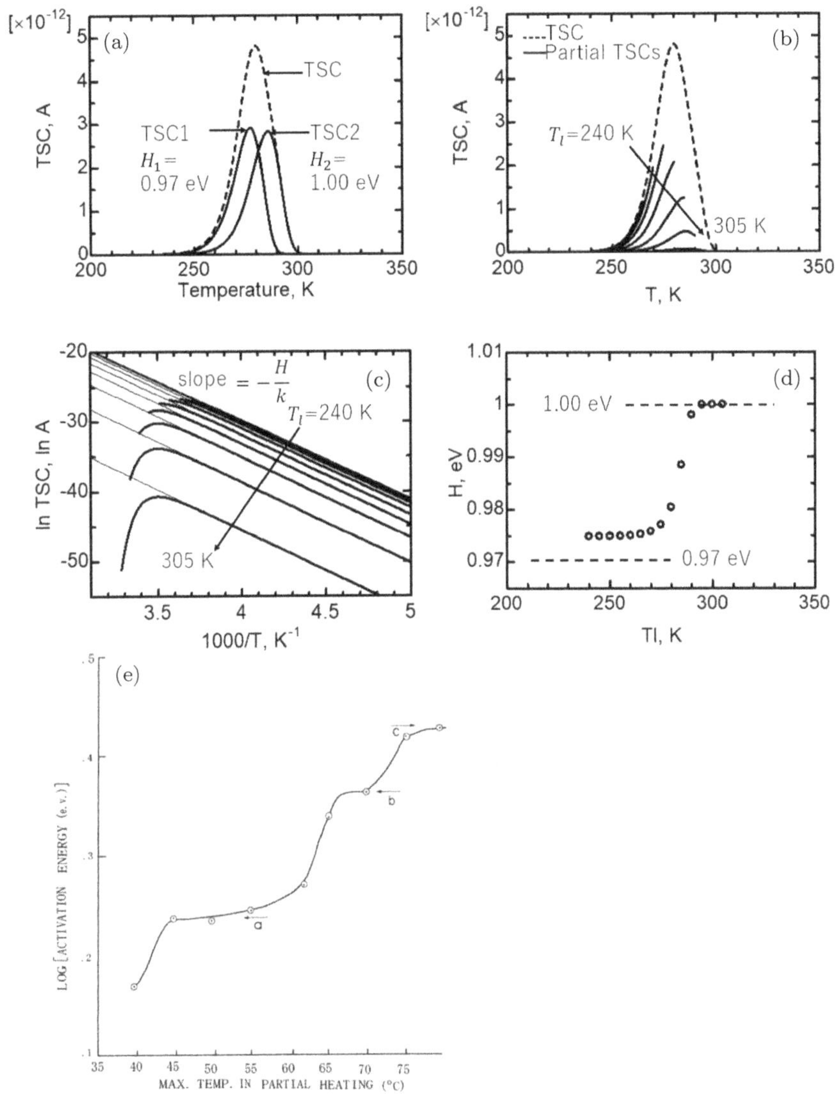

Figure 3.14. Partial heating method used in TSC analysis. Distribution of activation energy H is evaluated. (a) TSC of dipole with activation energy $H_1 = 0.97\,\text{eV}$ (TSC1) and $H_2 = 1.00\,\text{eV}$ (TSC2). (b) Partial heating with $T_l = 240, 245, \ldots, 300, 305\,\text{K}$. (c) Analysis of partial TSCs by initial rise method. (d) Activation energies extracted from partial TSCs. (e) Example of partial heating method. Reprinted from M. M. Perlman, J. Appl. Phys., Vol. 42, 2645, 1971, with the permission of AIP Publishing [19].

analyzing activation energies of Carnauba wax electrets in 0.25 inch (1 inch = 25.4 mm) thickness is shown in Fig. 3.14(e) [19, 20]. Noteworthy that, we can check whether there is interaction between TSC peaks or not. If there is no interaction between partial TSCs, the sum of partial TSCs returns to TSC measured with a single scan from the lowest temperature T_0 to T_∞.

For the discrimination of one TSC peak from overlapped TSC peaks, thermal sampling method can be employed, where the dipolar polarization formed at temperature between T and $T + \Delta T$ ($T_0 < T < T_\infty$) are selectively evaluated. After polarizing the TSC sample with $\tau < \tau(T + \Delta T) = t_b$ at temperature $T + \Delta T$, and the TSC sample is depolarized with $\tau < \tau(T) = t_s$ at temperature T, successively. In this way, polarization with relaxation time $t_s < \tau < t_b'$ is sampled among the total polarization (Fig. 3.15(a)). By increasing T step by step, and carry out TSC measurement for a series of the sampled polarizations, a set of TSC (TS in Fig. 3.15(b)), is obtained. By analyzing TS by using initial rise method, or other method, distribution of activation energy H is obtained. Figure 3.15(b) is an example of the thermal sampling method for analyzing overlapped TSC of polyethyleneterephthalate (PET) (thickness 16 μm) [21, 22]. TSC peak A and peak B is well analyzed by using thermal sampling method, and the distribution of activation energy H is determined by using initial rise method for a series of TS, as shown in Fig. 3.15(c).

3.4 TSC curves due to detrapped charges

The mechanism of TSC generation due to the detrapping of electrons (and holes) is quite different from that of TSC generation due to dipolar depolarization. The detrapping and the succeeding carrier transfer process are quite complex. It depends on the charging process of carrier traps, *i.e.*, trap filling process, which is carried out before TSC measurement. If traps are filled with carriers injected from electrodes by applying d.c. biasing voltage, trapped carriers serve as excess charges and they form a space charge field in film (see Fig. 3.16(a)). On the other hand, if traps are filled with excited electrons in films by radiation to the films such as by UV, visible light, γ-ray, *etc.*, no space charge field is formed in films (see Fig. 3.16(b)),

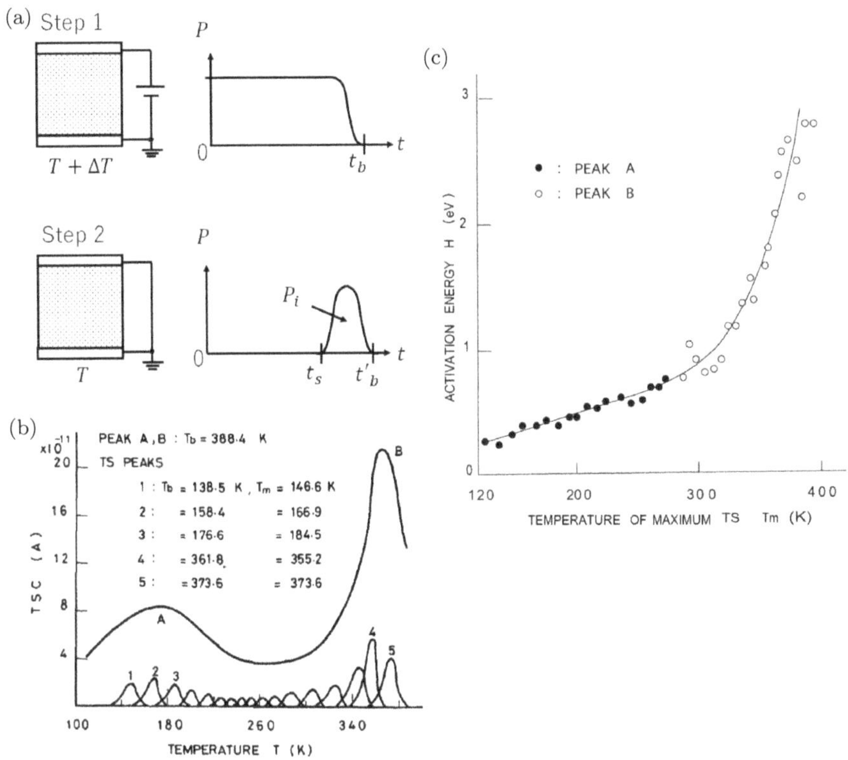

Figure 3.15. Thermal sampling (TS) method in TSC. Distribution of activation energy H is analyzed. (a) Experimental sequence in TS, that is used to take samplings of polarization P_i with single relaxation time τ_i ($t_s < \tau_i < t'_b$) among distributed polarizations: Step 1, under d.c. biasing at $T = T_b$ for time t_b, polarizations with distributed relaxation time $0 < \tau < t_b$ is formed; Step 2, by short circuiting two electrodes at $T = T_s$ ($< T_b$) for time t_s, polarizations with relaxation time $0 < \tau < t_s$ depolarize to zero. The polarization with a single relaxation time τ_i is sampled, and is analyzed by TSC. (b) Example of TS experiment of Polyethyleneterephthalate (PET) film, 16 μm in thickness, where TS is taken to analyze distribution of activation energy H. (c) The distribution of activation energy H analyzed by using TS. Reprinted from T. Hino, J. Appl. Phys., Vol. 46, 1956, 1975, with the permission of AIP Publishing [22].

because the same number of holes will remain inside of the film after the excitation of electrons.

If traps are filled with applying d.c. biasing, the space charge field formed by injected carriers significantly gives effect on the transport

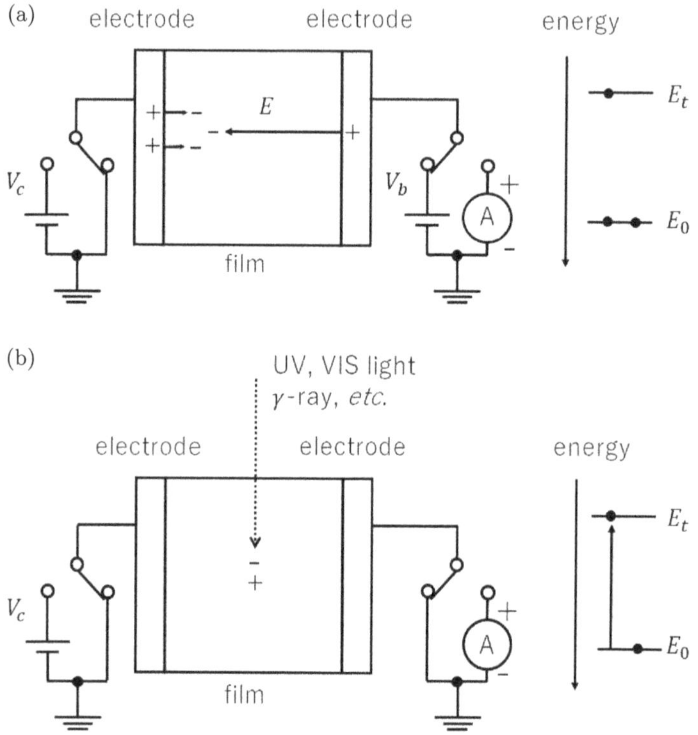

Figure 3.16. Trap filling process in TSC. (a) Carrier injection by using external d.c. biasing. Injected carriers are trapped as excess charges in film. In the figure, three electrons, indicated by "−" are trapped in a space between two electrodes, leaving three positive charges "+" on electrodes. (b) Excitation by using ultraviolet (UV) light and other light sources. A pair of positive and negative charges are generated in film. Subsequently, electron is trapped in film, but no space charge field is formed. In (a) and (b), the energetics of trapped electrons is illustrated in energy space, where E_t is trap energy level and E_0 is lower energy level. Trapped electron at E_t is transferred from electrode by applying external d.c. biasing ((a)), whereas trapped electron is supplied from lower energy level E_0 by excitation with UV, visible (VIS) light, γ-ray, *etc.* ((b)). After these trap filling processes, two electrodes are used to apply collecting voltage V_c during heating for TSC measurement.

of thermally released electrons (and holes), as described in Sec. 3.4(i). If traps are filled with excitation, we need to pay attention to carrier processes in energy space, as described in Sec. 3.4(ii). In the followings, we discuss these two cases of TSCs due to detrapped electrons.

Noteworthy that there is a possibility that excited electrons in metals move into the traps in films. In this case, space charge field is formed in the films by trap filling with excitation, but we do not consider such cases here.

(i) TSC due to detrapped electrons in the presence of space charge field

(i–1) Space charge field due to trapped electrons and TSC

Let us consider a case when a film is sandwiched between two identical parallel electrode 1 and electrode 2 as illustrated in Fig. 3.17(a), and empty traps in the film are filled with electrons injected from the electrode 1 by d.c. biasing. In other words, merely electrons are injected from the electrode 1, and then trapped in the film. After electrically shortening the film at low temperature, an electrostatic potential hill is built in the film at position $x = x^*$, as illustrated in Fig. 3.17(b), owing to the presence of excessive trapped charges. This position $x = x^*$ is called as zero-field position [23]. By solving the Poisson's equation $\nabla^2 V(x) = -\rho_t(x)/\varepsilon_s\varepsilon_0$ with the boundary condition $V(0) = V(d) = 0$, the potential $V(x)$ at $x = x^*$ is derived as [24]

$$V(x^*) = \frac{d - x^*}{d}\int_0^{x^*} x\frac{\rho_t(x)}{\varepsilon_s\varepsilon_0}dx + \frac{x^*}{d}\int_{d-x^*}^{d}(d - x)\frac{\rho_t(x)}{\varepsilon_s\varepsilon_0}dx. \quad (3.14)$$

$eV(x^*)$ (e: elementary charge 1.6×10^{-19} C) is the height of potential hill built in the film by trapped electrons. Released carriers from traps can move along the local electric field, and the direction of carrier transfer in film depends on the position of the carriers released from traps, with respect to the position of the potential hill ($x = x^*$). That is, electrons thermally released from traps located in the region between $x = 0$ and $x = x^*$ move in the direction to electrode 1 ($x = 0$), whereas electrons thermally released from traps located in the region between $x = x^*$ and $x = d$ move in the direction to electrode 2 (see the bottom panel of Fig. 3.17(b)). On electrodes 1 and 2, charges Q_1 and Q_2 are induced owing to the presence of trapped

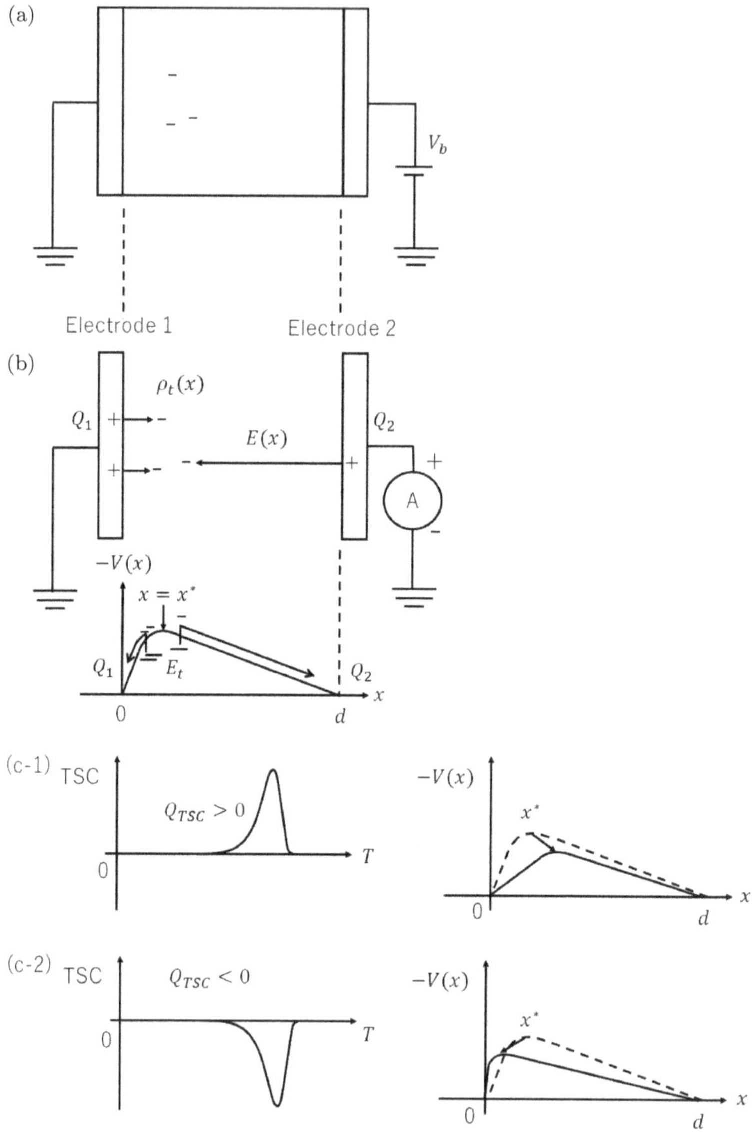

Figure 3.17. TSC of trapped charge after filling traps by using external voltage biasing. (a) As an example, electrons indicated by "−" are injected from electrode 1, and trapped, under d.c. biasing. (b) Electrostatic field and potential in TSC measurement. $x = x^*$ is zero-field position. (c–1) Positive TSC and (c–2) negative TSC, and the corresponding motion of zero-field position x^* during heating.

carriers in film. As an extension of Eq. (2.6), they are derived as

$$Q_1 = -\int_0^d \frac{d-x}{d}\rho_t(x)dx, \text{ and } Q_2 = -\int_0^d \frac{x}{d}\rho_t(x)dx. \tag{3.15}$$

Here, $\rho_t(x)$ is trapped charge density at $x = x$. As illustrated in Fig. 3.17(a), the direction of the space charge field in the region between $x = 0$ and $x = x^*$ is opposite to that in the region between $x = x^*$ and $x = d$. That is, the space charge field originating from electrode 1 converges on trapped charges located in the region between $x = 0$ and $x = x^*$, whereas the space charge field originating from electrode 2 converges on trapped charges located in the region between $x = x^*$ and $x = d$. Therefore, Eq. (3.15) can be rewritten as follows;

$$Q_1 = -\int_0^{x^*} \rho_t(x)dx \text{ and } Q_2 = -\int_{x^*}^d \rho_t(x)dx. \tag{3.16}$$

If the release of trapped electrons is from single-electron trap state with a trap depth of E_t, and this release is limited by thermal emission process, the number of trapped electrons changes on satisfying the following rate equation.

$$\frac{dn_t(x,T)}{dt} \approx -\frac{n_t(x,T)}{\tau_t}, \tag{3.17}$$

where $\rho_t = en_t$ and τ_t is detrapping time given by

$$\tau_t = \frac{1}{\nu}\exp\left(+\frac{E_t}{kT}\right)$$

with attempt to escape frequency ν. Noteworthy in the thermal emission limited process, once electrons are thermally released from the single trap with a time of τ_t, these electrons move to electrodes very quickly along the space charge field in time less than τ_t, and recombine with charges on electrodes, and finally disappear. Equation (3.17) is very similar to Eq. (3.2) for dipolar depolarization, and the following equation is derived for trapped electron density n_t.

$$n_t(x,T) = n_t(x,T_0)\exp\left[-\frac{\nu}{\beta}\int_{T_0}^T \exp\left(-\frac{E_t}{kT}\right)dT\right]. \tag{3.18}$$

Equation (3.18) suggests that the carrier density decays in the same way over the entire region between $x = 0$ and $x = d$. Consequently, the zero-field position x^* never moves during the release of electrons from traps, *i.e.*, $x^*(T) = x^*(T_0)$, and the potential $V(x^*)$ at temperature T given by Eq. (3.14) decays as follows:

$$V(x^*(T)) = V(x^*(T_0)) \exp\left[-\frac{\nu}{\beta}\int_{T_0}^{T} \exp\left(-\frac{E_t}{kT}\right) dT\right]. \tag{3.19}$$

MDC is the transient current that flows due to the change of induced charges Q_1 and Q_2 on electrodes with time. TSC is a special case of the MDC and it flows during heating with a heating rate β. In this case, the change of induced charges Q_1 and Q_2 on electrodes is due to recombination with electrons which are thermally released from traps and conveyed to electrodes 1 and 2. Therefore, using from Eqs. (3.3) and (3.16), we can get the following relation

$$TSC(T) = -\beta\frac{d}{dT}Q_1 = \beta\frac{d}{dT}Q_2 = \beta\rho_t(x^*(T))\frac{dx^*(T)}{dT}. \tag{3.20}$$

That is, TSC current flows according to the movement of zero-field position $x^*(T)$ (Fig. 3.18(a)), and no TSC flows if the zero-field position does not move during heating (Fig. 3.18(b)). Consequently, no TSC flows in short-circuit when the thermal emission process governs the detrapping of electrons in film (thermal emission limited process). To observe TSC generated due to thermal emission limited process by using short-circuit, complementary techniques are necessary (see Sec. 3.4(i–2)). On the other hand, it is found from Eq. (3.20) that total TSC charge Q_{TSC} recorded by the TSC measurement is given by

$$\begin{aligned} Q_{TSC} &= \frac{1}{\beta}\int_{T_0}^{T_\infty} TSC(T)dT = \int_{T_0}^{T_\infty} \rho_t(x^*(T))\frac{dx^*(T)}{dT}dT \\ &= \int_{x^*(T_0)}^{x^*(T_\infty)} \rho_t(x^*(T))dx^*(T) \leq Q_t. \end{aligned} \tag{3.21}$$

That is, Q_{TSC} is given by the small triangular area in the bottom panel of Fig. 3.18(a), and it is dependent on the movement of zero-field position in the films. The total trapped charge Q_t in the film is

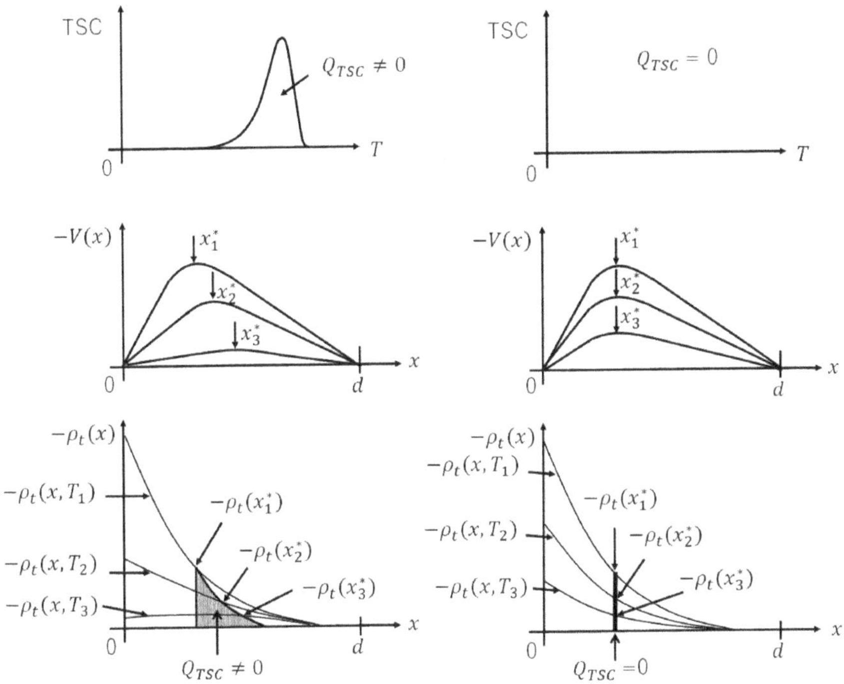

Figure 3.18. TSC of trapped electrons after filling traps by carrier injection from external electrode. Total TSC charge Q_{TSC} is given by an area swept by charge density $\rho_t(x^*)$ at zero-field position, and it is always smaller than total trapped charge Q_t. (a) Positive non-zero TSC flows meanwhile zero-field position x^* moves toward electrode 2, during heating. The shift of x^* is indicated by arrows labeled with x_1^*, x_2^*, and x_3^*, at temperature T_1, T_2, and T_3 ($T_1 < T_2 < T_3$), respectively. Negative TSC flows when the zero-field position moves toward electrode 1. (b) TSC current is zero during heating when the zero-field position x^* never moves. This is why no TSC flows when thermal emission governs charge detrapping process. Q_t is non-zero but Q_{TSC} is zero.

given by an area below $-\rho_t(x, T_1)$ in Fig. 3.18(a), and is expressed as

$$Q_t = \int_0^d \rho_t(x)dx.$$

Obviously, total TSC charge Q_{TSC} is always smaller than the total charge Q_t.

The movement of zero-field position is one principal reason of generation of negative TSC peaks, for example, peak C in Fig. 3.4.

In more detail, if the discharging process is limited by thermal emission process given by Eq. (3.18), detrapped electrons move to electrodes 1 and 2 immediately in the manner as we described earlier, and then recombine with charges on electrodes 1 and 2. That is, trapped charges and induced charges on electrodes decrease at the same rate. Consequently, no TSC flows because zero-field position $x^*(T)$ does not move, even when trapped charge distribution in film sandwiched between two electrodes is asymmetric in the film thickness direction. In other words, thermal detrapping mechanism that leads to the movement of zero-field position $x^*(T)$ is necessary to generate non-zero TSC. One of the possible mechanisms that gives non-zero TSC is the drift limited process. In this process, carriers are thermally detrapped according to Eq. (3.17) in the same way with the thermal emission limited process, but the detrapped carriers do not move to electrodes immediately. As that space charge field caused from detrapped charges in film directly give effect on the movement of detrapped carriers and the succeeding recombination process on electrodes. Accordingly, there is a possibility that the zero-field position $x^*(T)$ moves from side to side depending on the spatial distribution of trapped charges and detrapped charges in film, when trapped charge distribution is asymmetric in film thickness direction. Consequently, non-zero TSC is generated, but the direction of TSC flow is governed by the movement of the zero-field position (Eq. (3.20)). It is instructive to note here that asymmetric trapped charge distribution in film thickness direction is formed after d.c. stress biasing due to the difference of hole and electron injection from facing two electrodes, trapping of homo-charges around the injection electrode by electron-beam radiation, and so forth. Keeping into mind such carrier trapping in practical experiments, we can model asymmetric trapped charge distribution in film.

Noteworthy that in the case carrier trapping is energetically asymmetric in film direction, due to the presence of multi-electron trapping states, there is a possibility of the movement of zero-field position. For example, on one side of film with merely shallow trapping and on the other side of film with shallow and deep trapping. In such case, trapped electrons in film are thermally excited according to Eq. (3.17) from the shallow and deep trapping states, but the detrapping from the shallow traps completes at lower temperature in comparing with the detrapping from deep traps. Consequently, the movement of zero-field position x^* is anticipated in film, even when thermal emission process governs the carrier process in film.

As illustrated in Figs. 3.17(c–1) and (c–2), positive TSC flows when zero-field position x^* moves from electrode 1 to electrode 2, while negative TSC flows when zero-field position x^* moves from electrode 2 to electrode 1. This is a basic relationship between the generation of non-zero TSC and the movement of zero-field position x^*, and modeling of the movement of zero-field position x^* based on experimental evidence is one way for analyzing electrical transport properties of organic materials.

(i–2) TSC analysis

(i–2a) Estimation of total trapped charge Q_{total} and centroid of trapped charge distribution $\bar{x}$

In the TSC analysis due to carrier detrapping, the evaluation of total trapped charge Q_{total}, trapped charge distribution in film, determination of trap depth E_t, *etc.*, are important. And it is helpful if we can evaluate them even when thermal emission process governs the detrapping of electrons and no TSC flows in a closed short-circuit. To do this, it is necessary to apply a non-zero external d.c. electric field $E_{collect}$ to the film during heating in the TSC measurement by applying a collecting voltage $V_{collect}$. By applying this non-zero external electric field $E_{collect}$, the balance of the rate of discharging Q_1 and Q_2 on electrodes due to the recombination with transferred electrons is broken, because the zero-field position shifts from $x = x^*(E_{collect} = 0)$ to $x = x^*(E_{collect} \neq 0)$ by the application of the additional electric field $E_{collect}$ at temperature $T = T_0$, and electrons trapped in the region between $x = x^*(E_{collect} = 0)$ and $x = x^*(E_{collect} \neq 0)$ must change its direction to be conveyed after releasing from the traps. Figure 3.19 illustrates this situation. In short-circuit condition (see Fig. 3.19(a)), zero-field position is located at $x = x^*$ with $E_{collect} = 0$. As described in Sec. 3.4(i–1), trapped electrons in the region between $x = 0$ and x^* connect the diverging

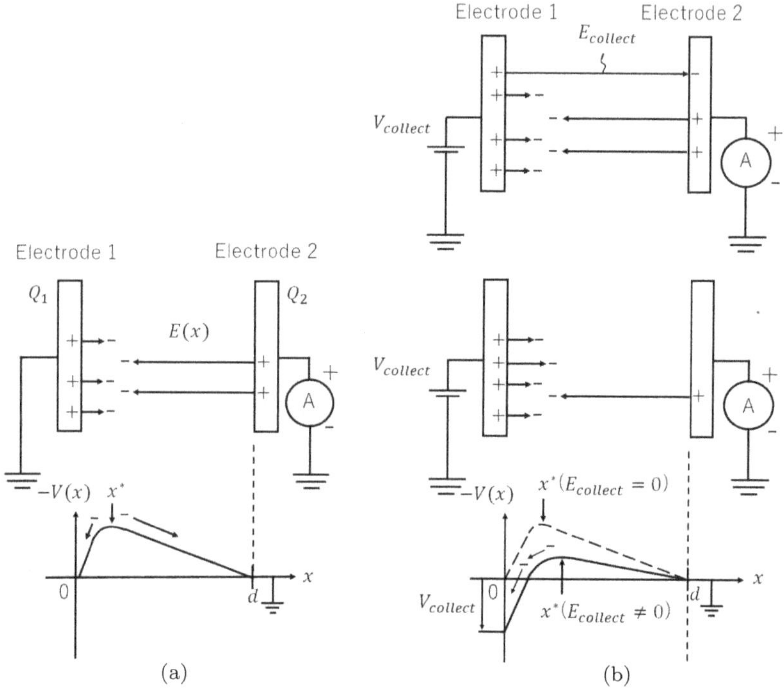

Figure 3.19. Analysis of trapped charge amount by using non-zero collecting electric field $E_{collect}$. (a) Detrapping of electrons with $E_{collect} = 0$ never generates TSC when thermal emission process proceeds. (b) Detrapping of electrons with nonzero collecting field $E_{collect} \neq 0$ is utilized to generate TSC to analyze trapped charge distribution. By shifting zero-field position from the original position in short-circuit condition, non-zero TSC flows in the closed circuit, even in the case where thermal emission governs carrier detrapping process.

electric field to electrode 1, while electrons between $x = x^*$ and d connect the diverging electric field to electrode 2. In more detail, modeling this situation using Fig. 3.19, where 5 electrons (indicated by "–" in Fig. 3.19) are assumed to be involved in the detrapping process during heating, we can illustrate the carrier motion. The carrier motion under the condition with $E_{collect} = 0$, illustrated in Fig. 3.19(a), is as follows: three negative charges (electrons) on the left-hand side in the figure can move to electrode 1 and two electrons on the right-hand-side can go to electrode 2, and they recombine with electrode charges (three positive charges on electrode 1 and two positive charges on electrode 2) and disappear on electrodes 1 and 2, respectively. As that, in this process, no TSC flows. By contrast, with non-zero collecting field $E_{collect} \neq 0$ (see Fig. 3.19(b)), positive collecting voltage ($V_{collect} > 0$) freshly induces positive charges on electrode 1 and negative charges on electrode 2, forming external field $E_{collect}$ pointing in the direction from electrode 1 to electrode 2 (top panel of Fig. 3.19(b)). In the way as we derived from Eq. (3.16), not from Eq. (3.15), this situation is illustrated as in a middle panel of Fig. 3.19(b). That is, applying collecting field $E_{collect}$ shifts the zero-field position from $x^*(E_{collect} = 0)$ to $x = x^*(E_{collect} \neq 0)$. In the figure, the zero-field position shifts in the direction from electrode 1 to electrode 2. Now four electrons on the left-hand-side connect the electric field to electrode 1, while one electron on the right-hand-side connects the field to electrode 2. That is, in comparing with the situation with $E_{collect} = 0$, one electron located between $x^*(E_{collect} = 0)$ and $x^*(E_{collect} \neq 0)$ changes its direction of the electric filed connection from electrode 2 to electrode 1. In this situation with $E_{collect} \neq 0$, during discharging process, four electrons go to electrode 1. One electron goes to electrode 2. Among the four electrons, three of them recombine with electrode charges (three positive charges) on electrode 1 in the same manner as discharging process with $E_{collect} = 0$, meanwhile remaining one electron flows through an ammeter, and recombines with induced charge on electrode 2. In other words, electrons located between $x^*(E_{collect} = 0)$ and $x^*(E_{collect} \neq 0)$ flows through the ammeter during a course of discharging. Consequently, non-zero charge given by Eq. (3.22) flows in the TSC measurement, though this charge is a part of total charge Q_{total} trapped in films.

$$Q_{TSC} = \int_{x^*(E_{collect}=0)}^{x^*(E_{collect}\neq 0)} \rho_t(x(T_0))dx < Q_{total}. \tag{3.22}$$

Noteworthy that when the $x = x^*(E_{collect} \neq 0)$ reaches the position of electrodes, *i.e.*, $x = x^*(E_{collect} \neq 0) = 0$ or $x = x^*(E_{collect} \neq 0) = d$, all charges released from traps must move to electrodes 2 (or 1). Accordingly, by choosing $E_{collect}$ appropriately, we can measure Q_1 and Q_2, respectively. Figure 3.20 shows an example of the simulated and experimental measurement of Q_1 and Q_2, by applying

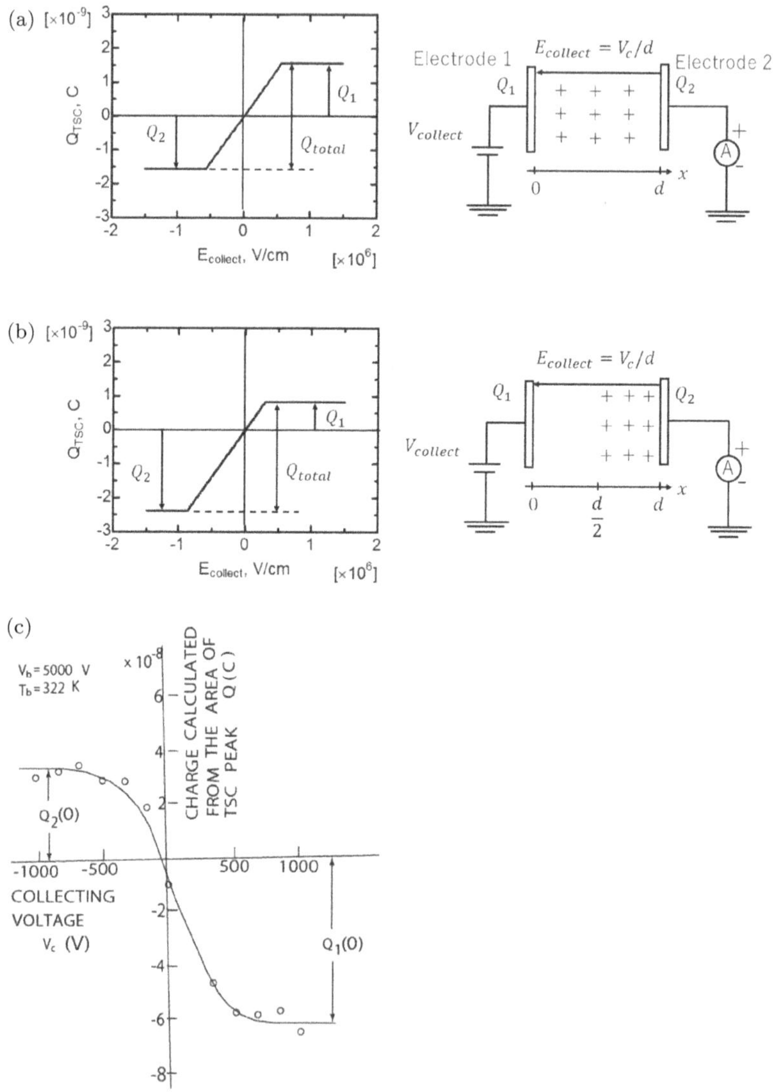

Figure 3.20. TSC charge at various collecting field $E_{collect}$. (a) Q_{total} is homogeneously trapped in the film between electrode 1 ($x = 0$) and electrode 2 ($x = d$). (b) Q_{total} is distributed in the region between $x = d/2$ to d. $\varepsilon_s = 1$, $Q_{total} = 10^{-7}\,\mathrm{C/cm^2}$, electrode area $3.1\,\mathrm{mm^2}$ are assumed. $E_{collect} = V_{collect}/d$ is -1.5 MV/cm to 1.5 MV/cm ($V_{collect}$ was -15 to 15 V applied to film with thickness $d = 100$ nm). (c) Example of TSC charge of polyethylene film. The film thickness $40\,\mu$m and $V_{collect} = 1000\,\mathrm{V}$, corresponding to collecting electric field $E_{collect} = 0.25\,\mathrm{MV/cm}$. Reprinted by permission from the Institute of Electrical Engineers of Japan [25a].

various $E_{collect}$, where we can find the saturation of TSC charge at Q_1 and Q_2. As the amount of Q_1 and Q_2 is governed by distribution of trapped charges, two examples are simulated in Figs. 3.20(a) and (b), where whole region of the film is filled with positive trapped charges and half region of the film near electrode 2 is filled with positive trapped charges, respectively. The total trapped charge Q_{total} is assumed the same, but saturated values Q_1 and Q_2 are different as seen in Figs. 3.20(a) and (b) of TSC measurement. Using Q_1 and Q_2, we can determine the total charge Q_{total} trapped in the film as follows (see Fig. 3.20):

$$Q_{total} = -(Q_1 + Q_2). \tag{3.23}$$

Also, the centroid of the trapped charge distribution $\bar{x}$ is determined using the experimental values of Q_1 and Q_2. That is, using the Eq. (3.15),

$$\bar{x} = \frac{\int_0^d x\rho_t(x)dx}{\int_0^d \rho_t(x)dx} = d\frac{Q_2}{Q_1 + Q_2}. \tag{3.24}$$

Figure 3.20(c) is an example of TSC carried out for polyethylene film (thickness: 40 μm) [25]. It is noteworthy that the application of $E_{collect}$ is mainly to change the direction of released electrons from traps, and not to convey the released electrons. In other words, the role of $E_{collect}$ is to alter the electric field distribution in film.

(i–2b) TSC analysis and the determination of trap parameters
In the TSC process limited by thermal emission, $\rho_t(x, T_0)$ decays on satisfying the relation Eq. (3.18), $\rho_t(x, T) = en_t(x, T)$. Therefore, TSC is derived with taking account of the discussion on Eq. (3.2) as follows

$$TSC(T) = \beta\frac{d}{dT}\left\{\int_{x^*(E_{collect}=0)}^{x^*(E_{collect}\neq 0)} \rho_t(x(T_0))dx \cdot \exp\left[-\frac{\nu}{\beta}\int_{T_0}^{T} \exp\left(-\frac{E_t}{kT}\right)dT\right]\right\} \tag{3.25}$$

$$= A\nu \exp\left(-\frac{E_t}{kT}\right) \exp\left[-\frac{\nu}{\beta}\int_{T_0}^{T} \exp\left(-\frac{E_t}{kT}\right) dT\right] \quad (3.25')$$

with $A \equiv \int_{x^*(E_{collect}=0)}^{x^*(E_{collect}\neq 0)} \rho_t(x(T_0))dx \leq Q_{total}.$

The form of Eq. (3.25) is very similar to Eq. (3.4) for dipolar depolarization, except that P_0 is replaced by A ($\leq Q_{total}$). Consequently, the trap depth E_t is evaluated by using the same method. Initial rise method, whole curve method, and $\ln T_m^2/\beta \sim 1/T_m$ plots are again typical ones. Experiments are carried out using these relations as shown in Figs. 3.21 and 3.22 [26, 27]. Noteworthy that in the derivation of TSC given by Eq. (3.25), we assumed that released carriers never go back to the initial trap states and they move to the electrodes along the space charge field.

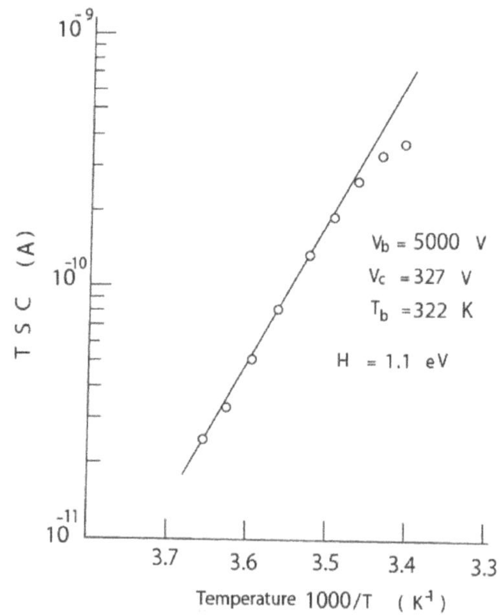

Figure 3.21. Example of TSC analysis using initial rise method for evaluating trap energy depth E_t. Reprinted by permission from the Institute of Electrical Engineers of Japan (T. Hino, K. Yamashita, IEEJ Trans. Fundam. Mater., Vol. 95, 79, 1975) [26]. Peak C of TSC of polyethylene in Fig. 3.4 is analyzed.

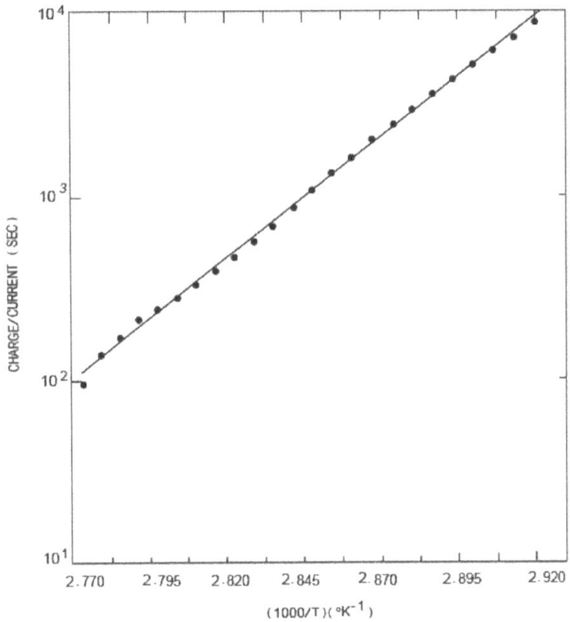

Figure 3.22. Example of TSC analysis using whole curve method for evaluating trap energy depth E_t. Reprinted from M. M. Perlman, R. A. Creswell, J. Appl. Phys., Vol. 42, 531, 1971, with the permission of AIP Publishing [27]. Corona-charged Mylar film is analyzed.

(ii) TSC due to detrapped electrons in the presence of no space charge field

In previous subsection (i), we discussed the case of TSC due to detrapped electrons in the presence of space charge field. In that case, the trapped charges are excess charges due to injected electrons from electrodes, and they go out from the film by thermal heating, and the film returns to the initial state before d.c. biasing. In other words, we can simply assume that only excess charges, which are injected and then trapped in films by d.c. biasing, contribute to the TSC. Based on the idea, we discussed the TSC under the assumption that thermally released carriers are conveyed swiftly along the space charge field to electrodes to recombine with charges on electrodes, while without retrapping during the carrier transfer. On the other hand, there are other ways to study carrier traps by filling them with light irradiation *etc.* (Fig. 3.16(b)). In this case electrons are excited and carrier traps are filled with excited electrons, but the same number of holes remains inside the film. As a result, no space charge field is formed in films but the state in film is non-equilibrium and the recombination of excited electrons with holes happens. In that situation, TSC measurement is carried out. In the measurement, we must apply non-zero external electric field E_c to convey detrapped electrons during heating, meanwhile the non-equilibrium state of the film will return to the initial state.

Therefore, we need to discuss the generation of TSC, with consideration of basic physical processes such as electron excitation, recombination, detrapping, retrapping, by using energy diagram. In the followings, after describing the fundamental carrier process in films on the basis of energy diagram, we discuss three-cases of TSCs which are limited by (a) slow retrapping process, (b) fast retrapping process, and (c) high external electric field, respectively.

(ii–1) Fundamental carrier process in films

Let us assume here that a film is modeled using a wide-gap semiconductor as illustrated in Fig. 3.23, where a single trap with a depth of E_t exists with a trap density of N_t, and trap is filled with electrons with density of n_t. E_c and E_v represent the bottom of conduction band and the top of valence band, respectively. In that situation, the fundamental rate equation of trapped carrier density n_t is described

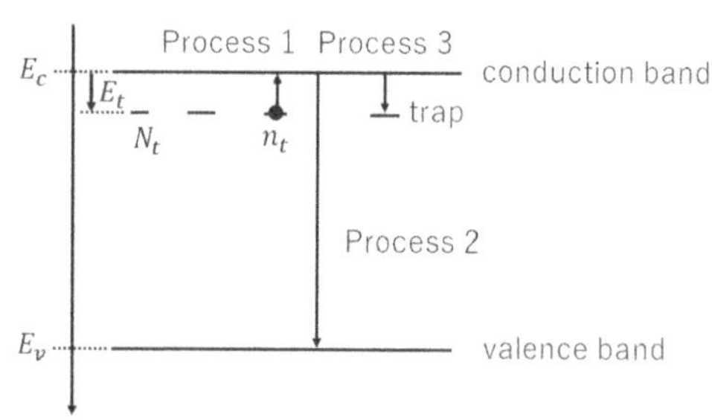

Figure 3.23. Fundamental carrier processes in energy band diagram. E_c: the bottom of conduction band, E_v: the top of valence band, E_t: trap energy level, N_t: trap density, n_t: trapped electron density.

as follows:

$$\frac{dn_t}{dt} = -e_t n_t + n_c(N_t - n_t)\sigma_t v_n. \tag{3.26}$$

Here, n_c represents free electron density in conduction band that is excited from the valence band. e_t is thermal emission rate, σ_t is electron capture cross-section, v_n is thermal velocity of electrons. The first term on the right-hand side represents the decrease of trapped electrons due to thermal emission (Process 1 in Fig. 3.23). On the other hand, the second term on the right-hand side represents the increase of trapped carrier density due to retrapping of excited electrons in conduction band (Process 3 in Fig. 3.23). Consequently, the rate equation of free carrier density in the conduction band is described as follows:

$$\frac{dn_c}{dt} = +e_t n_t - n_c(N_t - n_t)\sigma_t v_n - \frac{n_c}{\tau_l} + \frac{1}{e}\frac{\partial j}{\partial x}. \tag{3.27}$$

On the right-hand-side term, the first term represents the increase of free electron density due to thermal excitation of trapped carriers, the second term represents the decrease of free electron density due to retrapping, and the third term represents the decrease of free electron density due to recombination with holes in valence band (Process 2 in Fig. 3.23), and the last term represents increase of free electron density due to the flow of conductive current. Here τ_l is the lifetime of free electrons. Noteworthy that the conductive current density $j = enu$ (n: carrier density, u: velocity of carriers) is approximately given by $en\Delta x/\Delta t$, and the last term is rewritten as

$$\frac{1}{e}\frac{\partial j}{\partial x} \rightarrow \frac{1}{e}\frac{\Delta j}{\Delta x} = \frac{1}{e}\frac{\Delta\left(en\left(\frac{\Delta x}{\Delta t}\right)\right)}{\Delta x} \approx \frac{1}{e}\frac{\Delta en}{\Delta x}\frac{\Delta x}{\Delta t} = \frac{\Delta n}{\Delta t}.$$

That is, the last term represents the increase of free electron density due to supply of electrons by the conductive current flow.

(ii–1a) Slow retrapping ($\tau_l^{-1} \gg (N_t - n_t)\sigma_t v_n$)
In this process, most of the thermally released electrons from carrier traps prefer to recombine with holes in the valence band and docs not return to the carrier traps. Accordingly, from Eq. (3.26), the rate

equation for trapped electrons is approximately given by

$$\frac{dn_t}{dt} \approx -e_t n_t \tag{3.28a}$$

Therefore, as the solution of Eq. (3.28a), n_t is approximately given by

$$n_t(T) \approx n_t(T_0) \exp\left(-\beta^{-1} \int_{T_0}^{T} e_t dT\right) \tag{3.28b}$$

with $T = T_0 + \beta t$ (T_0: initial temperature). On the other hand, from Eq. (3.27), the rate equation of carrier density n_c in the conduction band is described as follows

$$\frac{dn_c}{dt} \approx +e_t n_t - \frac{n_c}{\tau_l} + \frac{1}{e}\frac{\partial j}{\partial x}. \tag{3.29}$$

The major contribution to the increase of carrier density dn_c/dt is due to the carrier flow that is described by the third term of the right-hand-side of Eq. (3.29). Accordingly, Eq. (3.29) can be approximately given as $dn_c/dt \approx 1/e \cdot \partial j/\partial x$. That is,

$$e_t n_t = \frac{n_c}{\tau_l}. \tag{3.30}$$

On the other hand, thermal emission rate from the trap is

$$e_t = \nu \exp\left(-\frac{E_t}{kT}\right).$$

Therefore, from Eqs. (3.28) and (3.30), we get the carrier density of n_c as follows:

$$\begin{aligned} n_c(T) &\approx n_t(T)\tau_l \nu \exp\left(-\frac{E_t}{kT}\right) \\ &\approx n_t(T_0)\tau_l \nu \exp\left(-\frac{E_t}{kT} - \frac{\nu}{\beta}\int_{T_0}^{T} \exp\left(-\frac{E_t}{kT}\right) dT\right). \end{aligned} \tag{3.31}$$

TSC flows due to the flow of $n_c(T)$ along the bottom of conduction band, and it is given as

$$TSC = en_c u = en_c \mu E_{ext}. \tag{3.32}$$

That is, TSC is observed as the change of effective conductivity, $\sigma = en_c\mu$ of film, where the carriers are conveyed by the external electric field, E_{ext}. In summary, TSC is described as follows:

$$TSC = en_c(T)\mu E_{ext} \approx en_t(T)\mu E_{ext}\tau_l \nu \exp\left(-\frac{E_t}{kT}\right)$$
$$\approx en_t(T_0)\mu E_{ext}\tau_l \nu \exp\left(-\frac{E_t}{kT} - \frac{\nu}{\beta}\int_{T_0}^{T} \exp\left(-\frac{E_t}{kT}\right) dT\right). \tag{3.33}$$

(ii–1b) Fast retrapping ($\tau_l^{-1} \ll (N_t - n_t)\sigma_t v_n$)
In this process, most of the thermally released electrons from carrier traps prefer to retrapping and they return to the carrier traps. Accordingly, from Eq. (3.26) and Eq. (3.27), the rate equation for trapped electrons is given by

$$\frac{dn_c}{dt} = -\frac{dn_t}{dt} - \frac{n_c}{\tau_l} + \frac{1}{e}\frac{\partial j}{\partial x}. \tag{3.34}$$

The major contribution to the increase of carrier density dn_c/dt is again due to the carrier flow that is described by the third term of the right-hand side of Eq. (3.34). Accordingly, (3.34) can be approximately given as $dn_c/dt \approx 1/e \cdot \partial j/\partial x$. That is, we obtain the following relation:

$$\frac{dn_t}{dt} = -\frac{n_c}{\tau_l}. \tag{3.35}$$

On the other hand, as thermally emitted electrons from the trap to the conduction band return to the carrier trap, it is considered that quasi-thermodynamic equilibrium is established between the carrier density of conduction band and the carrier density of trapped carriers. Therefore, the following relationship is satisfied

$$\frac{n_t}{n_c} \approx \frac{N_t}{N_c} \exp\left(+\frac{E_t}{kT}\right). \tag{3.36}$$

Substituting Eq. (3.36) into Eq. (3.35), the following rate equation on the trapped carrier density n_t is satisfied.

$$\frac{dn_t}{dt} = -\frac{n_t}{\tau_l \frac{N_t}{N_c} \exp\left(+\frac{E_t}{kT}\right)}. \tag{3.37}$$

Solving Eq. (3.37), we get the following equation on the trapped carrier density n_t,

$$n_t = n_t(T_0)\exp\left(-\frac{t}{\tau}\right), \text{where } \tau = \tau_l\frac{N_t}{N_c}\exp\left(+\frac{E_t}{kT}\right). \tag{3.38}$$

Therefore, in the TSC process with $T = T_0 + \beta t$ (T_0: initial temperature),

$$n_t(T) \approx n_t(T_0)\exp\left(-\frac{1}{\beta}\frac{N_c}{\tau_l N_t}\int_{T_0}^{T}\exp\left(-\frac{E_t}{kT}\right)dT\right). \tag{3.39}$$

Substituting Eq. (3.39) into Eq. (3.36), we get the carrier density $n_c(T)$ in the conduction band as follows.

$$n_c(T) \approx n_t(T_0)\frac{N_c}{N_t}\exp\left(-\frac{E_t}{kT} - \frac{1}{\beta}\frac{N_c}{\tau_l N_t}\int_{T_0}^{T}\exp\left(-\frac{E_t}{kT}\right)dT\right). \tag{3.40}$$

Using Eq. (3.40), we get the TSC as follows:

$$\begin{aligned} TSC &= en_c(T)\mu E_{ext} \\ &\approx en_t(T_0)\frac{N_c}{N_t}\mu E_{ext}\exp\left(-\frac{E_t}{kT} - \frac{1}{\beta}\frac{N_c}{\tau_l N_t}\int_{T_0}^{T}\exp\left(-\frac{E_t}{kT}\right)dT\right) \end{aligned} \tag{3.41}$$

Eq. (3.41) is very similar to Eq. (3.33). That is, shape of TSC curves is very similar, but the main difference appears in the TSC parameters. For the evaluation of TSC, we need to pay attention to the physical quantities evaluated from the experimental TSC parameters.

(ii–1c) Under high external electric field condition
This process corresponds to the process where detrapped electrons immediately move to electrodes without retrapping and recombination with holes in bulk. This process is similar with the case which we discussed previously on thermally stimulated current in the presence of space charge field. Main difference is that conductive current is always supplied from the electrodes and they are going out from the

counter electrode. The carrier behaviors at the electrodes is different, but we can carry TSC analysis in the same way.

Electrons detrapped from the carrier trap and transport to the electrode 1 move at a distance of $\bar{x}$ on the average, therefore TSC is described as follows:

$$TSC \approx \frac{\bar{x}}{d} e N_t^0 \nu \exp\left(-\frac{E_t}{kT} - \frac{\nu}{\beta}\int_{T_0}^{T} \exp\left(-\frac{E_t}{kT}\right) dT\right)$$

$$\text{where } eN_t^0 \equiv Q_t^0 = \int_0^d \rho_{t0}(x)dx, \text{ and } \bar{x} \equiv \frac{\int_0^d x\rho_{t0}(x)dx}{Q_t^0}. \qquad (3.42)$$

It is instructive here to note that Eq. (3.42) has no dependence on the external high electric field E_{ext}, though we utilize this electric field E_{ext} in the TSC measurement. In other words, the external high electric field E_{ext} is acting only to collect detrapped electrons at the electrode 1.

(ii–2) Estimation of physical quantities

TSC formula derived for the 3 cases, *i.e.*, Eq. (3.33), Eq. (3.41), and Eq. (3.42), are very similar, and they are also very similar to the TSC formula on dipolar depolarization, Eq. (3.4). Therefore, we can estimate the physical parameters corresponding to P_0, H and others, in the same way as that we carry in the analysis of TSC due to dipolar depolarization. For example, initial rise method, partial heating method, $\ln T_m^2/\beta \sim 1/T_m$ plots are available for the determination of trap depth E_t. Figure 3.24 shows an experimental example of TSC analysis for trapped carriers, where the carrier trap depth has been well determined [28].

3.5 TSC due to mobile ions

There are many other TSC sources. Among them are impurity ions, cations, anions and so on. The long-range movement of impurity ions gives rise to a TSC peak that depends on the applied field, while the behavior of cations and anions that compose materials is basically the short-range movement. TSC curves due to the short-range movement of ions is very similar with that of dipolar motion,

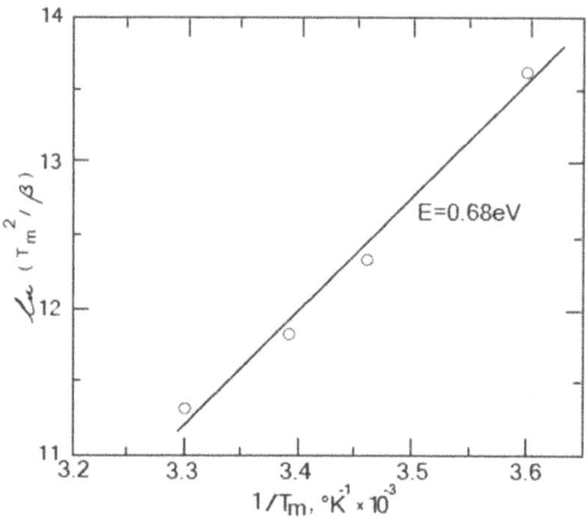

Figure 3.24. Example of TSC analysis using $\ln T_m^2/\beta \sim 1/T_m$ relationship for evaluating trap energy depth E_t. Reprinted from R. H. Bube, G. A. Dussel, Ching-Tao Ho, L. D. Miller, J. Appl. Phys., Vol. 37, 21, 1966, with the permission of AIP Publishing [28]. Anthracene single crystal with 2 mm thickness is analyzed by trap filling using laser excitation.

meanwhile TSCs due to the long-range movement of ions are quite different from that of dipolar motion. We here discuss briefly the TSC due to the impurity ions.

3.5.1 *Ion hopping model*

The movement of impurity ions in films is governed by diffusion and drift, and this movement is generally analyzed by solving the differential equation that describes the continuity of current comprised of the diffusion and the conduction drift currents, with consideration of the Poisson's equation of ions. However, it is not easy to solve the differential equation of ion motions in films. On the other hand, Ion Hopping Model is useful for analyzing ion motions in films. In the followings, we discuss this model, and show that this model describes the diffusion and drift motion of impurity ions.

Figure 3.25 illustrates the Ion Hopping Model used for analysis of ion behaviors in film, where a film is sandwiched between two electrodes 1 and 2 separated with a distance d. In the film, there are excessive impurity ions with the total number of N_t, and these ions

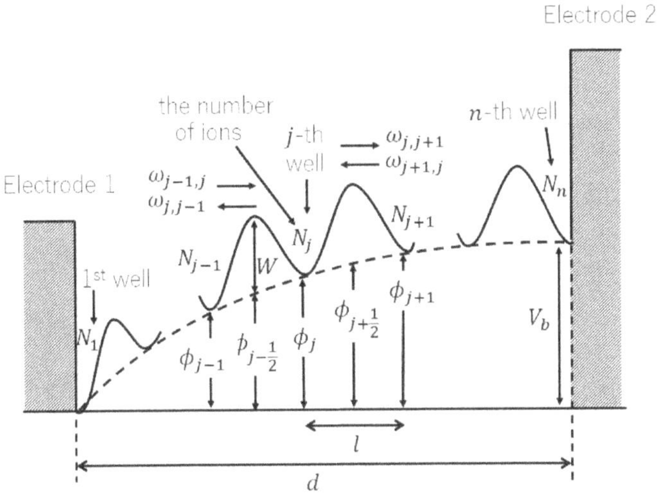

Figure 3.25. Ion hopping model. Ions (total number N_t) are distributed to n potential wells which are separated by distance l, located between electrodes 1 and 2 with the electrode spacing d. N_j is the number of ions in the j-th potential well ($1 \leq j \leq n$) where electrostatic potential is ϕ_j. The ions at j-th well is assumed to hop to $(j-1)$-th and $(j+1)$-th well at probability $\omega_{j,j-1}$ and $\omega_{j,j+1}$, respectively.

form a space charge field in the film. The two electrodes block ions to go out from the film, and there are n potential wells with a spacing of l, that is, $d = (n-1)l$. Between adjacent potential wells, there is a potential hill with a height of W. The N_t ions are distributed into the n potential wells, as illustrated in Fig. 3.25. Ions can move in the film by hopping from one potential well to the adjacent two potential wells. That is, ions in j-th well hops to $(j-1)$-th well and to $(j+1)$-th well. Consequently, this situation is described by the following rate equation.

$$\frac{dN_j}{dt} = \omega_{j-1,j}N_{j-1} - (\omega_{j,j-1} + \omega_{j,j+1})N_j + \omega_{j+1,j}N_{j+1}$$
$$(j = 1, 2, \ldots, n-1, n). \tag{3.43}$$

Here N_j represents ions that position in the j-th potential well, which is numbered from electrode 1 to electrode 2. $\omega_{j,j+1}$ and $\omega_{j+1,j}$ represent transfer rate of ions that hop from the j-th well to the $(j+1)$-th well, and the $(j+1)$-th well to the j-th well, respectively. ϕ_j and ϕ_{j+1} represent the potentials of the j-th well and $(j+1)$-th well, respectively, and $\phi_{j+\frac{1}{2}}$ represents the potential built between the j-th and the $(j+1)$-th wells. $\omega_{j,j+1}$ and $\omega_{j+1,j}$ are given by

$$\omega_{j,j+1} = \lambda \exp\left(-\frac{q\left(\phi_{j+\frac{1}{2}} - \phi_j\right)}{KT}\right),$$

$$\omega_{j+1,j} = \lambda \exp\left(-\frac{q\left(\phi_{j+\frac{1}{2}} - \phi_{j+1}\right)}{KT}\right) \tag{3.44}$$

with $\lambda = \nu \exp(-W/KT)$. Here, q is elementary charge. $\lambda^{-1} = \nu^{-1}\exp(+W/KT)$ represents the relaxation time and it corresponds to the dipolar depolarization time. As ions cannot go out from the film into electrodes 1 and 2, owing to the blocking property of electrodes, we set the boundary conditions in the way as $\omega_{0,1} = 0$, $\omega_{1,0} = 0$, and $\omega_{n+1,n} = 0$, $\omega_{n,n+1} = 0$, and $\sum_{j=1}^{n} N_j = N_t$. Noteworthy that this situation is observed for the case of impurity Na^+ and K^+ ions in an insulator layer in Metal–Insulator–Semiconductor (MIS) diodes [29]. If an interaction between ions and electrodes is active, the boundary condition, *i.e.*, $\omega_{0,1} = 0$, $\omega_{1,0} = 0$ and $\omega_{n+1,n} = 0$, $\omega_{n,n+1} = 0$, should be modified accordingly, with consideration of the interaction. This is the basis of ion hopping model, and diffusion constant D and ion mobility μ are given from the model as follows

$$D = \lambda l^2, \mu = \frac{q}{KT}\lambda l^2. \tag{3.45}$$

By solving Poisson's equation of charge distribution $\rho(x)$, we can calculate potential distribution ϕ_j in films as follows:

$$\phi_j = \frac{qN_t d}{\varepsilon_s \varepsilon_0}\left((n-j)\sum_{k=1}^{j}(k-1)\frac{N_k}{N_t} + (j-1)\sum_{k=j+1}^{n}(n-k)\frac{N_k}{N_t}\right) \cdot \frac{1}{(n-1)^2} + \frac{j-1}{n-1}V_b. \tag{3.46}$$

Here V_b is the d.c. applied voltage on electrode 2 with respect to electrode 1 grounded. Equation (3.46) suggests that the potential is well influenced by impurity ions in the case $(qN_td)/(\varepsilon_s\varepsilon_0) \gg V_b$, whereas the contribution of ions is minor in the case $(qN_td)/(\varepsilon_s\varepsilon_0) \ll V_b$. It is noteworthy that $(qN_td)/(\varepsilon_s\varepsilon_0) \gg V_b$ is rewritten as $qN_t \gg (\varepsilon_s\varepsilon_0)/d \cdot V_b$. That is, this situation appears when the total ion charge is much greater than capacitive charge on electrodes. In the same way, $(qN_td)/(\varepsilon_s\varepsilon_0) \ll V_b$ is rewritten as $qN_t \ll (\varepsilon_s\varepsilon_0)/d \cdot V_b$. That is, this situation appears when the total ion charge is much less than capacitive charge on electrodes.

The current density that flows from the j-th to the $(j+1)$-th potential well is given as

$$J_j = \omega_{j,j+1}qN_j - \omega_{j+1,j}qN_{j+1}. \tag{3.47}$$

Under the condition of $1 > q(\phi_{j+1/2} - \phi_j)/KT$, this equation is approximately written as

$$\begin{aligned} J_j &\approx \frac{q}{KT}\lambda l^2\left(\frac{-\phi_{j+1}+\phi_j}{l}\right)\frac{qN_j}{l} - \lambda l^2\left(\frac{qN_{j+1}}{l} - \frac{qN_j}{l}\right)\frac{1}{l} \\ &= \mu\left(\frac{-\phi_{j+1}+\phi_j}{l}\right)\frac{qN_j}{l} - D\left(\frac{qN_{j+1}}{l} - \frac{qN_j}{l}\right)\frac{1}{l}. \end{aligned} \tag{3.48}$$

In the limit $l \to 0$, we can assume

$$\frac{(-\phi_{j+1}+\phi_j)}{l} \to E(x), \quad \frac{qN_j}{l} \to \rho(x),$$

and

$$\left(\frac{qN_{j+1}}{l} - \frac{qN_j}{l}\right)\Big/ l \to \frac{d\rho(x)}{dx}.$$

Therefore, this equation is written as

$$J(x) \approx \mu E(x)\rho(x) + D\frac{d}{dx}\rho(x). \tag{3.49}$$

That is, the current given by Eq. (3.48) is sum of the drift current (the first term) and the diffusion current (the 2nd term). Further, the following relation is derived from Eq. (3.43) and Eq. (3.47):

$$\frac{d}{dt}\left(\frac{qN_j}{l}\right) = -\frac{J_{j+1} - J_j}{l}. \tag{3.50}$$

In the limit $l \to 0$, we get

$$\frac{d}{dt}\rho(x,t) = -\frac{\partial}{\partial x}J(x,t). \tag{3.51}$$

That is, Eq. (3.43) also satisfies the relation of the continuity of current. In other words, by using the Ion Hopping Model introduced here, we can analyze the long-range movement of ions in films, with consideration of ion drift motion and diffusion motion. It is instructive here to note the validity of ion hopping model to describe the electrostatic field in discrete fashion. There are a variety of numerical models that allow us to solve electrostatic problems. For example, forward, backward, and central differences are available in mathematics to approximate the first derivative terms of Eqs. (3.49) and (3.51). By contrast, ion hopping model is a fundamental model that can describe electrostatic field with consideration of discretely arranged ions with a spacing of l. Accordingly, the ion hopping model leads to Eq. (3.49) and Eq. (3.51) in the limit $l \to 0$.

3.5.2 *Ionic space charge polarization*

In Chap. 2, we showed that the displacement of point charge q with a distance of $\Delta x = a$ is identical to generate a dipole moment with a magnitude of $\mu = qa$, and this is the source of dipolar polarization. As an extension of this model, we can derive that ionic polarization is generated due to the change of distribution of ions. Assuming ion charge density at $t = 0$ is given by $\rho(x,0)$ and it changes into $\rho(x,t)$ at $t = t$ owing to the movement of ions, we can derive the polarization induced in film, due to the movement of ions as follows:

$$P(t)= \int_0^d \frac{x}{d} \left(\rho\left(x,t\right) - \rho(x,0)\right) dx. \tag{3.52}$$

Using the Ion Hopping Model illustrated in Fig. 3.25, this equation is rewritten as

$$P(t) = qN_t \sum_{j=1}^{n} \left(\frac{j-1}{n-1}\right) (f_j - f_j^0) \tag{3.53}$$

with $f_j = N_j/N_t$ $(j = 1, 2, \ldots, n)$. Here f_j^0 is the initial ion distribution at $t = 0$, and f_j is the ion distribution at $t = t$. $P(t)$ changes in accordance with ion movement, and Maxwell Displacement Current flows as follows:

$$I(t) = \frac{dP}{dt} = qN_t \sum_{j=1}^{n} \left(\frac{j-1}{n-1}\right) \frac{d}{dt}(f_j - f_j^0). \tag{3.54}$$

Therefore, if we carry this MDC measurement whilst films are being heated at a constant heating rate β, MDC $I(t)$ is recorded as $TSC(T)$ with temperature T. It is instructive here to note from viewpoint of analytical side that if we can make it clear the transition of f_j $(= N_j/N_t)$ with time, we can find the specific property of ion polarization.

In the initial state at $t = 0$, ions distribute in steady, that is,

$$\frac{d}{dt} f_j^0 = 0$$

and

$$J_j = 0 \text{ for } j = 1, 2, \ldots, n.$$

That is, ion distribution is in thermodynamic equilibrium state, and the following relation is satisfied between the ions in two adjacent wells.

$$\omega_{j,j+1} f_j^0 = \omega_{j+1,j} f_{j+1}^0 (j = 1, 2, \ldots, n-1). \tag{3.55}$$

Using a set of the $n-1$ equations for $j = 1, 2, \ldots, n-1$, we obtain

$$f_j^0 = \frac{\exp\left(-\frac{q\phi_j^0}{KT}\right)}{\sum_{j=1}^{n} \exp\left(-\frac{q\phi_j^0}{KT}\right)}. \tag{3.56}$$

Here ϕ_j^0 is the potential of the j-th well in the initial state. This equation indicates that ions distribute on satisfying the Boltzmann statistics. By the application of voltage, the ion distribution changes with time, but finally the steady state is established. That is,

$$\omega_{j,j+1} f_j^l = \omega_{j+1,j} f_{j+1}^l \tag{3.57}$$

and

$$f_j^l = \frac{\exp\left(-\frac{q\phi_j^l}{KT}\right)}{\sum_{j=1}^n \exp\left(-\frac{q\phi_j^l}{KT}\right)}. \tag{3.58}$$

Here f_j^l $(j = 1, 2, \ldots, n)$ represents the final state of ion distribution at $t \to \infty$. It is interesting to note that the ion distributes in the film on satisfying the Boltzmann statistics. Therefore, the relaxation time of the ion polarization specifies the repose from the initial state to the final state, which is ruled by the Boltzmann statistics. It is interesting here to note that the response time of ion hopping between two adjacent potential wells is given by $\tau = \lambda^{-1}$. Meanwhile, Ion Hopping Model suggests that ions can hop many potential hills successively. Accordingly, the response time should differ from $\tau = \lambda^{-1}$, as shown below.

To estimate the relaxation time of ion movement illustrated by Hopping model, we need to analyze the transition of $(f_1, f_2, \ldots, f_n)$ at the same time. Here we define a vector $\vec{f}^T \equiv (f_1, f_2, \ldots, f_n)$ and rewrite the Eq. (3.43) as follows:

$$\frac{d}{dt_b}\vec{f} = -L\vec{f}, \text{ with } t_b = \lambda t. \tag{3.59}$$

Here $\vec{f}^T$ is transpose vector of $\vec{f}$ and L is $n \times n$ matrix. Solving Eq. (3.59), we obtain

$$\vec{f}(t_b) = \exp\left(-Lt_b\right)\vec{f}(0) = \sum_{k=1}^n \exp(Q_k t_b) e_k \cdot e^k \vec{f}(0). \tag{3.60}$$

Here Q_k $(k = 1, 2, \ldots, n)$ is eigenvalue of Matrix L, and e_k is its eigenvector. e^k is transpose eigenvector e_k. Therefore, $-Q_k^{-1}\lambda^{-1}$ represents the relaxation time of the polarization.

For $qN_td/\varepsilon_s\varepsilon_0 \gg V_b$, the influence of ionic space charge is strong and it is difficult to derive the solution, whereas for $qN_td/\varepsilon_s\varepsilon_0 \ll V_b$, we can determine $-Q_k\lambda$, and $-Q_k\lambda = 4\lambda\sin^2(k\pi/2n)$ ($k = 1, 2, \ldots, n-1$), $-Q_k\lambda = 0$ ($k = n$) are obtained. Therefore,

$$P(t) = P(\infty) + \sum_{k=1}^{n-1} p_k \exp\left(-4\lambda\sin^2\left(\frac{k\pi}{2n}\right)\cdot t\right) \tag{3.61a}$$

$$I(t) = \sum_{k=1}^{n-1} 4\lambda\sin^2\left(\frac{k\pi}{2n}\right) p_k \exp\left(-4\lambda\sin^2\left(\frac{k\pi}{2}n\right)\cdot t\right). \tag{3.61b}$$

For $n \gg 1$, $-Q_k\lambda \approx \lambda(k\pi/n)^2$ ($k = 1, 2, \ldots$) is derived. That is, the polarization and MDC are comprised of the number of n relaxations. The longest relaxation time is given by $\lambda^{-1}n^2/\pi^2$, indicating that it takes much longer time than the response time $\tau = \lambda^{-1}$ to establish the thermodynamic equilibrium state. Actually, the relaxation time of ionic polarization is very long and the influence appears in the low frequency region. Here, p_k ($k = 1, 2, \ldots, n-1$) is determined by ion distribution in film at $t = 0$ (or $t \to \infty$). At $t = 0$,

$$P(0) = P(\infty) + \sum_{k=1}^{n-1} p_k. \tag{3.62}$$

As mentioned above, the relaxation is composed of the number of n relaxations, but these relaxations have relations each other.

TSC is derived by substituting $T = \beta t + T_0$ in the derivation of $I(t)$ using $P(t)$.

$$TSC(T) = \sum_{k=1}^{n} 4\sin^2\left(\frac{k\pi}{2}n\right) p_k(T_0)\nu \cdot \exp\left(-\frac{W}{KT} - \frac{\nu}{\beta}4\sin^2\left(\frac{k\pi}{2n}\right)\int_{T_0}^{T}\exp\left(-\frac{W}{KT}\right)dT\right). \tag{3.63}$$

Therefore, the TSC peak temperature is estimated as follows:

$$\frac{\beta W}{KT_m^2} = 4\nu\sin^2\left(\frac{k\pi}{2n}\right)\exp\left(-\frac{W}{KT_m}\right), \quad (k = 1, 2, \ldots, n-1). \tag{3.64}$$

The maximum temperature T_m in TSC curve shifts to higher temperature as the long-range movement of ions involves the relaxation as in the case of $k = 1$.

In actual TSCs, for the case of $(qN_t d)/(\varepsilon_s \varepsilon_0) > V_b$, the relaxation becomes more complex. Figure 3.26 shows a simulating result based on ion hopping model and transition of ion distribution from initial state f_j^0 to final state f_j^l, and Fig. 3.27 shows TSC calculated from the simulation. As qN_t increases, T_m increases. The shift of T_m depending on total ion density N_t is characteristics of TSC of long-range motion of ions, which differs from TSC of short-range motion of ions and dipolar depolarization process [30].

The total density of ions $Q_i = qN_t$ is determined from TSC of ions. The maximum polarization is limited at high biasing voltage V_b if the interaction between ions and electrodes is minor. That is, ions are mostly distributed nearby electrode at sufficiently high bias voltage, and TSC charge Q_{TSC} saturates. In Fig. 3.26(a), from Eq. (3.56), positive ions N_t is in thermal equilibrium state with positive biasing voltage on electrode 2 in reference to electrode 1 grounded. Most of ions occupy the 1st well next to electrode 1 in this situation, *i.e.*, $f_1^0 \approx 1$ and $f_j^0 \approx 0$ $(j = 2, 3, \ldots, n)$ at $t = 0$. These ions are redistributed to a new thermal equilibrium in short-circuit condition by heating, and reaches distribution f_j^l at $t \to \infty$. The change of polarization due to movement of ion impurity is, from Eq. (3.53),

$$Q_A = P(\infty) = qN_t \sum_{j=1}^{n} \left(\frac{j-1}{n-1}\right)(f_j^l - f_j^0) = qN_t \sum_{j=1}^{n} \left(\frac{j-1}{n-1}\right) f_j^l. \tag{3.65}$$

In a similar way, if sufficiently large negative biasing is applied to the electrode 2, positive ions are mostly located in n-th well, $f_n^0 \approx 1$ and $f_j^0 \approx 0$ $(j = 1, 2, \ldots, n-1)$ at $t = 0$. The change of polarization at $t \to \infty$ is

$$\begin{aligned} Q_B = P(\infty) &= qN_t \sum_{j=1}^{n} \left(\frac{j-1}{n-1}\right)\left(f_j^l - f_j^0\right) \\ &= qN_t \sum_{j=1}^{n} \left(\frac{j-1}{n-1}\right) f_j^l - qN_t \end{aligned} \tag{3.66}$$

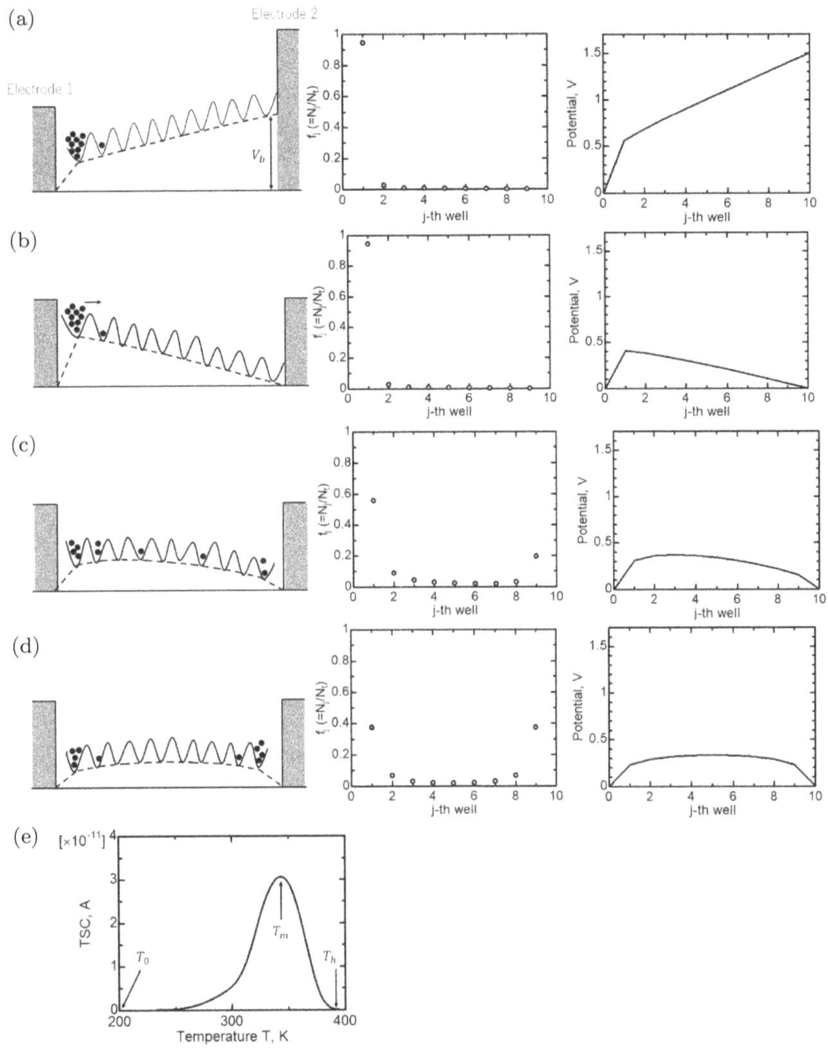

Figure 3.26. Ion hopping during TSC measurement. Ion distribution and electrostatic potential in the film (a) under thermal equilibrium after applying biasing voltage $V_b = +1.5$ V at $T_b = 370$ K for 10 s. Total ion density $N_t = 0.5 \times 10^{13}$ cm^{-2}, $\nu = 10^7$ 1/s, $W = 0.56$ eV, $l = 2$ nm, $\varepsilon_r = 3.9$, $d = 20$ nm, the number of wells $n = 9$ are assumed, (b) after short circuit electrodes 1 and 2 ($V_c = 0$ V) at initial temperature $T_0 = 200$ K. (c) During temperature increase ($\beta = 10$ K/min), ion distribution is changing to equilibrium distribution at $V_c = 0$. The distribution at $T_m = 343$ K ($t = 859$ s) was illustrated. (d) Equilibrium at $V_c = 0$ is reached at $T_h = 392$ K. (e) TSC during ion hopping where electrode area 3.1 mm^2 was assumed.

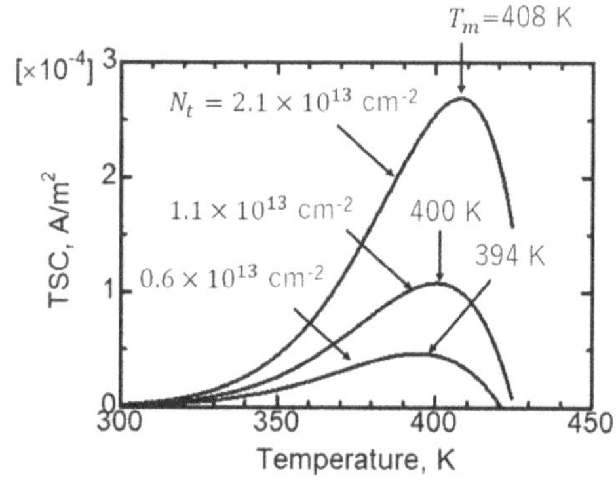

Figure 3.27. Numerical simulation of TSC based on ion hopping model. The total ion density was a parameter ($N_t = 0.6$, 1.1, 2.1×10^{13} cm^{-2}). As predicted by using Ion Hopping model, maximum temperature T_m of TSC shifts to higher temperature with increasing ion density N_t, which is characteristics of TSC of long-range motion of ions.

From Q_A and Q_B, total density of ions Q_i is

$$Q_i = Q_A - Q_B = qN_t. \tag{3.67}$$

Figures 3.28(a) and (b) simulates bias voltage V_b dependence on TSC charge Q_{TSC} with the same numerical parameter used for the ion hopping model in Figs. 3.26 and 3.27. The simulated TSC charge Q_{TSC} saturates at positive and negative biasing voltage, and total density of ions Q_i is determined by using the relation Eq. (3.67). Actually, in experimental TSC measurement of SiO_2 film with intentionally introduced sodium ion impurity, Na^+, the TSC charge saturates at the positive and negative biasing voltage, as shown in Fig. 3.28(c) [31]. Experimental TSC of SiO_2 film without the intentional ion introduction exhibits the same saturation behavior, with much smaller ion charge density (Fig. 3.29(b)) [30].

Collecting voltage V_c changes carrier transport behaviors in TSC measurement. That is, movement of ions follows drift and diffusion processes, as described by Eq. (3.43), and collecting voltage V_c changes electric field in the film. Basically, the electric field drifts ions faster and slower depending on the polarity of collecting voltage. But the effect of collecting voltage and resulting TSC curve differ

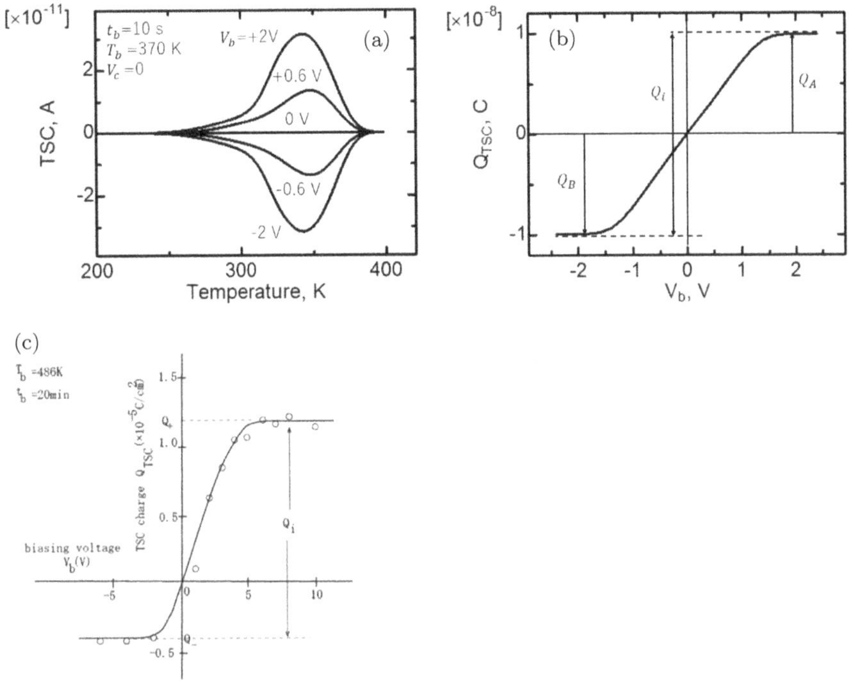

Figure 3.28. (a) Bias voltage dependence of TSC curves calculated by using Ion Hopping model. (b) Saturation of TSC charge Q_{TSC} at high biasing voltage. (c) Example of TSC experiment showing saturation of Q_{TSC} at high biasing voltage (150 nm thick SiO_2 where Na^+ impurity is intentionally introduced). Reprinted by permission from the Institute of Electrical Engineers of Japan (K. Yamashita, T. Hino, IEEJ Trans. Fundam. Mater., Vol. 97, 537, 1977) [31]. (a)(b) was calculated by using parameters, $N_t = 0.5 \times 10^{13}/\text{cm}^2$, $\nu = 10^7$ /s, $W = 0.56$ eV, $H = \infty$, $l = 2$ nm, $\varepsilon_s = 3.9$, electrode area 3.1 mm^2, $n = 9$, $\beta = 10$ K/min, $V_c = 0$ V. As an initial distribution of ions, biasing voltage $V_b = -2.4$ to $+2.4$ V was applied at $T_b = 370$ K for 10 s for homogeneously distributed ions from 1st well to n-th well.

from the case of TSC due to detrapping. That is, the role of collecting voltage is very simple in the TSC of detrapping, as described in Sec. 3.4, whereas in the case of TSC of ions, the effect of collecting voltage is not so simple owing to long-range motion of ions. The collecting voltage directly gives effect on the drift motion of ions, but the change of drift motion impacts on the change in diffusion due to the resulting ion distribution. This point is quite different from the

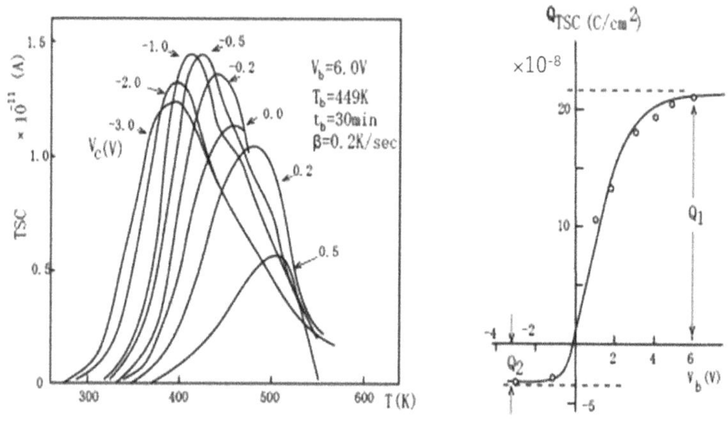

(a) (b)

Figure 3.29. TSC of 100 nm thick SiO_2 thermally grown on n-type Si. (a) TSC with various collecting voltage V_c. (b) TSC charge as a function of bias voltage V_b. Reprinted by permission from the Institute of Electrical Engineers of Japan (M. Iwamoto, K. Yamashita, T. Hino, IEEJ Trans. Fundam. Mater., Vol. 101, 445, 1981) [30].

case of TSC due to detrapping, where detrapping process from trapping sites goes independently. Figure 3.29(a) is experimental result of TSC of ions in thermally grown SiO_2 film (thickness 100 nm) with non-zero collecting voltage application [30]. TSCs at larger collecting voltage (for example $V_c = -3.0$ V) shows TSC maximum temperature T_m is shifted to lower temperature. The result indicates that the collecting voltage makes an impact on the drift motion of ions, and transition from initial distribution of ions to final distribution proceeds in shorter time with non-zero collecting voltage.

3.6 Recent TSC study trend

As mentioned in earlier sections, TSC measurement is very simple, but we can obtain useful information on carrier trapping, polarization, and others, in a variety of materials and electronic devices. The materials include dielectrics, ferroelectrics, semiconductors, ceramics, plastics, and so on. Recent developments of new advanced materials and technologies have motivated researchers to utilize TSC measurement as a tool for characterizing new materials and new devices, in terms of the electrical properties of materials, device performance, and their reliability. Among them are the evaluation of defect states of advanced electronic materials, the structure of nano-enabled materials, and the performance of organic and inorganic devices such as functional oxide devices, C60 FETs, organic LEDs, and emerging photovoltaic cells. As a result, we can see many TSC reports in scientific journals and magazines. In the following, we focus on the TSC study trend in Japan and discuss briefly.

3.6.1 *Analysis of dipolar TSC for development of advanced materials*

The number of TSC studies is increasing with emergence of new and advanced materials. Water is very important for our daily life and for production in industries. Accordingly, much attention has been paid to the behavior of water itself, *i.e.*, water molecules in biomaterials, polymers, and so forth. The fundamental TSC study of water has been carried out by using ice made from water, ion exchanged water, and so on [32]. These TSCs of water show dipolar behavior that follows Eq. (3.4), and their peak positions clearly reflect the difference between the structures of pure water and heavy water, and between the contributions of ions such as K^+, Na^+, Li^+, Ca^{++}, and Mg^{++}. Also, the contribution of water molecules in polymers such as Nylon 66 has been intensively studied, in terms of the molecular structure of polymers, for further development of advanced functional polymers [33]. It is noteworthy that from the viewpoint of new TSC experimental technique, we can say that TSC spectra of ice of water can be evaluated with high sensitivity, though it was difficult in the past.

With the development of advanced ionic ceramics, TSC has attracted much attention in the field of ceramics. Ionic ceramics such as hydroxyapatite ($Ca_{10}\,[PO_4]_6\,[OH]_2$) are being studied using TSC, in terms of the polarization originating from protons [34]. The dipolar motion of protons greatly affects TSC, and four types of polarization have been identified in a wide range of temperatures. The successful finding of these four polarizations is mainly due to the progress in the TSC experimental technique and its analytical method.

In the field of telecommunication technologies, TSC has attracted much attention as a tool for evaluating the degradation and deterioration of electro-optic (EO) polymers. In this research field, the development of new EO polymers that are available in the long term is an urgent task. Generally, it takes a much longer time to evaluate degradation by using isothermal measurement, because the relaxation time of polarized dipoles in EO polymers is much longer at room temperature. The TSC method has the potential to overcome this situation. By using EO polymers, *e.g.*, Disperse Red 1 (DR1)/poly(methyl methacrylate) (PMMA) guest/host polymer, side-chain copolymer DR1–PMMA (PMMA–*co*–DR1) and so forth, dipolar depolarization in EO polymers has been examined by TSC measurement, and the characteristic time τ given by $\tau = \tau_0 \exp(+H/KT)$ was determined as an index of the thermal stability of EO polymers [15, 35]. The time required for the determination of τ is much shorter by TSC measurement than by isothermal measurement such as the potential decay measurement at room temperature. Consequently, information on relaxation time is promptly available for the improvement and redesign of new EO polymers. The determination of the relaxation time τ by TSC measurement is now getting available as a new tool for the development of advanced thermally stable EO polymers. As mentioned above, many TSC experiments are being carried out in a variety of new research fields focusing on dipolar depolarization.

3.6.2 *Analysis of traps and defects by TSC*

3.6.2.1 *Organic devices*

The presence of carrier traps, defects, and interface states is a serious problem in electronic devices, sensors, and so forth. To analyze these traps and defects in terms of a device life and materials degradation, TSC is employed in R&D sections along with the development of new organic semiconductor materials, new devices, and so forth. Among them are TSC studies on new organic materials [36, 37], carrier traps, and defects of organic light-emitting diodes (OLEDs) [38, 39], Pentacene Schottky diodes [40], organic field-effect transistors (OFETs) [41], organic solar cells (OSCs) [42, 43], and so forth. In these TSC studies, Eq. (3.25′) is basically used in the analysis under the assumption that thermal emission governs TSC. TSC measurements well identify the traps and defects of these devices, and account for the effects of water, oxygen, photo-radiation, and high-energy particle radiation on the device performance. That is, TSC is well utilized to identify the trap depth and changes in trap density. For example, for the tris(8–hydroxyquinolinato)aluminum(III) (Alq3) layer in OLEDs, it was shown that the density of deep traps increases with degradation, and that electron traps in the N,N'–diphenyl–N,N'–bis(1–naphthalenyl)–1,1′–biphenyl–4,4′–diamine (α–NPD) layer and of electron and hole traps in the Alq3 layer have been identified [38]. Moreover, for the application of new materials to devices, the impurity of the materials and carrier traps greatly affect device performance. For this purpose, TSC has been employed for analyzing influence of impurities on the efficiency of OSCs using C60, and the relationship between trap density and the impurities of C60 has been studied. The OSC efficiency is shown to be considerably improved to about 22% owing to the increase in the purification level from 99.95 to 99.999%, but the trap density of C60 films evaluated by TSC shows no significant dependence on the impurities of C60 in this high-purity region. This result suggested that the improvement of the interfacial device structure such as flatness is a key parameter for further increase in the efficiency of OSCs with C60, and no further purification is necessary [42]. Perovskite is one of the hot-topic materials in the field of solar batteries, but its degradation is a very serious problem for practical use. In this research field, TSC has been employed for studying the degradation mechanism of perovskite, and it has been demonstrated that residual water molecules in perovskite are the origin of hole traps that cause the degradation [43]. In this way, for the development of new organic devices using new functional materials such as Alq3, C60, and perovskite, TSC is being well utilized.

TSC measurement has been widely used to study the interfacial states of MIS diodes and also field-effect transistors. In earlier days, the main devices were Si-based MOS diodes, transistors, and other inorganic devices, and TSC was employed to study the interface states of these devices [44, 45]. The main reason is that the interface property governs device performance, and thus the analysis of the semiconductor–insulator interface is very important for the improvement of device performance. Nowadays, TSC measurement is being used to evaluate organic diodes, OFETs, and so on [37, 41, 46]. Again, the interface is a key issue in organic devices. Generally, the MIS structure is used to study the interface states between an insulator and a semiconductor, but for organic semiconductor devices the discrimination of hole and electron traps is a difficult task, and the carrier transport of OFETs greatly depends on the quality of the interface. That is, owing to the ambiguity of the polarity of carrier species, *e.g.*, holes or electrons, and to the device performance of organic devices being different from those of conventional Si-based devices, we need to carry out TSC experiments with consideration of the specific properties of organic semiconductor devices, although the TSC technique employed for analyzing organic devices is very similar to the technique used for analyzing conventional Si-based devices. Accordingly, besides its use for organic MIS diodes, TSC measurement is also directly employed for OFETs [41]. For the TSC measurement of OFETs, by choosing the polarity of the applied gate voltage, we can selectively investigate electron and hole behaviors in devices, and also study electron and hole traps and defects in the channel region of OFETs. As organic semiconductor materials, pentacene [41], 7,7,8,8–tetracyano–*p*–quinodimethane (TCNQ) [37], and so on are used, and insulator–semiconductor interface properties are analyzed by using TSC, and interfacial energy states, the density of interface states, and the density of defects are well evaluated.

3.6.2.2 *Inorganic devices*

TSC measurement is also used for inorganic semiconductors. Wide-gap semiconductors such as GaN, SiC, and ZnO are anticipated for use in high-speed devices, high-power devices, and optoelectronic devices, and many investigations are being extensively carried out. However, defects and radiation effects are key issues for their practical applications. Accordingly, TSC measurement is employed to evaluate the energy levels of traps, trap density, and defects induced by photo-radiation and particle radiation in the bulk of these semiconductors [47–49]. The technique used for TSC measurement of inorganic devices is basically the same as other TSC measurement techniques, but the TSC system is designed so that photo-radiation and particle radiation can be carried out at low temperatures. The samples used for experiments are semiconductors, and the energy level is evaluated from the slope of $1/T_m - \ln(T_m^4/\beta)$ plots by carrying out experiments at various heating rates β, on the basis of the relation called Look's relation [3]. It is noteworthy that the energy level of traps in insulator is generally evaluated from the slope of $1/T_m - \ln(T_m^2/\beta)$ plots (see Fig. 3.13(c) and Fig. 3.24), where the temperature dependence of the effective carrier density N_c is presumably constant. As mentioned above, the TSC measurement is well used for analyzing wide-gap semiconductors.

Lead zirconate titanate (PZT) is a ferroelectric material and it is used for memory devices. To study the defects of such ferroelectric material, TSC measurement has been carried out. In the TSC measurement, TSC peaks due to spontaneous polarization appear and also leakage current flows across PZT materials. These contributions observed in PZT are removed experimentally from the TSC spectra by the thermal cleaning method, and carrier traps are extensively studied [51–53].

In the application of materials to electronic components, they are subjected to an applied d.c. electric field and external environment stimuli such as radiation, water, and light. Accordingly, defects and carrier traps have direct effects on the material's electrical and electronic properties and device performance. TSC measurement is well utilized to characterize carrier traps in materials and devices.

3.6.3 *Development of TSC technology*

TSC is very small current, generally less than 1 pA. Therefore, it is quite important to use instruments and circuits that can measure such a small current. Measuring-instruments and circuits must be designed to be free of mechanical noise and electrical noise. Highly insulating materials are used in electric cables for the electrical circuit, and many noise sources such as charges induced by friction and capacitance changes induced by vibrations are removed. Nowadays, we can measure a very small current with high reproducibility by setting samples in a Faraday cage and connecting the sample to an electro-ammeter. In other words, knowledge of electrostatic phenomena is quite important to establish TSC measurement, and the TSC experimental technology is advancing to measure very small currents by removing noises [54–56].

The analytical technique of TSC is also advancing. Basically, TSC analysis is carried out using Eq. (3.4), but in the actual case of TSCs, many TSC peaks overlap with each other. Consequently, techniques that can discriminate TSC peaks from one original TSC curve are essential for analyzing TSCs (see Fig. 3.14 and Fig. 3.15). Experimentally, it can be done using a partial heating method as well as a thermal cleaning method (see Sec. 3.3.2(iii)), which are widely used. From the viewpoint of the analytical method, many techniques have been introduced including the curve fitting method, and Bucci's plots (whole curve method). These techniques have been used for analyzing TSC curves. Nevertheless, these are still insufficient to discriminate overlapped TSC peaks from the original TSC curve. On the basis of advanced computer technology, new analytical methods have been developed in Japan including the asymptotic estimation method (AEM) [57] and maximum entropy method (MEM) [58]. By using these methods, activation energy and other TSC parameters are accurately determined. By employing MEM for analyzing MIS capacitors with the structure $Al/Si_aN_b/Si$, TSC due to single-electron detrapping from a single trap has been well discriminated [59]. In this way, TSC technology is advancing experimentally and analytically on the basis of advances in computer technology.

TSC is basically displacement current. This current is generated owing to the carrier motion in samples, such as the orientational motion of dipoles and carrier detrapping, and it flows through an external ammeter. Consequently, we can speculate on the carrier motion in samples from the trace of recorded current. In other words, we merely probe carrier motion indirectly from the recorded TSC curve. Accordingly, it is difficult to distinguish electron movement and hole movement in a single film sandwiched between two identical parallel electrodes. It is of great help to develop a TSC technique that can directly probe carrier motion by coupling with other experimental techniques [60–62]. The TSC technique coupled with the optical second-harmonic generation (SHG) method has been developed, and it has been used to study dipolar motions in thin films [63]. Recently, it has been applied to the study of the interfacial states of OLED and organic MIS diodes by coupling the TSC measurement with the electric-field-induced SHG (EFISHG) method [60, 61] (see also Sec. 9.3.1).

3.6.4 *Future scope*

Recent progress in TSC measurement has been briefly reviewed. We can see TSC studies in various research fields, as well as experimental techniques and analytical methods that have developed accordingly. TSC is short-circuit current; it can be measured using one electro-ammeter. Many new materials will be continuously discovered and developed in the future. We therefore believe that the TSC technique will serve as one of the key techniques for characterizing these materials in the future. Considering this situation, we anticipate that TSC should be standardized so that every researcher can use it easily and reproducibly. The TSC measurement method for plastic films has been standardized as JIS K7131 in Japan [64], but it is basically for detecting TSCs due to dipolar depolarization and is no longer sufficient for characterizing traps and defects. It will be necessary to standardize the method of measuring TSC due to carrier detrapping. Along with globalization, activity for the standardization of TSC is ongoing, focusing on the evaluation of carrier traps of electronic devices using nano-enabled materials in Japan [65, 66].

References

1. Hino, T. (1985). IEEJ Technical Report, No. 2–194 (in Japanese).
2. Vanderschueren, J. and Gasiot, J. (1979). *Field-induced Thermally Stimulated Currents*, in *Topics in Applied Physics* 37, ed. Bräunlich, P., Chapter 4, "Thermally Stimulated Relaxation in Solids," (Springer-Verlag, Berlin Heidelberg New York), https://doi.org/10.1007/3540095950_10. pp. 135–223.
3. Look, D. C. (1983). *Semiconductors and Semimetals*, 19, eds. Robertson, R. K. and Beer, A. C., Chapter 2 "The electrical and photoelectronic properties of semi-insulating GaAs," (Academic, New York), https://doi.org/10.1016/S0080-8784(08)60275-6.
4. Engström, O. (2014). *The MOS System* (Cambridge University Press), https://doi.org/10.1017/CBO9780511794490. Section 7.1 Thermally stimulated current method. pp. 168–173.
5. Chen, R. and Kirsh, Y. (1979). *Analysis of Thermally Stimulated Processes* (Pergamon Press, Oxford U.K.).
6. Turnhout, J. V. (1975). *Thermally Stimulated Discharge of Polymer Electrets* (Elsevier, Amsterdam).
7. Sessler, G. M. (1979). *Electrets* in *Topics in Applied Physics*, 33 (Springer, Berlin).
8. Meier, H. (1974). *Organic Semiconductors* (Verlag Chemie, Berlin).
9. Mori, T. (2013). *Electronic Properties of Organic Conductors* (Springer Japan, Tokyo).
10. Schmechel, R. and von Seggern, H. (2005). *Physics of Organic Semiconductors*, ed. Brütting, W., Chapter 10 "Electronic traps in organic transport layers," (Weinheim, Wiley-VCH). pp. 273–304, https://doi.org/10.1002/3527606637.
11. Iwamoto, M. and Taguchi, D. (2014). *Thermally Stimulated Current in Electric and Electronic Materials — Analysis and Application* (Corona Publishing, Tokyo) (in Japanese).
12. Iwamoto, M. ed. (2016). *Latest Research Trend in Thermally Stimulated Current Method for Development of Materials and Devices* (CMC, Tokyo) (in Japanese).
13. Iwamoto, M. and Taguchi, D. (2018). *Jpn. J. Appl. Phys.*, 57, 03EA04, https://doi.org/10.7567/JJAP.57.03EA04.
14. Hino, T. and Yamashita, K. (1975). *Trans. IEEJ A*, 95, 79 (in Japanese), https://doi.org/10.1541/ieejfms1972.95.79.
15. Mo, X., Kinemura, K., Yamada, T., Otomo, A., Taguchi, D., Manaka T. and Iwamoto, M. (2013). *Jpn. J. Appl. Phys.*, 53, 01AD04, https://doi.org/10.7567/JJAP.53.01AD04.
16. Bucci, C., Fieschi, R. and Guidi, G. (1966). *Phys. Rev.*, 148, 816, https://doi.org/10.1103/PhysRev.148.816.
17. Hino, T., Suzuki, K. and Yamashita, K. (1973). *Jpn. J. Appl. Phys.*, 12, 651–656, https://doi.org/10.1143/JJAP.12.651.
18. Hino, T. (1973). *IEEJ Trans. Fundam. Mater.*, 93-A, pp. 99–106 (in Japanese), https://doi.org/10.1541/ieejfms1972.93.99.

19. Perlman, M. M. (1971). *J. Appl. Phys.*, 42, 2645, https://doi.org/10.1063/1.1660602.
20. Takamatsu, T. (1972). *Rep. Inst. Phys. Chem. Res.*, 48, 109–118.
21. Hino, T. (1979). Oyo Buturi. (*Applied Physics*), 48, 1186–1193 (in Japanese), https://doi.org/10.11470/oubutsu1932.48.1186.
22. Hino, T. (1975). *J. Appl. Phys.*, 46, 1956, https://doi.org/10.1063/1.321873.
23. Lindmayer, J. (1965). *J. Appl. Phys.*, 36, 196, https://doi.org/10.1063/1.1713874.
24. Iwamoto, M. and Taguchi, D. (2014). *Thermally Stimulated Current in Electric and Electronic Materials — Analysis and Application* (Corona Publishing, Tokyo). (in Japanese). pp. 174–175. Answer to problem 3 of Chap. 3.
25. (a) Hino, T., Yamashita, K. and Otsuka, T. (1974). Netsu Shigeki Denryuu ni Yoru polyethylene Maku he no Chuunyuu Kuukanndenka to sono Chuunyuu Fukasa no Sokutei (Thermally stimulated current measurement of space charge injected into polyethylene film and its charge centroid), *Dai 7 Kai Denki Zetsuen Zairyou Shinpojiumu Yokousyuu (Proceedings of the Seventh Symposium on Electrical Insulating Materials)*, I–3, pp. 55–58 (in Japanese). (b) Hino, T. (1975). *J. IEEJ*, 95, 109 (in Japanese), https://doi.org/10.11526/ieejjournal1888.95.109.
26. Hino, T. and Yamashita, K. (1975). *IEEJ Trans. Fundam. Mater.*, 95, 79 (in Japanese), https://doi.org/10.1541/ieejfms1972.95.79.
27. Perlman, M. M. and Creswell, R. A. (1971). *J. Appl. Phys.*, 42, 531, https://doi.org/10.1063/1.1660057.
28. Bube, R. H., Dussel, G. A., Ho, C. T. and Miller, L. D. (1966). *J. Appl. Phys.*, 37, 21, https://doi.org/10.1063/1.1707811.
29. Nauta, P. K. and Hillen, M. W. (1978). *J. Appl. Phys.*, 49, 2862, https://doi.org/10.1063/1.325168.
30. Iwamoto, M., Yamashita, K. and Hino, T. (1981). *IEEJ Trans. Fundam. Mater.*, 101, 445 (in Japanese), https://doi.org/10.1541/ieejfms1972.101.445.
31. Yamashita, K. and Hino, T. (1977). *IEEJ Trans. Fundam. Mater.*, 97, 537 (in Japanese), https://doi.org/10.1541/ieejfms1972.97.537.
32. Sato, K., Koizumi, T. and Okajima, K. (2005). *Anal. Sci.*, 21, 331, https://doi.org/10.2116/analsci.21.331.
33. Sato, K., Koizumi, T. and Okajima, K. (2005). *Polym. J.*, 37, 137, https://doi.org/10.1295/polymj.37.137.
34. Tanaka, Y., Iwasaki, T., Nakamura, M., Nagai, A., Katayama, K. and Yamashita, K. (2010). *J. Appl. Phys.*, 107, 014107, https://doi.org/10.1063/1.3265429.
35. Ikemoto, R., Taguchi, D., Manaka, T., Iwamoto, M., Yamada, T. and Otomo, A. (2016). Evaluation of thermal stability of organic electro-optic device by using thermally stimulated current, *J. Nanosci. Nanotech.*, 16, pp. 3378–3382, https://doi.org/10.1166/jnn.2016.12282.
36. Iwamoto, M. and Nakano, K. (1988). *Jpn. J. Appl. Phys.*, 27, 1224, https://doi.org/10.1143/JJAP.27.1224.
37. Sakai, M., Hanada, M., Kuniyoshi, S., Yamauchi, H., Nakamura, M. and Kudo, K. (2012). Crystals, 2, 730, https://doi.org/10.3390/cryst2030730.

38. Matsushima, T. and Adachi, C. (2008). *Jpn. J. Appl. Phys.*, 47, 1748, https://doi.org/10.1143/JJAP.47.1748.
39. Mori, T., Miyake, S. and Mizutani, T. (1995). *Jpn. J. Appl. Phys.*, 34, pp. 4120–4121, https://doi.org/10.1143/JJAP.34.4120.
40. Kuniyoshi, S., Watanabe, M., Iizuka, M., Nakamura, M., Kudo, K. and Tanaka, K. (2003). *Synth. Met.*, 137, 895, https://doi.org/10.1016/S0379-6779(02)01130-X.
41. Fujii, T., Matsui, H., Hasegawa, T., Kuniyoshi, S., Sakai, M., Kudo, K. and Nakamura, M. (2010). *IEICE Tech. Rep.*, OME2009–76, 51 (in Japanese).
42. Ohashi, N. and Yoshida, Y. (2016). *Yuuki Taiyou Denti* (*Organic Photovoltaics*) in *Latest Research Trend in Thermally Stimulated Current Method for Development of Materials and Devices*, M. Iwamoto ed. (CMC, Tokyo) Chap. 3.3 (in Japanese).
43. Qin, C., Matsushima, T., Fujihara, T., Potscavage, W. J. and Adachi, C. (2016). *Adv. Mater.*, 28, 466, https://doi.org/10.1002/adma.201502610.
44. Yamashita, K., Iwamoto, M. and Hino, T. (1978). *J. Appl. Phys.*, 49, 2866, https://doi.org/10.1063/1.325169.
45. Kuniyoshi, S., Tanaka, K. and Ohki, H. (1983). *IEEJ Trans. Fundam. Mater.*, 103, 222 (in Japanese), https://doi.org/10.1541/ieejfms1972.103.222.
46. Kuniyoshi, S., Ozeki, T., Tubuki, Y., Iizuka, M., Sakai, M. and Kudo, K. (2013). *Tech. Rep. IEEJ*, DEI–13–054, 27 (in Japanese).
47. Kuriyama, K., Ooi, M., Onoue, A., Kushida, K., Okada, M. and Xu, Q. (2006). *Appl. Phys. Lett.*, 88, 132109, https://doi.org/10.1063/1.2190446.
48. Kuriyama, K., Ooi, M., Matsumoto, K. and Kushida, K. (2006). *Appl. Phys. Lett.*, 89, 242113, https://doi.org/10.1063/1.2405867.
49. Yano, H., Hatayama, T. and Fuyuki, T. (2014). Hyomen Kagaku (*Journal of the Surface Science Sosiety of Japan*), 35, 90 (in Japanese), https://doi.org/10.1380/jsssj.35.90.
50. Nishida, T., Matsuoka, M., Okamura, S. and Shiosaki, T. (2003). *Jpn. J. Appl. Phys.*, 42, 5947, https://doi.org/10.1143/JJAP.42.5947.
51. Okino, H., Toyoda, Y., Shimizu, M., Horiuchi, T., Shiosaki, T. and Matsushige, K. (1998). *Jpn. J. Appl. Phys.*, 37, 5137, https://doi.org/10.1143/JJAP.37.5137.
52. Fukami, T., Kusunoki, M. and Tsuchiya, H. (1987). *Jpn. J. Appl. Phys.*, 26 (Suppl. 26–2), 46, https://doi.org/10.7567/JJAPS.26S2.46.
53. Fukami, T. and Karasawa, J. (1993). *J. Ceram. Soc. Jpn.*, 101, 394 (in Japanese), https://doi.org/10.2109/jcersj.101.394.
54. Hirayama, T. (2009). Rigaku J., 40, 1 (in Japanese).
55. Iwamoto, M. ed. (2016). *Latest Research Trend in Thermally Stimulated Current Method for Development of Materials and Devices* (CMC, Tokyo) Chapter 9 (in Japanese).
56. *Koukando Sokutei Handobukku* (*Low Level Measurements Handbook*) (Keithley Instruments) 6th Ed. (in Japanese).
57. Yoshida, F. and Maeta, S. (1998). *IEEJ Trans. Fundam. Mater.*, 118, 1035 (in Japanese), https://doi.org/10.1541/ieejfms1990.118.9_1035.
58. Yonamoto, Y. (2013). *Entropy*, 15, 1663, https://doi.org/10.3390/e15051663.

59. Yonamoto, Y. (2015). *J. Therm. Anal. Calorimetry*, 122, 1299, https://doi.org/10.1007/s10973-015-5097-8.
60. Taguchi, D., Manaka, T. and Iwamoto, M. (2014). *Int. Conf. Solid State Devices and Materials (SSDM2014)*, K–6–2.
61. Iwamoto, M. ed. (2016). *Latest Research Trend in Thermally Stimulated Current Method for Development of Materials and Devices* (CMC, Tokyo) Chapter 10.3 (in Japanese).
62. Arita, Y., Shiratori, S. S. and Ikezaki, K. (2003). *J. Electrostat.*, 57, 263, https://doi.org/10.1016/S0304-3886(02)00166-3.
63. Köhler, W., Robello, D. R., Dao, P. T., Willand, C. S. and Williams, D. J. (1990). *J. Chem. Phys.*, 93, 9157, https://doi.org/10.1063/1.459205.
64. Japanese Industrial Standards, JIS K 7131, *Plastic Film No Netsu Shigeki Denryuu Shiken Houhou* (*Testing Method for Thermally Stimulated Current of Plastic Films*) (1994), https://www.jisc.go.jp/.
65. International Standard, Technical specification, IEC TS 62607–8–1:2020, *Nano-enabled Metal–Oxide Interfacial Devices — Test Method for Defect States by Thermally Stimulated Current*, 9th April 2020, https://www.iec.ch/.
66. International Standard, Technical specification, IEC TS 62607–8–2: 2021, *Nano-enabled Metal–Oxide Interfacial Devices — Test Method for the Polarization Properties by Thermally Stimulated Depolarization Current*, 5th January 2021, https://www.iec.ch/.

Chapter 4

Maxwell–Wagner Effect and Interface Phenomena

The properties of *Surface and Interface of materials* are quite different from those of the bulk. The transport of electrons, holes and ions, and the rotational motion of dipoles are uniquely restricted at the surface and interface. Understanding the role of *Surface and Interface of materials* is very important for deeply understanding the generation of Maxwell displacement current and Optical second-harmonic generation. Also, it is key to make use well of *Surface and Interface* in electronics and electrical engineering. In this chapter, we mainly focus on the electronic charge accumulation and trapping phenomena at the interface, in terms of the Maxwell Displacement Current and Optical Second-Harmonic Generation. In the next Chap. 5, we focus on dipolar motion on the surface and interfaces. In the following sections, we begin with discussion on the Maxwell–Wagner effect that can macroscopically account for charge accumulation at the interface. Then we discuss the interface state of materials, and argue how the interface state is evaluated by using Maxwell Displacement Current (MDC) and Thermally Stimulated Current (TSC).

4.1 Maxwell–Wagner effect

Macroscopic electrical properties of materials are characterized using two key material parameters. These are dielectric constant ε and conductivity κ, and they specify the macroscopic bulk property of materials. The ratio $\tau = \varepsilon/\kappa$ gives a relaxation time and stands for spreading time of excess carriers in the materials. Steady-state charge distribution is established in materials with a relaxation time $\tau = \varepsilon/\kappa$. That is, this value $\tau = \varepsilon/\kappa$ is also a parameter that specifies the electrical bulk properties of materials (see, for example, [1, 2], and Appendix B). Charge accumulation happens at the interface between two materials with different relaxation times when conductive current flows across the interface [3–5]. This is nothing but the effect called the Maxwell–Wagner (MW) effect, and the charge accumulation due to this effect appears at any kinds of two-material interface, such as metal–insulator interface, metal–semiconductor interface, insulator–semiconductor interface, insulator–insulator interface, and others.

According to the Maxwell's electro-magnetic theory, a total current flowing across materials is the sum of the conduction current and the Maxwell's displacement current. That is, there are two routes for the current flow. As described in Chap. 1, this situation is modeled using equivalent circuit of parallel R and C, as illustrated in Fig. 4.1(a), where the R is resistance and it represents a path for the conduction current, and C is capacitance and it represents a path for the Maxwell's displacement current. For materials sandwiched between two parallel electrodes, R is given by $R = l/A\kappa$, and C is given by $C = \varepsilon A/l$, (A : the area of the electrode, l: the thickness of the materials, *i.e.*, spacing between the two electrodes). Consequently, we can see that the relaxation time $\tau = \varepsilon/\kappa = RC$ represents electrical response time of materials, and it equals to the response time of equivalent RC circuit.

We here discuss a case of steady-state current flowing across the interface of the two-material system, as illustrated in Fig. 4.1(b). The two materials are described using a set of two parameters, dielectric constant ε and conductivity κ, and constant steady-state current with a current density i is flowing across the interface. The current flow is given by $\vec{i} = \kappa\vec{E}$, where $\vec{E}$ represents the electric field and κ is conductivity of materials. In steady-state, $\nabla\cdot\vec{i} = 0$ is satisfied over the entire region of the two-material system. The electric field $\vec{E}$ satisfies the relation $\vec{E} = \vec{D}/\varepsilon$, where $\vec{D}$ is electric flux density. Hence, we obtain the following relation concerning electric flux density $\vec{D}$ at

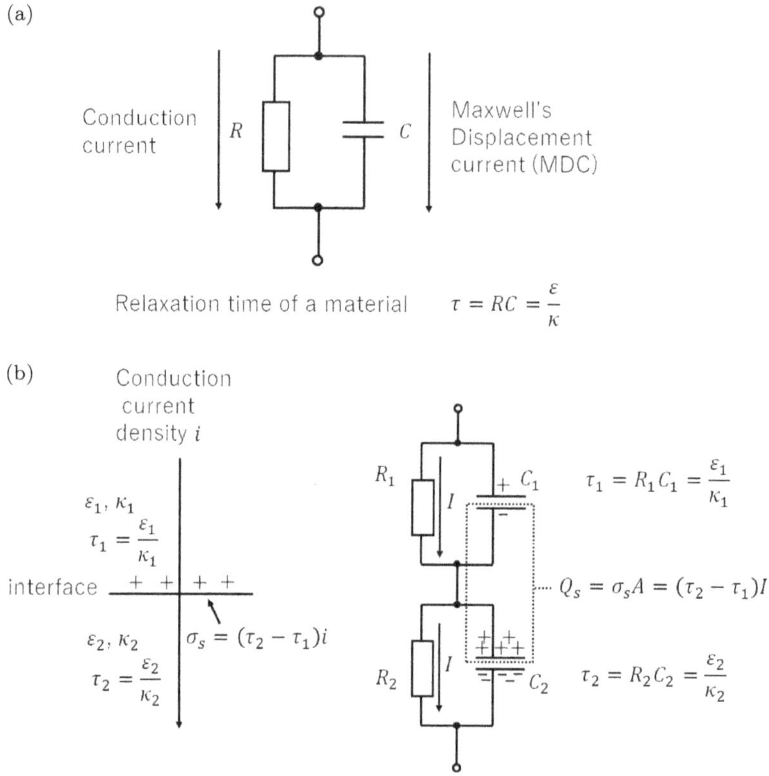

Figure 4.1. (a) Current flowing solids is the sum of conduction current and Maxwell displacement current. The route of the two currents is represented by a branch of R (resistance) and C (capacitance). (b) Current flowing across the interface causes charge accumulation due to different relaxation time. This is the Maxwell–Wagner effect.

the two-material interface as

$$\nabla \cdot \vec{D} = \nabla \cdot \varepsilon \vec{E} = \nabla \cdot \left(\frac{\varepsilon}{\kappa}\right) \vec{i} = \tau \nabla \cdot \vec{i} + \nabla \tau \cdot \vec{i}$$
$$= \nabla \tau \cdot \vec{i} = q_s (\neq 0) \tag{4.1}$$

In more detail, from Eq. (4.1) we obtain the following relation

$$(\tau_2 - \tau_1) i = \sigma_s (\neq 0) \tag{4.2}$$

Here $\tau_1 = \varepsilon_1/\kappa_1$ and $\tau_2 = \varepsilon_2/\kappa_2$ are relaxation times of materials 1 and 2, respectively. i is steady-state current per unit area, which flows normal to the surface of the interface, and σ_s is charge accumulated at the interface per area. Equations (4.1) and (4.2) show that charge is accumulated at the interface between two materials with different relaxation times, *i.e.*, $\tau_1 \neq \tau_2$. In addition, accumulated charge σ_s is given by the product of relaxation time difference $\Delta\tau = \tau_2 - \tau_1$ and steady state current density i. This is the Maxwell–Wagner effect that accounts for charge accumulation that is observed at all kinds of two-material interfaces. The charge σ_s is accumulated at the interface, and it is net charge accumulated due to the conductive current flow through the resistance. This accumulated interface charge is excess charge, and is acting as a source of space charge field in the system.

4.2 Interface charge in Metal–Insulator–Metal system and double layer system due to the Maxwell–Wagner effect

a. Metal–Insulator–Metal (MIM) system

Figure 4.2(a) shows a structure of metal–insulator–metal (MIM) system, in which we can find a metal–insulator interface. The conductivity of metal, κ_m, is high in comparison with insulator. On the other hand, the conductivity of insulator, κ_{ins}, is extremely low and it is nearly zero. In that situation, relaxation times of the metal and the insulator become $\tau_m \to 0$ and $\tau_{ins} \to \infty$, respectively. As a result, current flowing across the insulator becomes nearly zero in steady state, $i \to 0$, in the presence of external electric field $E = V/L$ in insulator, where V is applied voltage across the insulator and L is the distance between the two electrodes. Substituting these relations into Eq. (4.2), we obtain the induced charge on metals σ_M ($= \sigma_s$) as follows:

$$\begin{aligned}\sigma_M = (\tau_{ins} - \tau_m)i &= \left(\frac{\varepsilon_{ins}}{\kappa_{ins}}\right)\kappa_{ins}E_{ins} - \left(\frac{\varepsilon_m}{\kappa_m}\right)\kappa_m E_m \\ &= \varepsilon_{ins}E_{ins} - 0 = +\varepsilon_{ins}\frac{V}{L}. \end{aligned} \tag{4.3}$$

ε_m and ε_{ins} arc dielectric constant of metal and insulator, respectively, and E_m and E_{ins} are electric field in metal and insulator.

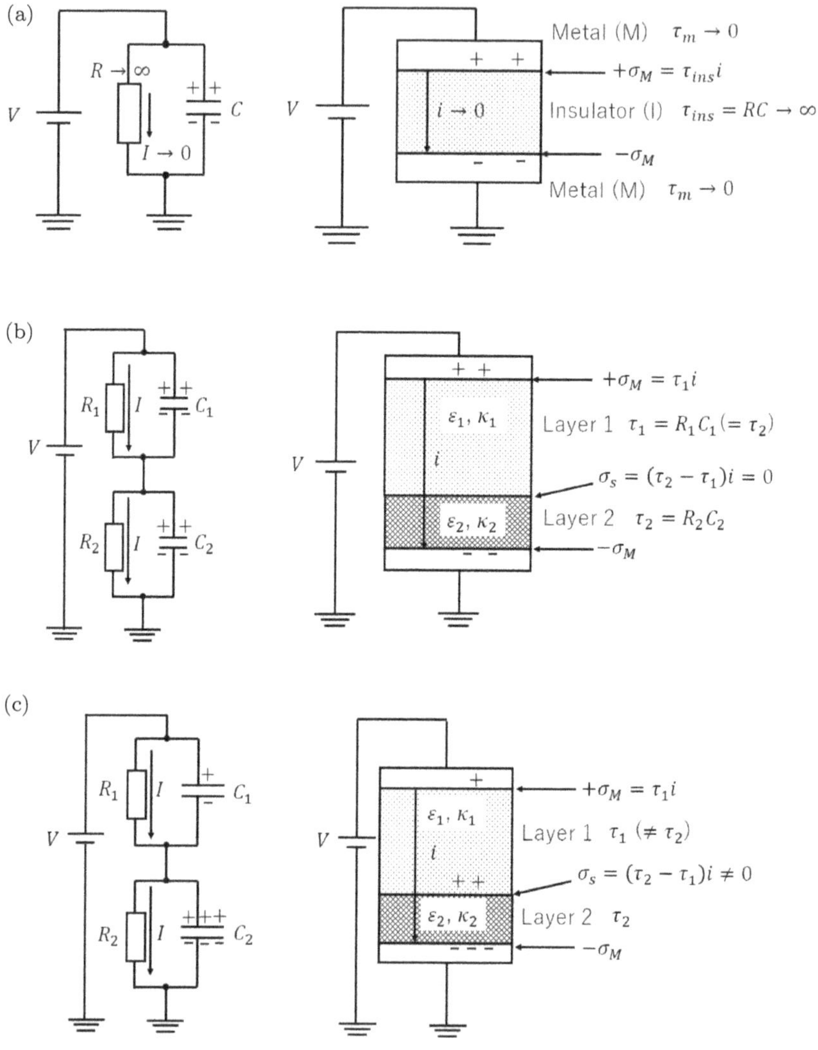

Figure 4.2. Interface charge accumulation due to Maxwell–Wagner effect. (a) MIM structure. (b) Double layer structure where the two layers have identical relaxation time $\tau_1 = \tau_2$. (c) Double layer structure where the two layers have different relaxation time $\tau_1 \neq \tau_2$.

In the derivation of Eq. (4.3), we assumed $E_m = 0$ with consideration of the fundamental electrical property of electrical conductors, *i.e.*, there is no potential difference inside of the metal electrodes [6]. Equation (4.3) is the general equation of capacitor that describes induced charge on metals of MIM structure. That is, induced charge $\pm\sigma_M$ on electrodes is proportional to the applied voltage V, and gives a capacitor element C in electrical circuit analysis as

$$C = \frac{\varepsilon_{ins}}{L} \quad \text{and} \quad \sigma_M = CV. \tag{4.3$'$}$$

b. Metal–Double-layer–Metal system (2-electrode system)
Figures 4.2(b) and 4.2(c) show metal–double-layer–metal structures. Typical examples are seen in double-layer dielectric (insulator) structures, double-layer organic EL devices, double-layer solar cells, *etc.* In these structures, we can see an interface between the adjacent two layers. As Eq. (4.1) suggests that no charge is accumulated at the interface, when the relaxation times of the two layers are the same as $\tau_1 = \tau_2$, *i.e.*, $q_s = 0$ and $\sigma_s = 0$ (see Eq. (4.1) and Eq. (4.2)), even though a steady-state current i flows across the interface. In that situation, the current i and electric flux density D is the same over the entire region of the both two-materials, and the following relations are satisfied for electric field in layer 1 E_1 and layer 2 E_2:

$$\kappa_1 E_1 = \kappa_2 E_2 (= i) \quad \text{and} \quad \varepsilon_1 E_1 = \varepsilon_2 E_2 (= D) \tag{4.4}$$

Equation (4.4) suggests that the ratio of the electric field formed in the two materials is given by

$$\frac{E_1}{E_2} = \frac{\varepsilon_2}{\varepsilon_1} = \frac{\kappa_2}{\kappa_1}$$

and it depends on the ratio of the dielectric constant and conductivity of the two materials. It is noteworthy that σ_M accumulates at metal–material interface in the same way as expressed by Eq. (4.3) as

$$\sigma_M = (\tau_1 - \tau_m)i = \tau_1 i - 0 = +\varepsilon_1 \frac{V}{L^*} \left(= \frac{C_1 C_2}{C_1 + C_2} V \right)$$

where $L^* = L_1 + \varepsilon_1/\varepsilon_2 \cdot L_2$ (L_1: layer 1 thickness, L_2: layer 2 thickness) is effective thickness.

On the other hand, in a case of double-layer system with different dielectric relaxation times, $\tau_1 \neq \tau_2$, as illustrated in Fig. 4.2(c), charge is accumulated at the interface. The electric field distribution in the two-material system depends on the accumulated charges at the interface. Assuming charge $Q_s = A\sigma_s$ (A: electrode area) is accumulated at the interface, we can derive the electric fields in materials 1 and 2, respectively, as follows:

$$E_1L_1 = \frac{Q_M}{C_1} - \frac{Q_s}{C_1 + C_2} \tag{4.5a}$$

$$E_2L_2 = \frac{Q_M}{C_2} + \frac{Q_s}{C_1 + C_2} \tag{4.5b}$$

Here $Q_M = \sigma_M A$ is the charge induced on metal electrodes with $Q_M = C^*V$, where $C^* = C_1C_2/(C_1 + C_2)$ is a total capacitance of the double layer system with $C_1 = \varepsilon_1 A/L_1$ and $C_2 = \varepsilon_2 A/L_2$. That is, the electric field formed in the double layer system changes depending on the accumulated charge at the interface Q_s. In steady state, Eq. (4.5a) and Eq. (4.5b) are given by

$$E_1L_1 = V\frac{R_1}{R_1 + R_2} \text{ and } E_2L_2 = V\frac{R_2}{R_1 + R_2} \tag{4.6}$$

by substituting the Eq. (4.2). That is, the general relation (Eq. (4.6)) derived from the RC circuit in steady state is obtained. As discussed above, the second terms of Eq. (4.5a) and Eq. (4.5b) represent the contribution of interface charge Q_s to the electric fields E_1 and E_2. For understanding the role of double-layer, information on the interface charge Q_s and the electric field distribution in each layer is important.

4.3 Discharging of interface charge accumulated due to the Maxwell–Wagner effect

As mentioned in Sec. 4.1, charge Q_s ($= A\sigma_s$) is accumulated at the interface according to the Maxwell–Wagner effect. For the case of double layer system, under voltage application $V = V_g$ and steady current I ($=Ai$) flowing the resistance, interface charge Q_s is described as follows:

$$Q_s = I\Delta\tau = I(\tau_2 - \tau_1) = \frac{G_1G_2}{G_1 + G_2}V_g\Delta\tau \tag{4.7}$$

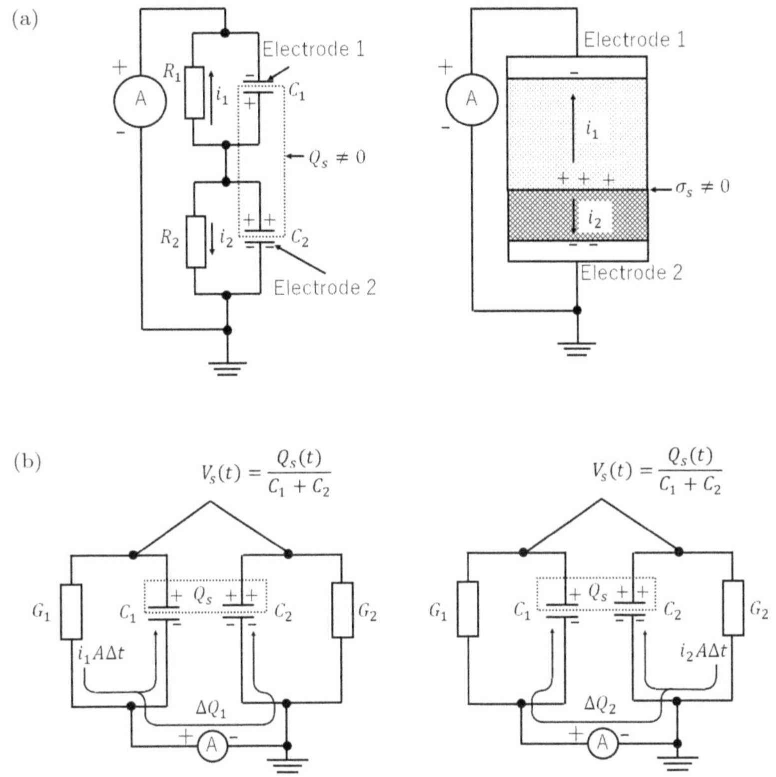

Figure 4.3. (a) Discharging of interface charge accumulated by the Maxwell–Wagner effect. (b) Equivalent circuit model after the shortening, where electrodes 1 and 2 are grounded. The charge $\Delta Q = \Delta Q_1 - \Delta Q_2$ flows through an ammeter, where ΔQ_1 and ΔQ_2 are a part of $i_1 A\Delta t$ and $i_2 A\Delta t$ flowing through conductance G_1 and G_2, respectively.

with $\tau_1 = \varepsilon_1/\kappa_1$, $\tau_2 = \varepsilon_2/\kappa_2$. In this situation, let us discuss how accumulated charge discharges under short-circuit condition. Figure 4.3 shows the equivalent circuit, which represents the situation just after the shortening. The interface charge Q_s will be discharged with a response time of $\tau_{MW} = (C_1 + C_2)/(G_1 + G_2)$, where $C_1 + C_2$ and $G_1 + G_2$ represent the parallel capacitance and conductance, respectively, and they are defined between the two-material interface and the ground (see Fig. 4.3(b)). The physical meaning of taking a parallel capacitance and conductance is clear in Fig. 4.3(b) where the interface charge Q_s discharges through two parallel branches of circuits C_1:G_1 and C_2:G_2. That is, the interface charge Q_s decays in the manner as

$$Q_s(t) = Q_s \exp\left(-\frac{t}{\tau_{MW}}\right), \tag{4.8}$$

where charge decays through both resistance $R_1 = 1/G_1$ and $R_2 = 1/G_2$. That means interface charge Q_s separates and moves to electrodes 1 and 2, respectively, where Ohmic conduction in resistance R_1 and R_2 governs discharging current, *i.e.*, carrier transport limited process. In the electrical circuit, the potential built at the interface is given by

$$V_s(t) = \frac{Q_s(t)}{C_1 + C_2} \tag{4.9}$$

with $C_1 = \varepsilon_1 A/d_1$ and $C_2 = \varepsilon_2 A/d_2$. Here $V_g \geq V_s(t)$. Owing to the establishment of potential $V_s(t)$ caused by the interface charge Q_s, Ohmic current i_1 and i_2 flow in opposite directions each other, through G_1 and G_2, respectively, and they are given by

$$i_1 = \kappa_1 E_1 = \kappa_1 \frac{V_s(t)}{d_1} = \frac{\kappa_1}{\varepsilon_1} \frac{C_1}{A} \frac{Q_s(t)}{C_1 + C_2}$$

and

$$i_2 = \kappa_2 E_2 = \kappa_2 \frac{V_s(t)}{d_2} = \frac{\kappa_2}{\varepsilon_2} \frac{C_2}{A} \frac{Q_s(t)}{C_1 + C_2} \tag{4.10}$$

The amount of charge that can arrive electrodes 1 and 2 in time Δt are $i_1 A \Delta t$ and $i_2 A \Delta t$, respectively. Consequently, the amount of charge ΔQ that flows through an external ammeter in time Δt is calculated as follows:

$$\begin{aligned} \Delta Q &= \Delta Q_1 - \Delta Q_2 \\ &= \frac{C_2}{C_1 + C_2} i_1 A \Delta t - \frac{C_1}{C_1 + C_2} i_2 A \Delta t \\ &= \Delta t \left(\frac{\kappa_1}{\varepsilon_1} - \frac{\kappa_2}{\varepsilon_2}\right) \frac{C_1 C_2}{C_1 + C_2} \frac{Q_s(t)}{C_1 + C_2} \end{aligned} \tag{4.11}$$

where ΔQ_1 and ΔQ_2 are a part of $i_1 A \Delta t$ and $i_2 A \Delta t$ illustrated in Fig. 4.3(b). By integrating ΔQ between the time $t = 0$ and $t = \infty$, we get

$$\begin{aligned} Q &= \left(\frac{\kappa_1}{\varepsilon_1} - \frac{\kappa_2}{\varepsilon_2}\right) \frac{C_1 C_2}{C_1 + C_2} \frac{Q_s}{C_1 + C_2} \tau_{MW} \\ &= \frac{G_1 G_2}{G_1 + G_2} \frac{Q_s}{C_1 + C_2} \Delta\tau = \frac{G_1 G_2}{G_1 + G_2} V_s \Delta\tau \end{aligned} \tag{4.12}$$

As $V_g \geq V_s$ $(= Q_s/(C_1 + C_2))$, we get the following relation using Eq. (4.7) and Eq. (4.12),

$$Q_s \geq Q. \tag{4.13}$$

That is, the amount of total charge Q recorded by an ammeter is less than the amount of charge Q_s accumulated at the interface. This situation is the same with that we discussed on the charging in Sec. 2.1 (see Fig. 2.3), where the charge q' $(= +qx/L)$ observed in ammeter is always smaller than the actual charge q, *i.e.*, $q \geq q'$. Further we should note that in this discharging process accumulated charge is conveyed to the both electrodes as Ohmic conduction current, and the discharging of Q_s is limited by the carrier transport process across the resistance.

4.4 Interface states and carrier traps

As we discussed above, charges can be accumulated at the interface when the relaxation times of the two adjacent layers are different, *i.e.*, $\tau_1 \neq \tau_2$, but the interface charge can be discharged with a response time $\tau_{MW} = (C_1 + C_2)/(G_1 + G_2)$. Noteworthy that τ_{MW} is geometry dependent, *i.e.*, thickness of two layers, and above discussion on charging and discharging is based on the bulk property of the two layers. In practical double layer system, the interface is not ideal, owing to the discontinuity of materials. For example, there are lattice mismatching between the adjacent two materials, presence of interface traps, presence of depletion layer, and so on. Among them, the presence of carrier traps at the interface makes a significant contribution to the long-term charge storage even after the shortening the circuit, and the charge storage continues much longer time than the Maxwell's relaxation time τ_{MW}. And once carriers are trapped at the interface traps, they remain as trapped charge for long time, and works as a source of space charge field. It is very important to get information on such interface states for carrier traps for understanding materials interface. Generally, the interface trap-states are distributed energetically, and many characterization methods have been developed. Among them are Deep-Level Transient Spectroscopy (DLTS) [7–10], Thermally Stimulated Current measurement (TSC) (see Chap. 3), Capacitance–frequency (C–f) measurement [11, 12], Conductance–frequency (G–f) measurement [13], Maxwell Displacement Current (MDC) measurement [14–16], and others. The principles of these measurements are similar in that the behavior of carriers released from traps is explored. In Sec. 4.5, we discuss two cases. One is the discharging of interface charges trapped at the interface between two dielectric materials. The other is the discharging of charges trapped at the interface states of insulator–semiconductor interface of Metal–Insulator–Semiconductor (MIS) diodes. We discuss the evaluation of interface states by using TSC, keeping in mind that TSC method is nothing but the MDC method whilst heating. Noteworthy that release of trapped carriers at the interface for discharging is mainly limited by the detrapping process, and this is totally different from the discharging of accumulated interface charge due to the Maxwell–Wagner effect. That is, it takes much longer time to release trapped carriers from the traps in comparison with the transport of released carriers across the dielectric layer, but once the carriers are released from the interface states via detrapping, they can leave very quickly from the interface to the electrode.

4.5 Discharging of trapped carriers at interface

(a) Interface charges trapped at interface between two dielectric materials

Figure 4.4(a) illustrates the carriers trapped at the interface between two dielectrics, where the double layer is electrically shorted. Q_{ss} represents the interface trapped carriers, and a potential V_{ss} is built

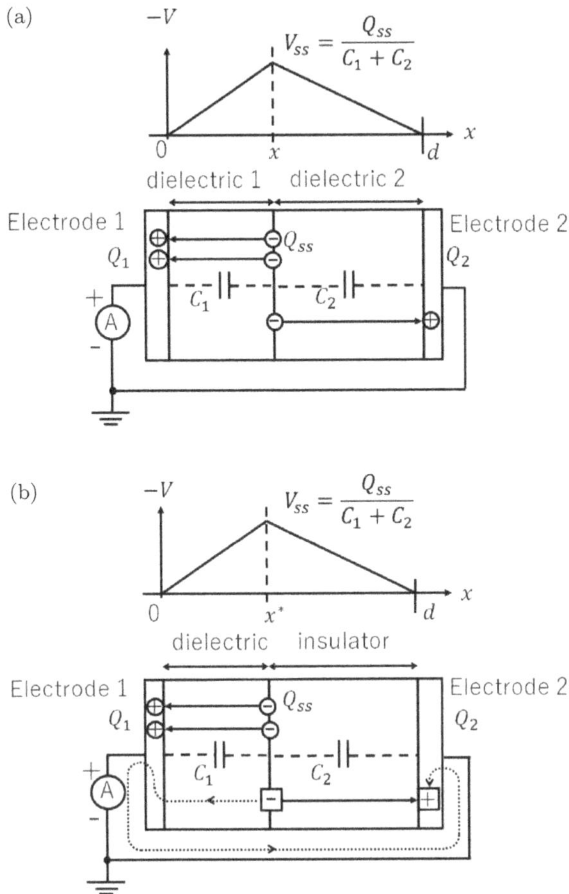

Figure 4.4. Discharging of interfacial trapped charges where the adjacent two layers are (a) two dielectric layers, and (b) dielectric layer and insulating layer where insulating layer functions as the blocking layer. Solid arrows represent the electric field originating from trapped electrons that is connected to positive charges induced on electrodes. In (a), among three electrons, two electrons connect the electric field to two positive charges induced on electrode 1 and the other one electron connects the electric field to one positive charge on electrode 2. After electrons are detrapped, two electrons move to electrode 1 while one electron moves to electrode 2, and recombine with positive charges. No current flows during this process. In (b), the electric formed in the layer is the same as in (a). But the electron connecting electric field to electrode 2 cannot move across the insulator. The electron moves to electrode 1, flows through the external circuit, and recombines with the positive charge on electrode 2. In this process, one electron flows through the ammeter, and non-zero current flows.

at the interface. This potential is given by

$$V_{ss} = \frac{Q_{ss}}{C_1 + C_2}. \tag{4.14}$$

Here C_1 and C_2 are capacitance of the two layers. As a result, internal electric field is formed in the capacitor in the direction from the two-layer interface to the electrode as follows:

$$E_1 = \frac{V_{ss}}{d_1} \quad \text{and} \quad E_2 = \frac{V_{ss}}{d_2}. \tag{4.15}$$

Consequently, on electrodes 1 and 2, charges Q_1 and Q_2 are induced, respectively, and they are given as

$$Q_1 = -C_1 V_{ss} = -C_1 \frac{Q_{ss}}{C_1 + C_2} \quad \text{and} \quad Q_2 = -C_2 V_{ss} = -C_2 \frac{Q_{ss}}{C_1 + C_2} \tag{4.16}$$

Now we consider the situation that electrons are released from the interface trap and then move to electrodes 1 and 2 very quickly under the potential difference between at the interface and at electrodes, and disappear due to the recombination with positive charges induced on electrodes 1 and 2. That is, the time for detrapping is much longer than the time required for carrier transport from the interface to electrodes 1 and 2. This is so-called electron emission limited process from traps, and this process is different from carrier transport limited process described in Sec. 4.3. We here assume that ΔQ_{ss} is released from the trap, and then charges Q_1 and Q_2 induced on electrodes 1 and 2 immediately lose the charge ΔQ_1 and ΔQ_2, respectively, owing to the recombination with charges conveyed through dielectrics 1 and 2. Then we have the following relation

$$Q_1 - \Delta Q_1 = -C_1 \frac{Q_{ss} - \Delta Q_{ss}}{C_1 + C_2} \quad \text{and} \quad Q_2 - \Delta Q_2 = -C_2 \frac{Q_{ss} - \Delta Q_{ss}}{C_1 + C_2}. \tag{4.17}$$

That is, charge $\Delta Q_1 = -C_1 \cdot \Delta Q_{ss}/(C_1 + C_2)$ and $\Delta Q_2 = -C_2 \cdot \Delta Q_{ss}/(C_1 + C_2)$ are lost on electrodes 1 and 2, respectively, owing to the detrapping of interface charge ΔQ_{ss}. In this process, as we discussed in Chap. 2, displacement current I flows and the charge ΔQ flows through an external closed circuit. They are given as follows.

$$\Delta Q = \frac{C_2}{C_1 + C_2}\Delta Q_1 - \frac{C_1}{C_1 + C_2}\Delta Q_2 \tag{4.18}$$

and

$$I = -\frac{\Delta Q}{\Delta t} = -\frac{C_2}{C_1 + C_2}\frac{\Delta Q_1}{\Delta t} + \frac{C_1}{C_1 + C_2}\frac{\Delta Q_2}{\Delta t} \tag{4.19}$$

Substituting ΔQ_1 and ΔQ_2 obtained from Eq. (4.17) into Eq. (4.18) and Eq. (4.19), we get

$$\Delta Q = 0 \tag{4.20}$$

$$I = 0. \tag{4.21}$$

That is, no MDC current flows and the charge recorded by an ammeter is zero. That is, if the carrier behavior of detrapped carriers is ruled by the emission limited process, we cannot see any MDC currents in the short circuit.

It is instructive here to discuss the detrapping of trapped carriers themselves. Assuming that Q_{ss} is being trapped at a single-electron trap state with an energy depth of E_t (>0), the lifetime τ_t of the trapped charge is described as

$$\tau_t = \nu^{-1} \exp\left(+\frac{E_t}{kT}\right) \tag{4.22}$$

Here, ν is attempt to escape frequency, and k is Boltzmann constant and T is absolute temperature. Q_{ss} decays due to emission from traps, on satisfying the following rate equation.

$$\frac{dQ_{ss}(t)}{dt} = -\frac{Q_{ss}(t)}{\tau_t} \tag{4.23}$$

Consequently, in the TSC process where the temperature T increases with a constant heating rate β, *i.e.*, $T = \beta t + T_0$ (T_0: initial temperature), Q_{ss} is given as follows:

$$Q_{ss}(T) = Q_{ss}(0) \exp\left(-\frac{\nu}{\beta}\int_{T_0}^{T} \exp\left(-\frac{E_t}{kT}\right) dT\right). \tag{4.24}$$

That is, trapped charges are thermally released and the interface charge decreases on satisfying Eq. (4.24). In the above discussion, we assumed that carriers emitted from traps can leave the interface and then move to electrodes 1 and 2 in the same way. However, in actual situation, owing to the difference of barrier height of dielectric materials 1 and 2 for electrons emitted from traps, non-zero TSC current will flow. For the evaluation of traps at the interface, it is desirable that we can force to deliver all detrapped carriers to one of the two electrodes, *i.e.*, electrode 1 or electrode 2. Using blocking layer is one of ideas for this purpose, where very high and excellent insulating layer is chosen as one of dielectric layers [17, 18].

We here discuss the case where dielectric material 2 is acting as blocking layer (Fig. 4.4(b)). Obviously, the basic relations on induced charge on electrodes 1 and 2 are the same with Eq. (4.16). Also, the emission of trapped carriers is given by the rate equation Eq. (4.23). However, after the detrapping of trapped carriers, all detrapped electrons ΔQ_{ss} can move only in the direction to electrode 1. On the electrode 1, recombination happens, but only the amount of ΔQ_1 can recombine with the charge Q_1 on electrode and the charge $\Delta Q_2 = \Delta Q_{ss} - \Delta Q_1$ goes through the electrode 1. ΔQ_2 flows through the ammeter installed in the external circuit, and finally reaches electrode 2. And on electrode 2 recombination with the charge Q_2 happens. Though the route of ΔQ_2 from interface to electrode 2 is not open inside of blocking layer, ΔQ_1 and ΔQ_2 must be given using the same equations as $\Delta Q_1 = -C_1 \cdot \Delta Q_{ss}/(C_1 + C_2)$ and $\Delta Q_2 = -C_2 \cdot \Delta Q_{ss}/(C_1 + C_2)$, respectively, to satisfy Eq. (4.17). In summary, by using blocking layer, ΔQ_2 are conveyed to the electrode 2 via external circuit, *i.e.*, through electrode 1 and ammeter, successively. As a result, the MDC current given by the following equations flows through external circuit, and non-zero MDC is recorded:

$$I = +\frac{dQ_2}{dt} \tag{4.25a}$$

$$I = +\frac{dQ_2}{dt} = -C_2\frac{d}{dt}\left(\frac{Q_{ss}}{C_1 + C_2}\right) = -C_2\frac{d}{dt}V_{ss} \tag{4.25b}$$

$$I = +\frac{dQ_2}{dt} = -C_2\frac{d}{dt}\left(\frac{Q_{ss}}{C_1 + C_2}\right) = -\frac{C_2}{C_1 + C_2}\frac{dQ_{ss}}{dt} \tag{4.25c}$$

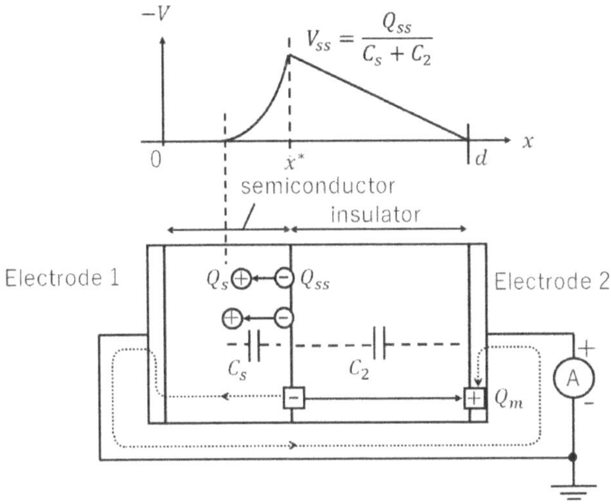

Figure 4.5. Discharging of interfacial trapped charges where the adjacent two layers are n-type semiconductor and insulator.

Equations (4.25a), (4.25b), and (4.25c) express the MDC in terms of electrode charge, interface potential, and interface charge, respectively. Therefore, using blocking layer is a good idea for studying interface states by measuring the MDC and TSC.

(b) Interface charges trapped at insulator–semiconductor interface in MIS diode

Figure 4.5 shows the metal–insulator–n-type semiconductor (MIS) diodes, where interface charges Q_{ss} exist at the insulator–semiconductor interface, and charge Q_m is induced on metal electrode, and Q_s is charge in semiconductor. Q_s is spatially distributed, and it is comprised of charge in depletion layer, inversion layer, and so on. Among these charges, there is a relation of charge neutrality as follows

$$Q_{ss} + Q_s(\psi_s) + Q_m = 0, \tag{4.26}$$

under assumption that there are no trapped charges in insulators, and the work function differencc between semiconductor and metals is negligible.

Here, ψ_s is surface potential of semiconductor and it changes depending on the semiconductor charge Q_s.

According to the semiconductor device physics, space charge Q_s is given as a function of ψ_s as follows [19, 20]:

$$Q_s = [2\varepsilon_s kTn_i\{\exp(+\beta_e\psi_b)(\exp(-\beta_e\psi_s) + \beta_e\psi_s - 1) + \exp(-\beta_e\psi_b)(\exp(+\beta_e\psi_s) - \beta_e\psi_s - 1)\}]^{\frac{1}{2}}. \quad (4.27)$$

Here, $\exp(+\beta_e\psi_b) = n_i/N_D$ and $\beta_e = q/kT$. n_i is intrinsic carrier density, N_D is donor density, and q is electron charge.

Noteworthy that insulator works as blocking layer for thermally emitted electrons from the interface traps, and these electrons cannot cross over the insulator. On the other hand, thermally emitted electros can enter the semiconductor region. Owing to the release of electrons from the interface into semiconductor, Q_{ss}, Q_s and Q_m change accordingly. In the final stage, thermodynamic equilibrium is established and we can see the following relation

$$Q_{ss}^f + Q_s(\psi_s^f) + Q_m^f = 0. \quad (4.28)$$

Here, ψ_s^f, Q_{ss}^f, $Q_s(\psi_s^f)$, and Q_m^f, represents the final state of surface potential, interface charge, semiconductor charge, and charge on metal, respectively.

By using Eq. (4.25), we can derive the thermally stimulated current (TSC) as follows:

$$TSC = -\beta\frac{dQ_m}{dT} \quad (4.29a)$$

and

$$TSC = C_2\beta\frac{d}{dT}\left(\frac{Q_{ss}}{C_s + C_2}\right). \quad (4.29b)$$

$$TSC = \beta\frac{dQ_{ss}}{dT} \text{ with condition } C_2 \gg C_s. \quad (4.29b')$$

Note that polarity of Eq. (4.29) is opposite to Eq. (4.25), reflecting the connection of an ammeter to electrode 2 in Fig. 4.5. Here, $C_s = dQ_s(\psi_s)/d\psi_s$ and this is a function of ψ_s that changes with thermal emission of Q_{ss} at the interface. Accordingly, relation corresponding to Eq. (4.25c) is not the case. Instead, by substituting Eq. (4.26) into Eq. (4.25).

$$TSC(T) = \beta\frac{dQ_{ss}}{dT} + \beta\frac{dQ_s}{dT}. \tag{4.30}$$

In summary, TSC is comprised of thermally emission process of trapped charges Q_{ss} and change of Q_s, but in the case $C_2 \gg C_s$, TSC is approximately given by only the first term of Eq. (4.30) (see Eq. (4.29b′)).

(c) Estimation of interface states in MIS structure

(c–1) Thermally Stimulated Current

TSC provides a powerful way to evaluate interface states [21, 22]. Figure 4.6(a) shows the energy band diagram of MIS diode after d.c. biasing the diode by applying V_b at temperature T_b. Electrons are captured with interface states, where the interface states with energies lower than the Fermi energy E_f of semiconductor are filled with electrons. Figure 4.6(b) shows the energy diagram just after applying voltage V_c to release trapped electrons at low temperature T_c. As the temperature T_c is much lower than T_b, electrons are mostly as trapped while detrapping gradually proceeds. In the semiconductor region near the insulator–semiconductor interface, there is a depleted region. Figure 4.6(c) shows the situation where trapped electrons are thermally emitted from the traps at the interface. As we can see in the figure, insulating layer is acting as electron blocking layer and electrons emitted from the interface traps only enter the semiconductor layer along the depleted region. Figure 4.6(d) shows the energy diagram after the completion of the thermal emission of electrons from the interface traps, where the thermal equilibrium condition is established. As mentioned above, electrons trapped at the interface are thermally emitted from the interface and all of them go into the semiconductor. As that, TSC current flows in a closed circuit, and we can evaluate the distribution of interface energy states. In the followings we show the analysis of interface states by using TSC.

Assuming that, at the interface, electrons are trapped at the energy states between the bottom of conduction energy band, E_c (<0), and the middle of the energy band, and the density of

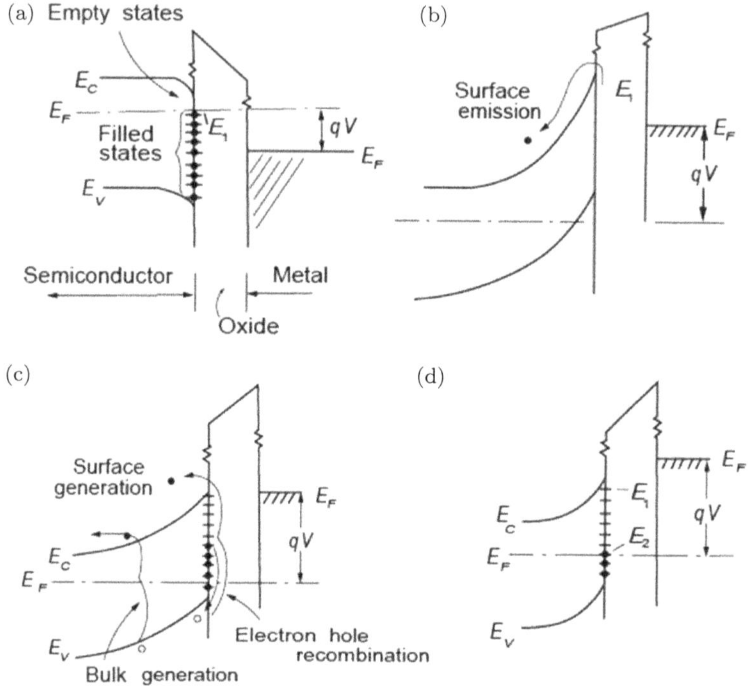

Figure 4.6. Energy diagram of metal–insulator–n-type semiconductor (MIS) diode during a course of TSC measurement. (a) Under d.c. voltage biasing V_b at temperature T_b. (b) Just after applying voltage V_c to release trapped electrons at low temperature T_c. (c) Detrapping from interface states and hole–electron recombination during heating. (d) After the completion of the thermal emission of electrons from the interface traps, and the thermal equilibrium condition is established. Reprinted from Solid-State Electronics, Vol. 17, J. G. Simmons, G. W. Taylor, Theory of non-steady-state interfacial thermal currents in MOS devices, and the direct determination of interfacial trap parameters, p. 125, Copyright (1974), with permission from Elsevier [21].

$\delta n_{ss} = n_{ss} dE'$ electrons are being trapped in the region of energy dE', we can rewrite the rate equation of the emission given by Eq. (4.23) as follows,

$$\frac{d(\delta n_{ss})}{dt} = -e_n \delta n_{ss} \tag{4.31}$$

$$\text{with } e_n = \nu \exp\left(\frac{E' - E_c}{kT}\right)$$

e_n is the thermal emission rate and E' (< 0) is the energy level of interface states measured in reference to the vacuum level. Equation (4.31) represents that the process is limited by thermal emission process, because electrons never return to the trap states due to the re-trapping process. E' represents the energy level, and the trap depth E_t (> 0) is given as $E_c - E'$. From Eq. (4.31), we get the following relation

$$\delta n_{ss} = N_{ss}(E') \exp\left(-\beta^{-1} \int_{T_0}^{T} e_n dT\right) dE' \tag{4.32}$$

where $N_{ss}(E')$ is trap density per unit energy at energy level E'. Using the relation $\delta n_{ss} = n_{ss} dE'$, we get the following relation

$$n_{ss}(E') = N_{ss}(E') f(E', T) \text{ with } f(E', T) \equiv \exp\left(-\beta^{-1} \int_{T_0}^{T} e_n dT\right). \tag{4.33}$$

For $E_c - E' \geq kT$, Eq. (4.33) can be approximately written as

$$f = \exp\left(-\beta^{-1} \int_{T_0}^{T} e_n dT\right) \approx \exp\left(-\frac{e_n k T^2}{\beta(E_c - E')}\right). \tag{4.34}$$

It is instructive here to note that form of Eq. (4.34) is very similar to the Fermi–Dirac distribution function (see Eq. (4.39)), as shown in Fig. 4.7.

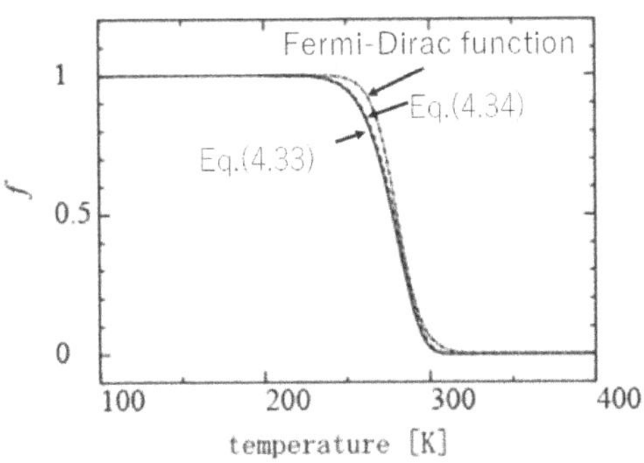

Figure 4.7. Similarity of functions $f(E', T)$ (Eq. (4.33)), f (Eq. (4.34)), and Fermi–Dirac function (Eq. (4.39)). Example shown here is calculated with parameters: $\beta = 10\,\text{K/min}$, $\nu = 10^7$ Hz, $E_c = -3.0\,\text{eV}$, and $E' = -3.5\,\text{eV}$.

Assuming that the thermal emission from the interface is the most at the energy state $E' = E^*$, we can consider that the Fermi energy level E^* is just the energy state $E' = E^*$. At the enery state corresponding to the Fermi energy level, half of the trapped electrons are released, we can have the following relation

$$\exp\left(-\beta^{-1}\int_{T_0}^{T} e_n(E^*)dT\right) \approx \exp\left(-\frac{e_n(E^*)kT^2}{\beta(E_c - E^*)}\right) = \frac{1}{2}. \quad (4.35)$$

Therefore, we may consider that all of the electrons trapped at the energy level $E' > E^* + 2kT$ are released, whereas electrons trapped at the energy level $E' < E^* - 2kT$ are as trapped. With consideration of these, we may conclude that TSC curve gives a peak at the energy state $E' = E^*$ at a temperature of T_m. In this situation, we can assume that, at $T = T_m$ and $E' = E^*$, we have the following relation (see the relation Eq. (3.5))

$$e_n(E^*)^{-1} = \nu^{-1}\exp\left(+\frac{E_c - E^*}{kT_m}\right) = \frac{kT_m^2}{\beta(E_c - E^*)} \quad (4.36)$$

If we substitute this relation into Eq. (4.35), we get the following relation

$$\exp\left(-\beta^{-1}\int_{T_0}^{T_m} e_n(E^*)dT\right) \approx \exp\left(-\frac{e_n(E^*)kT_m{}^2}{\beta(E_c - E)}\right) = e^{-1} = 0.3679. \quad (4.37)$$

Equation (4.37) suggests that the relation given by Eq. (4.35) is mostly satisfied at $T = T_m$. Therefore, the derivation of the relationship between $E_c - E^*$ and T_m from Eq. (4.36) is important. However it is difficult to find analytical solution from the relation Eq. (4.36), and we can only estimate the following relation that satisfies Eq. (4.36) as follows:

$$E_c - E^* = \left(1.92\log\frac{\nu}{\beta} + 3.2\right) \times 10^{-4}T_m - 0.0155(\text{eV}) \quad (4.38)$$

In Fig. 4.8, we plot the result of Eq. (4.38) and the results of numerical simulation that satisfies Eq. (4.36). We can find that relation Eq. (4.38) well agrees with the numerical simulation. With energy level $E = E^*$ corresponding to TSC temperature $T = T_m$,

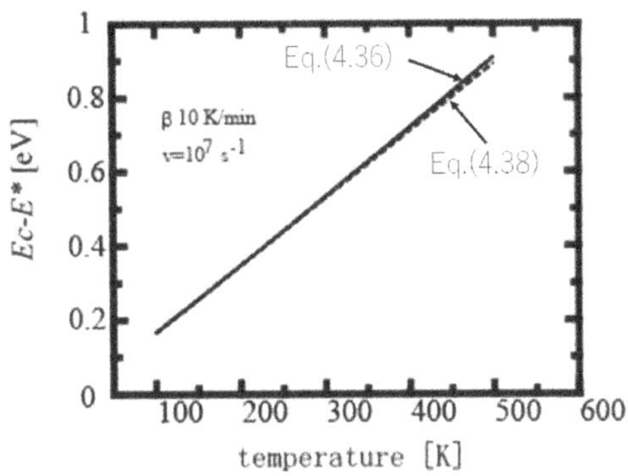

Figure 4.8. Relationship between evaluated trap depth $E_c - E^*$ in eV unit and peak temperature T_m in TSC, estimated from computer simulation of rigorous equation (Eq. (4.36)) and the approximation equation of Eq. (4.38).

the function f (Eq. (4.34)) is represented by Fermi–Dirac-type distribution,

$$f \approx \frac{1}{1 + 1.7\exp\left(2\frac{E'-E^*}{kT}\right)}. \tag{4.39}$$

Eq. (4.39) is $f \cong 1$ for energy level $E' < E^* - 2kT$, $f \cong 0$ for energy level $E' > E^* + 2kT$, and $f = e^{-1}$ at $E' = E^*$ satisfying Eq. (4.37). Fermi–Dirac function (Eq.(4.39)) is calculated by substituting $E' = E^*$ at $T = T_m$ using the relation Eq. (4.38), and is shown in Fig. 4.7. We can see the Fermi–Dirac function agrees to calculation results from the theoretical expression (Eq.(4.33)) and approximated expression (Eq. (4.34)).

On the basis of discussions above, we now extend our analysis to the case where interface states are energetically distributed between the bottom of conduction band and the top of valence band. Assuming the energy level at the center of the energy gap is E_{mi}, the following equation is derived from Eq. (4.31) and Eq.(4.32).

$$\frac{dn_{ss}}{dt} = \int_{E_{mi}}^{E_c} N_{ss}(E')e_n(E')\exp\left(-\beta^{-1}\int_{T_0}^{T} e_n dT\right) dE'. \tag{4.40}$$

In the thermally stimulated process, electrons will be thermally emitted from the interface state gradually from electrons trapped at the shallow states until electrons trapped at the middle energy state E_{mi} is emitted, and the electrons will be conveyed in the direction to semiconductor. In the thermally stimulated current measurement, the temperature increases at a constant heating rate β, Eq. (4.40) is rewritten as

$$\beta\frac{dn_{ss}}{dT} = \int_{E_{mi}}^{E_c} N_{ss}(E')e_n(E')f(E',T)dE'. \tag{4.41}$$

Substituting Eq. (4.41) into Eq. (4.29), we get the TSC equation for analyzing interface states as follows:

$$\begin{aligned} TSC &\approx \beta\frac{qC_i}{C_i + C_d}\frac{dn_{ss}}{dT} \\ &= \frac{qC_i}{C_i + C_d}\int_{E_{mi}}^{E_c} N_{ss}(E')e_n(E')\exp\left(-\beta^{-1}\int_{T_0}^{T} e_n dT\right)dE'. \end{aligned} \tag{4.42}$$

Here C_i is the capacitance of insulating layer (I), and C_d is the capacitance of depletion layer of semiconductor layer (S). Now we here assume that $f(E',T)$ is approximately written in the form of Fermi–Dirac distribution, and freshly define function $P(E',T)$ as

$$P(E',T) \equiv e_n \exp\left(-\beta^{-1}\int_{T_0}^{T} e_n dT\right) \tag{4.43}$$

and substitute in the TSC formula Eq. (4.42). We can find that this function $P(E',T)$ is approximated using δ-function that distributes around $E' = E^*$ with a width of about $2kT$ (see Fig. 4.9). Noteworthy the following relation is satisfied from Eq. (4.43) (see Appendix C).

$$\frac{T}{\beta}\int_{T_0}^{T}\frac{e_n(E^*)}{T}dT = 1 \tag{4.44}$$

In the TSC measurement, E^* changes gradually, but linearly with temperature $T = T_m$, according to the Eq. (4.38), we can conclude that TSC curve itself traces the interface density of states of the MIS diode. In more detail, taking account of similarity between $f(E',T)$

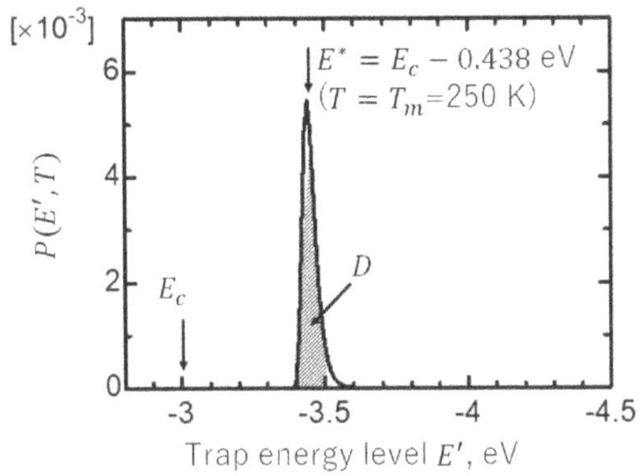

Figure 4.9. The function $P(E',T)$ that is similar to δ-function. D is the area of $P(E',T)$ (see Eq. (4.46) and Appendix D), and $P(E',T) \approx D\delta(E'-E^*)$. Example shown here is $P(E',T)$ for $T = T_m = 250$ K, corresponding to $E^* = E_c - 0.438$ eV. Calculation is carried out for $E_c = -3.0$ eV, $\beta = 10$ K/min, $\nu = 10^7$ s^{-1}, $T_0 = 100$ K. The width of $P(E',T)$ is $2kT = 0.043$ eV, which is narrow compared with the location of trap energy $E_t = E_c - E' = 0.438$ eV.

and Fermi–Dirac distribution function, Eq. (4.43) is approximately written using δ-function as

$$D\delta(E' - E^*) \approx e_n \exp\left(-\beta^{-1} \int_{T_0}^{T} e_n dT\right), \tag{4.45}$$

and we get the TSC formula (Eq. (4.42)) as follows

$$TSC \approx \frac{qC_i}{C_i + C_d} \int_{E_{mi}}^{E_c} N_{ss}(E') D\delta(E' - E^*) dE' \approx \frac{qC_i}{C_i + C_d} DN_{ss}(E^*) \tag{4.46}$$

with $D = 1.2\beta(E_c - E^*)/T$ (see Appendix D). From Eq. (4.46), we find that TSC curve traces the interface states, with Eq. (4.38). If the assumption $C_i \gg C_d$ is valid, Eq. (4.46) is

$$TSC \approx q \int_{E_{mi}}^{E_c} N_{ss}(E') D\delta(E' - E^*) dE' \approx qDN_{ss}(E^*). \tag{4.46$'$}$$

Keeping in mind that temperature T of TSC is linear function of trap energy level E^*, TSC directly traces the interface density of states $N_{ss}(E^*)$. Noteworthy that insulating layer is acting as blocking layer for electrons emitted from the interface states, and all of electrons thermally excited electrons,

$$Q_{ss} = q \int_{E_{mi}}^{E_c} N_{ss}(E')D\delta(E' - E^*)dE',$$

are conveyed to the semiconductor layer. The coefficient $C_i/(C_i + C_d)$ in Eq. (4.46) means this situation. Using the relationship in this section, we can evaluate the interface energy states. Figure 4.10 shows an example of the evaluation of interface states of metal–nitride–oxide–semiconductor (MNOS) diodes by using the TSC measurement [22].

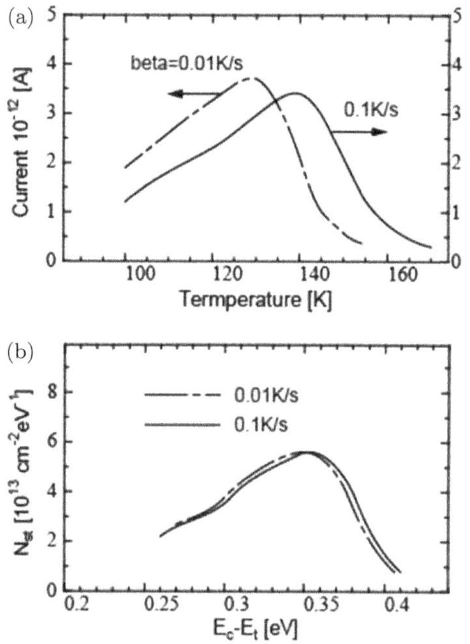

Figure 4.10. Example of tracing interface state distribution in energy space by using TSC. MNOS device is analyzed where the device structure is nitride 850 Å/oxide 18 Å/n-type silicon. (a) TSC curves at two heating rates $\beta = 0.01$ K/s and 0.1 K/s. (b) Evaluated interface states. Reprinted from Solid-State Electronics, Vol. 17, H. A. Mar, J. G. Simmons, Determination of the energy distribution of interface traps in MIS system using non-steady-state techniques, p. 131, Copyright (1974), with permission from Elsevier [22].

(c–2) Thermally Stimulated Surface Potential (TSSP)

In previous sections we show the analysis of interface states by using TSC measurement, where insulating layer (I) works well as electron blocking layer, and all of thermally emitted electrons Q_{ss} at the insulator–semiconductor interface moves toward the semiconductor layer. Nevertheless, semiconductor charge Q_s changes depending on the interface potential ψ_s, on satisfying the relation Eq. (4.27). As a result, TSC depends on the change of Q_{ss} as well as Q_s. In order to overcome this situation, open circuit measurement is very helpful [23, 24]. In open circuit, the amount of charge Q_m on metal electrode never changes during the measurement, even when the temperature of MIS diode increases in the measurement.

On keeping MDC zero, *i.e.*, on measuring surface potential at open circuit condition, zero-field position never move in a course of heating [25, 26] (see Sec. 3.4(i–1), Eq. (3.20)). In other words, no charges cross the insulator nor conveyed through external circuit. According to the charge neutrality condition of the MIS diode, the following relationship is satisfied,

$$Q_{ss}(T) + Q_s(T) + Q_m = 0, \tag{4.47}$$

under the assumption that there are no trapped charges inside of insulator and the work function difference between metals and semiconductor is discarded. In the open circuit, charges induced on metal electrode never move across the insulator, and it is constant. That is, after we put charge $Q_m = Q_m^0$ on metal electrode at $T = T_0$ by applying collecting voltage V_{c0}, the circuit is open. Afterwards, the amount of this charge $Q_m = Q_m^0$ never changes at temperatures $T = T\ (> T_0)$, *i.e.*, during a course of heating. Consequently, during the thermal emission process, Q_{ss} changes, and Q_s changes accordingly, but Q_m is constant, and the following relationship is satisfied.

$$-\frac{dQ_{ss}}{dT} = +\frac{dQ_s}{dT} \tag{4.48}$$

That is, the decrease of interface charge Q_{ss} leads to the increase of semiconductor charge Q_s only. It is instructive here to note that

Q_m^0 is described as follows

$$Q_m^0 = C_i V_i = C_i (V_{TSSP} - \psi_s), \tag{4.49}$$

where V_i is the potential difference across the insulator, V_{TSSP} is the potential recorded in the TSSP measurement, and ψ_s is the surface potential of the semiconductor layer in reference to the semiconductor bulk grounded. From Eq. (4.49), we get the following relation

$$V_{TSSP}(T) = \psi_s(T) + \frac{Q_m^0}{C_i} \tag{4.50}$$

That is, during the heating process in the TSSP, thermal emission of interface charge Q_{ss} leads to the change of surface potential $\psi_s(T)$. Obviously, the thermal emission process in the MIS diodes is the same between in the TSC measurement and in the TSSP measurement, and the Q_{ss} changes with increase of temperature as follows:

$$Q_{ss}(T) = \int_{E_v}^{E_c} qN_s(E') \exp\left(-\beta^{-1} \int_{T_0}^{T} e_n dT\right) dE', \tag{4.51}$$

with $e_n = \nu \exp(E' - E_c)/kT$ with $E' < 0$.

E_c and E_v are energy level of the bottom of conduction band and the top of valence band, respectively. Looking at relation Eq. (4.48), we find that Q_s changes according to the change of Q_{ss}, leading to the change of $\psi_s(T)$, and results in the change of $V_{TSSP}(T)$. The experimental procedure of TSSP measurement is similar to that of TSC measurement. The way of TSSP is illustrated in Fig. 4.11. At first, by applying d.c. biasing voltage V_b to the gate metal electrode at temperature T_b in the same way as we have done for the TSC measurement, the interface trap-states $N_{ss}(E')$ of MIS diode are filled with electrons (see Fig. 4.11(a)). After cooling down the MIS diode to the temperature T_0, we remove the biasing voltage V_b and apply a collecting voltage V_{c0} very shortly, by switching S1 from on-state to off-state, successively. After that we gradually increase the temperature of MIS diode at a constant heating rate β, and measure the electrostatic potential $V(T)$ of metal electrode, in reference to semiconductor grounded. This potential $V(T)$ measured with increasing temperature is the thermally stimulated surface

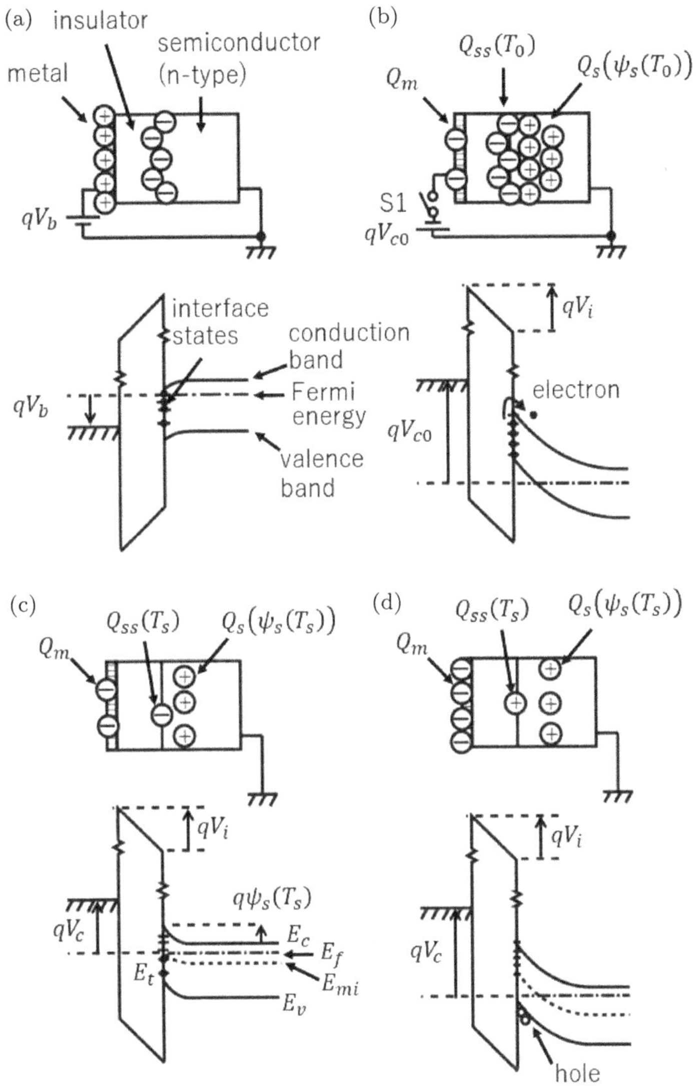

Figure 4.11. (*Continued*)

potential $V_{TSSP}(T) = V(T)$. The distribution of interface trap states can be analyzed in the same way with that of the TSC analysis, by using $V_{TSSP}(T)$. One of the most useful ways is to analyze the distribution of interface states on focusing the thermally equilibrium states of MIS diodes. In the TSSP measurement, Q_{ss} is gradually released from the interface with increase of temperature by the electron emission limited process. Eventually at $T = T_s$ ($\gg T_0$), $Q_s(T_s)$ reaches the thermal equilibrium state, where the following relation is satisfied.

$$Q_{ss}(T_s) + Q_s(\psi_s(T_s)) + Q_m = 0 \tag{4.52}$$

As illustrated in Fig. 4.11(c), at this situation all electrons that have been trapped at the energy states higher than the Fermi–energy of semiconductor is released from the interface trap states and the corresponding interface states are empty. That is, electrons trapped in non-equilibrium state at low temperature T_0 is totally released. The surface potential of semiconductor is $\psi_s(T_s)$ at temperature $T = T_s$. This temperature T_s is easily found in a curve of $V_{TSSP}(T)$ where the tendency of potential change is flattened as shown in Fig. 4.12 (see the dotted line T_s in Fig. 4.12). That is, during a heating from temperature T_0 to T_l ($T_0 < T_s < T_l$), trapped

←

Figure 4.11. TSSP measurement of MIS diode. (a) At first, by applying d.c. biasing voltage V_b to the gate metal electrode at temperature T_b in the same way as in the TSC measurement, the interface trap-states $N_{ss}(E')$ of MIS diode are filled with electrons. (b) After that, cooling down the MIS diode to the temperature T_0, we remove the biasing voltage V_b and apply a collecting voltage V_{c0} very shortly, by switching S1 from on-state to off-state, continuously. During heating process, V_{TSSP} changes and this $V_{TSSP}(T)$ is used for analyzing interface trap distribution. (c)(d) At temperature $T_s \gg T_0$, all of trapped charge located above Fermi–energy in thermal equilibrium state is detrapped. Further heating increase temperature from T_s to T_l only slightly change a location of Fermi–energy, resulting in mostly flat $V_{TSSP}(T)$ in temperature $T_s < T < T_l$. (c) and (d) illustrate equilibrium state where Fermi–energy at semiconductor surface is above and below middle of band-gap E_{mi}, respectively. The location of Fermi–energy after establishing thermal equilibrium depends on the collecting voltage V_{c0}. Reprinted from K. Yamashita, M. Iwamoto, T. Hino, J. Appl. Phys., Vol. 49, 2866, 1978, with the permission of AIP Publishing [24].

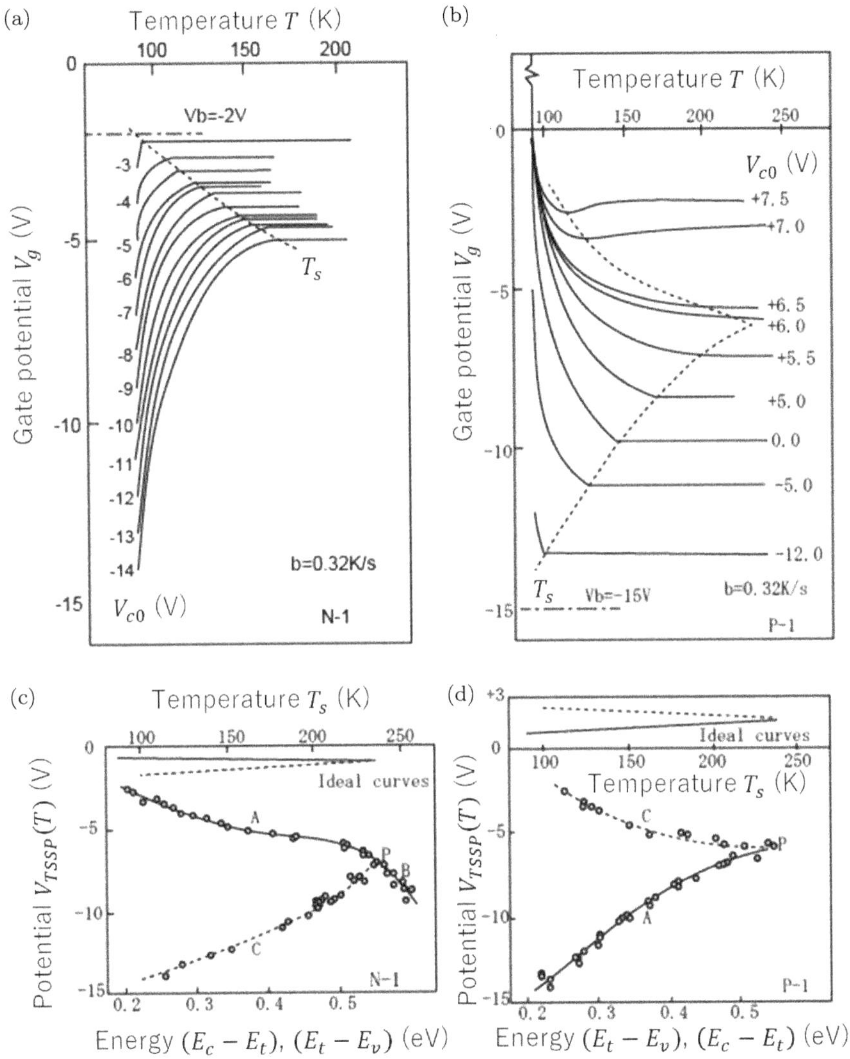

Figure 4.12. (*Continued*)

charge in non-equilibrium state smoothly goes out from trapped state from T_0 to T_s, and arrives at thermal equilibrium state at T_s. This process results in monotonous change of $V_{TSSP}(T)$. For example, in Fig. 4.12(a), $T = 90 \sim 150$ K of the curve starting at $V_{c0} = -11$ V corresponds to this detrapping process from non-equilibrium to equilibrium state. Further heating results in charge distribution change due to a shift of Fermi–energy level as the temperature increases. Assuming that the energy of semiconductor conduction band is E_c, the effective electron density of states at the conduction band is N_c, and the impurity donor density is N_D, the Fermi–level of semiconductor $E_F(T_s)$ at $T = T_s$ is described as:

$$E_F(T_s) = E_c + q\psi_s(T_s) - kT_s \ln \frac{N_c}{N_D}. \tag{4.53}$$

In the temperature region $T_s < T < T_l$, occupation of Q_{ss} changes in accordance with shift of Fermi–energy $E_F(T)$, but this change is small. Consequently, $V_{TSSP}(T)$ from T_s to T_l looks flat. In Fig. 4.12, $T > 150$ K ($V_{c0} = -11$ V) is this flat region. $T_s = 150$ K and corresponding $V_{TSSP}(T_s) = -5$ V is easily defined in experiment. This $V_{TSSP}(T_s) - T_s$ relation is utilized to analyze distribution of interface state $N_{ss}(E_t)$ as described in the following.

Figure 4.12. TSSP measurement of MIS diode (N–1: Al/SiO$_2$ 150 nm/n-type Si, P–1: Al/SiO$_2$ 150 nm/p-type Si). TSSP as a function of temperature where trace of T_s is indicated by a dotted curve for (a) n-type Si and (b) p-type Si MIS diode. $V_{TSSP}(T_s) - T_s$ relationship for (c) n-type Si and (d) p-type Si MIS diode. $V_{TSSP}(T_s) - T_s$ relation is used for analyzing interfacial state. Ideal curves in (c) and (d) are calculated results of Eq. (4.55). Solid curve is ideal curve corresponding to experimental $V_{TSSP}(T_s) - T_s$ relation indicated by A and B. These curves (A, B) reflect detrapping process of electrons from interface states to conduction band. Dotted curve is corresponding to curve C in experiment, which corresponds to detrapping of electrons through recombination with holes in valence band. Accordingly, T_s correlates with $E_c - E_t$ and $E_t - E_v$ for solid curve and dotted curve, respectively. Reprinted from K. Yamashita, M. Iwamoto, T. Hino, J. Appl. Phys., Vol. 49, 2866, 1978, with the permission of AIP Publishing [24].

Using Eq. (4.50) and Eq. (4.52), we get the following relation for the TSSP at $T = T_s$, *i.e.*, $V_{TSSP}(T_s)$ in Fig. 4.12(c)(d).

$$V_{TSSP}(T_s) = \psi_s(T_s) - \frac{Q_s(\psi_s(T_s))}{C_i} - \frac{Q_{ss}(\psi_s(T_s))}{C_i} \tag{4.54}$$

On the other hand, in the case of ideal MIS diodes with no interface states, if the surface potential of semiconductor is $\psi_s(T_s)$, TSSP $V^0_{TSSP}(T_s)$ at $T = T_s$ is derived as follows,

$$V^0_{TSSP}(T_s) = \psi_s(T_s) - \frac{Q_s(\psi_s(T_s))}{C_i}. \tag{4.55}$$

We can calculate the value of $V^0_{TSSP}(T_s)$ with assuming material parameter of semiconductor. For example, $Q_s(\psi_s(T_s))$ is calculated by assuming ψ_s in Eq. (4.27), and ψ_s is estimated by using Eq. (4.53) with assuming $E_t = E_F$ in Eq. (4.53), and $E_t = E^*$ and $T_s \approx T_m$ in Eq. (4.38). The calculated ideal curve of surface potential $V^0_{TSSP}(T_s)$ is shown in Fig. 4.12(c)(d). From Eq. (4.54) and Eq. (4.55), trapped charge $Q_{ss}(\psi_s(T_s))$ is calculated following the relation:

$$Q_{ss}(\psi_s(T_s)) = C_i(V^0_{TSSP}(T_s) - V_{TSSP}(T_s)). \tag{4.56}$$

That is, once we determine the $V_{TSSP}(T_s)$ using the TSSP experimentally, we can determine the $Q_{ss}(\psi_s(T_s))$. If $V_{TSSP}(T_s) - T_s$ relation is experimentally obtained in temperature range $T_{s1} \sim T_{s2}$, corresponding to energy range $E_{t1}(T_{s1}) \sim E_{t2}(T_{s2})$, $Q_{ss}(\psi_s(T_{s1})) - Q_{ss}(\psi_s(T_{s2}))$ is trapped charge in this energy region. $E_{t1}(T_{s1})$ and $E_{t2}(T_{s2})$ are, for example, calculated from Eq. (4.38) with $T_m \to T_{s1}$, T_{s2}. Interface state distribution is derived from Eq. (4.56). Using Eq. (4.56), we get the following relation.

$$-qN_{ss}(E_t) = \frac{\Delta Q_{ss}(\psi_s(T_s))}{\Delta E_t}\frac{\Delta E_t}{\Delta T_s}. \tag{4.57}$$

Here, $\Delta E_t/\Delta T_s$ is obtained, for example, by using Eq. (4.38) with $T_m \to T_s$. In the derivation of Eq. (4.57), we assume that

$$f(E', T) \equiv \exp\left(-\beta^{-1}\int_{T_0}^{T} e_n dT\right)$$

is approximately written as

$$f(E', T) = [1 + \exp((E' - E_F(T))/kT)]^{-1},$$

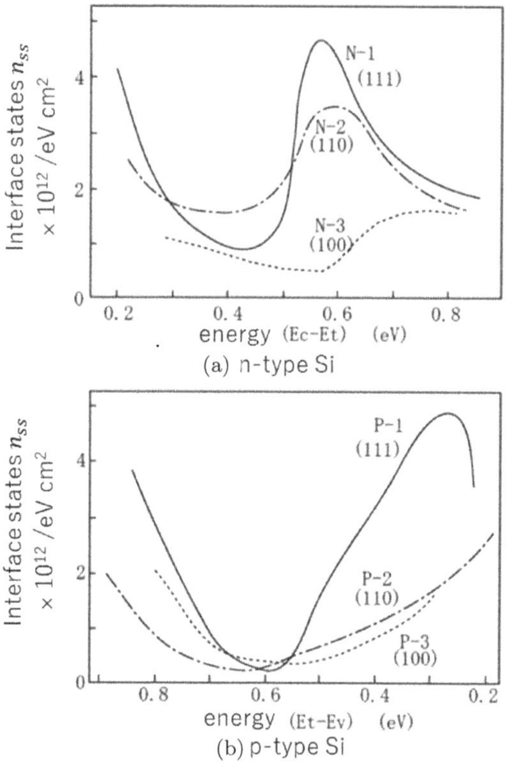

Figure 4.13. Evaluation of interfacial states by TSSP. MIS diode. (a) Al/SiO_2/n-type Si. (b) Al/SiO_2/p-type Si. SiO_2 thickness is 100 – 150 nm. SiO_2 is thermally grown on (111), (110), (100) crystallographic face of Si, which are polished and chemically etched. The growth was carried out in wet oxygen atmosphere. Reprinted from K. Yamashita, M. Iwamoto, T. Hino, J. Appl. Phys., Vol. 49, 2866, 1978, with the permission of AIP Publishing [24].

and carriers trapped at energies $E' = E_t$ around the Fermi energy $E_F(T)$ is thermally released from the interface. Figure 4.12 shows an example of TSSP, $V_{TSSP}(T)$. From $V_{TSSP}(T)$ curves with different biasing voltage V_b and collecting voltage V_{c0}, a set of T_s and $V_{TSSP}(T_s)$ is obtained and plotted in Fig. 4.12(c)(d). The computed $V^0_{TSSP}(T_s)$ is also shown in the figure, which are almost linear to T_s. Using these relations with Eq. (4.57), we can plot the distribution of interface states. Figure 4.13 shows an example of the evaluation of the interface states $N_{ss}(E_t)$ of Si–SiO_2 interface for n-type and p-type Si, where SiO_2 is thermally grown on (111), (110), (100)

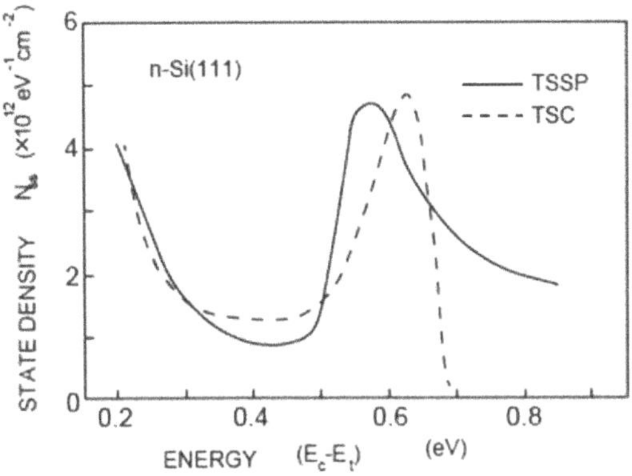

Figure 4.14. Evaluation of interfacial states $N_{ss}(E_t)$ by using TSC and TSSP. These analyzed results are similar in general, but detailed structure is different depending on the physical processes that are responsible to TSC and TSSP. Reprinted from K. Yamashita, M. Iwamoto, T. Hino, J. Appl. Phys., Vol. 49, 2866, 1978, with the permission of AIP Publishing [24].

crystallographic face polished and chemically etched. Oxide growth was carried out in wet oxygen atmosphere. Results well reflected general tendency that interface state density decreases in the order (111), (110), (100). Note that, by increasing collecting voltage step-by-step as $V_{c0}, V_{c1}, V_{c2}, \ldots,$ during the course of heating, quick experimental way to obtain $V_{TSSP}(T_s) - T_s$ relation for electron detrapping to conduction band is developed [23, 24].

Figure 4.14 shows the comparison of the analysis by TSSP and TSC. TSSP result shown here is the interface states $N_{ss}(E_t)$ of n-type Si (111)–SiO_2 interface in Fig. 4.13(a). TSC result is that measured for the same sample used for the TSSP measurement, and was analyzed in the way described in Sec. 4.5(c–1), with biasing voltage $V_b = 0$ V and constant collecting voltage $V_c = -13$ V. General tendency given by TSSP and TSC analysis looks similar but detailed structure of interface states are different. One reason is that TSSP uses $V_{TSSP}(T_s)$ at T_s where Fermi–energy at thermal equilibrium is used as a trap energy depth while TSC uses maximum temperature T_m where changing from non-equilibrium to equilibrium state is the most. Accordingly, interface state distribution $N_{ss}(E_t)$ of TSC steeply decreases to zero near 0.65 eV that is limited by the applied collecting voltage. Another reason is the fact that TSSP is open circuit measurement while TSC is short circuit measurement. As a result, zero-field position is fixed in TSSP but is possible to move in TSC measurement. In TSC measurement, changing of carrier behavior due to motion of zero-field position can result in additional interface states apparently.

4.6 Summary

In this chapter, accumulation and trapping of charge at interface was described. Maxwell–Wagner effect and carrier trapping states are introduced as an origin of interface charge accumulation phenomena. As a conventional way to analyze interface states, TSC and TSSP measurements were described. By using these experimental methods, on keeping in mind fundamental physical process behind the measurement and analysis, high-quality and stable characterization are carried out, and is helpful for development of new electrical and electronic materials and devices. The concept of Maxwell–Wagner effect is well applied to analyze carrier processes in organic devices and introduced in Chap. 7 and Chap. 8. TSC is well adopted to new research field such as organic electronics as described in Chap. 3. Further, analysis of interface states using TSC is connected to new experimental method of optical second-harmonic generation (SHG), which are described in Chap. 6 and Chap. 9.

References

1. Purcell, E. M. (1963). *Electricity and Magnetism*, Berkley physics course volume 2 (McGraw–Hill, New York London) Sec. 4.11. pp. 138–141.
2. Maxwell, J. C. (1954). *A Treatise on Electricity & Magnetism*, vol. 1 (Dover, New York) Art. 325. pp. 450–451.
3. Maxwell, J. C. (1954). *A Treatise on Electricity & Magnetism*, vol. 1 (Dover, New York) Art. 328–329. pp. 452–456.
4. Wagner, K. W. (1914). Archiv für Elektrotechnik, 2, 371–387, https://doi.org/10.1007/BF01657322.
5. Oka, S. and Nakata, O. (1960). *Kotai Yudentai Ron* (*Theory of Condensed Dielectric Materials*) (Iwanami, Tokyo) Chap.5. pp. 80–105 (in Japanese), https://doi.org/10.11501/1378002.

6. Fynmann, R. P., Leighton, R. B. and Sands, M. (1964). *The Fynmann Lectures on Physics*, vol. II (Addison-Wesley, Massachusetts) Sec. 5–9 "The fields of a conductor".
7. Lang, D. V. (1974). Deep-level transient spectroscopy: A new method to characterize traps in semiconductors, *Appl. Phys. Lett.*, 45, 3023–3032, https://doi.org/10.1063/1.1663719.
8. Usami, A. and Tokuda, Y. (1990). *Handoutai Debaisu Koutei Hyouka Gijutsu: Raifutaimu DLTS Hyouka O Chuushin Toshite* (*Evaluation Techniques for Semiconductor Devices: Lifetime Evaluation by using Deep Level Transient Spectroscopy*) (Realize, Tokyo) (in Japanese); Usami, A. and Tokuda, Y. (2006). *Handoutai Debaisu Koutei Hyouka Gijutsu: Raifutaimu DLTS Hyouka O Chuushin Toshite* (*Evaluation Techniques for Semiconductor Devices: Lifetime Evaluation by using Deep Level Transient Spectroscopy*) (Realize AT, Tokyo) (in Japanese).
9. Crowell, C. R. and Alipanahi, S. (1981). Transient distortion and nth order filtering in deep level transient spectroscopy (DnLTS), *Solid-State Electronics*, 24, 25–36, https://doi.org/10.1016/0038-1101(81)90209-4.
10. Thurzo, I., Pogány, D. and Gmucová, K. (1992). *Solid-State Electronics*, 35, 1737–1743, https://doi.org/10.1016/0038-1101(92)90254-A.
11. Torres, I. and Taylor, D. M. (2005). Interface states in polymer metal–insulator–semiconductor devices, *J. Appl. Phys.*, 98, 073710, https://doi.org/10.1063/1.2081109.
12. Takagi, K., Nagase, T., Kobayashi, T. and Naito, H. (2016). Determination of deep trapping lifetime in organic semiconductors using impedance spectroscopy, *Appl. Phys. Lett.*, 108, 053305, https://doi.org/10.1063/1.4941235.
13. Nicollian, E. H. and Goetzberger, A. (1967). The Si–SiO_2 interface-electrical properties as determined by the metal–insulator–silicon conductance technique, *The Bell System Technical Journal*, 46, pp. 1055–1133, https://doi.org/10.1002/j.1538-7305.1967.tb01727.x.
14. Egusa, S., Gemma, N., Miura, A., Mizushima, K. and Azuma, M. (1992). Carrier injection characteristics of the metal/organic junctions of organic thin-film devices, *J. Appl. Phys.*, 71, 2042–2044, https://doi.org/10.1063/1.351150.
15. Noguchi, Y., Miyazaki, Y., Tanaka, Y., Sato, N., Nakayama, Y., Schmidt, T. D., Brütting, W. and Ishii, H. (2012). Charge accumulation at organic semiconductor interfaces due to a permanent dipole moment and its orientational order in bilayer devices, *J. Appl. Phys.*, 111, 114508, https://doi.org/10.1063/1.4724349.
16. Tanaka, Y., Noguchi, Y., Kraus, M., Brütting, W. and Ishii, H. (2013). Three-terminal capacitance–voltage measurements of pentacene field-effect transistors during operation, *Org. Electron.*, 14, 2491–2496, https://doi.org/10.1016/j.orgel.2013.06.013.
17. Suzuki, T., Kudo, K. and Tanaka, S. (1980). Thermally stimulated current from polymethyl-methacrylate (PMMA) electrets with a needle-plane electrodes geometry, *IEEJ Trans. Fundam. Mater.*, 100, 95–102 (in Japanese), https://doi.org/10.1541/ieejfms1972.100.95.

18. Giacomettii, J. A., Malmonge, J. A. and Neto, J. M. G. (1986). Open-circuit TSD method and anomalous air gap current in Teflon®FEP, *IEEE Trans. Electr. Insul.*, EI–21, 383–387, https://doi.org/10.1109/TEI.1986.349080.
19. Sze, S. M. and Ng, Kwok K. (2007). *Physics of Semiconductor Devices*, 3rd Ed. (New York, John Wiley & Sons), https://doi.org/10.1002/0470068329 p. 202, Eq. (10); Sze, S. M. (1981). *Physics of Semiconductor Devices*, 2nd Ed. (New York, John Wiley & Sons) p. 367, Eq. (11).
20. Iwamoto, M. and Taguchi, D. (2014). *Thermally Stimulated Current in Electrical and Electronic Devices — Analysis and Application* (Tokyo, Corona Publishing) p. 117 and p. 176, Chap. 5, Exercise 4 (in Japanese).
21. Simmons, J. G. and Taylor, G. W. (1974). Theory of non-steady state interfacial thermal currents in MOS devices, and the direct determination of interfacial trap parameters, *Solid-State Electronics*, 17, 125–130, https://doi.org/10.1016/0038-1101(74)90060-4.
22. Mar, H. A. and Simmons, J. G. (1974). Determination of the energy distribution of interface traps in MIS system using non-steady-state techniques, *Solid-State Electronics*, 17, 131–135, https://doi.org/10.1016/0038-1101(74)90061-6.
23. Yamashita, K., Iwamoto, M. and Hino, T. (1978). A method for determining MIS interface states by thermally stimulated surface potential, *IEEJ Trans. Fundam. Mater.*, 98, 93–100 (in Japanese), https://doi.org/10.1541/ieejfms1972.98.93.
24. Yamashita, K., Iwamoto, M. and Hino, T. (1978). A method for determining MIS interface states by thermally stimulated surface potential, *J. Appl. Phys.*, 49, 2866–2875, https://doi.org/10.1063/1.325169.
25. Kato, K., Iwamoto, M. and Hino, T. (1986). Measurement of the position of zero-field plane of space charge field in insulators by temperature gradient thermally stimulated surface potential, *IEEJ Trans. Fundam. Mater.*, 106, 251–258 (in Japanese), https://doi.org/10.1541/ieejfms1972.106.251.
26. Iwamoto, M. and Kato, K. (1987). Determination of the zero-field position in an insulating film by thermally stimulated surface potential measurement in the presence of a temperature gradient, *IEEE Trans. Electr. Insul.*, EI–22, 413–417, https://doi.org/10.1109/TEI.1987.298901.

Chapter 5

Interface Dielectric Polarization and its Characterization Using Optical Methods

The properties of *Surface and Interface of Materials* are quite different from those of the bulk. In this chapter, we focus on the specific feature of the surface and interface, where dipolar orientation and dipolar motion are restricted. With consideration of the restriction, we discuss the dielectric polarization property of monolayers on the surface. The status of positional arrangement and orientational alignment of constituent molecules on the surface is key to understand the interface dielectric polarization property. The concept of orientational order parameter is adopted to analyze the dielectric property of organic monolayers. It is shown that spontaneous polarization, linear polarization, and nonlinear polarization are expressed by using orientational order parameters $S_n (n = 1, 2, 3, \ldots)$. We use optical and electrical experimental methods to characterize the dielectric polarization of surface monolayers. Optical second-harmonic generation (SHG) measurement is utilized to probe nonlinear polarization. Ellipsometry, Brewster–Angle Microscopy (BAM), and Brewster–Angle reflectivity (BAR) measurement are used to probe the linear polarization of organic monolayers. Electrical Maxwell displacement current (MDC) measurement is used to probe spontaneous polarization. In this chapter, we discuss how the dielectric property of monolayers is characterized by using these experimental techniques, in terms of the orientation order parameters.

5.1 Surface layers and dielectric property of monolayers

On account of symmetry broken structure of monolayers on the material surface, molecular films on solid and water surface show interesting behaviors. The behaviors are restricted to two-dimensional (2D) ones, and quite different from those of 3D bulk materials. In other words, material is the same but shows different behavior at the surface due to symmetry breaking. Among surface films, floating films on the water surface are one of the most popular ones. We can easily handle their two-dimensional behaviors by compression and expansion using a floating movable-barrier. We therefore can get important information on the specific dielectric properties of surface ultrathin films through the study of floating monolayers on the water surface. We here focus on organic monolayers on the water surface, but the knowledge from this study is useful for the understanding of the properties of monomolecular films on any kind of material surfaces.

Since the discovery of chemical and physical force of floating monomolecular films on the water surface by Langmuir [1, 2], monolayers at the air–water interface become a research topic for physicists, chemists and biologists. The floating monomolecular films are called as Langmuir–films. The study of Langmuir films advanced when the technique of surface pressure–area ($\pi - A$) isotherm measurement was established. Step by step, many instruments equipped with the $\pi - A$ isotherm measuring system have been developed. Among them are optical instruments for the measurement of the thickness and the refractivity, e.g., ellipsometer, mechanical instruments for the measurement of the viscosity of floating monolayer, electrical instruments for the measurement of the potential difference across the monolayer, *etc.* [2].

Noteworthy that for the past several decades, many new experimental techniques and analytical methods have been developed for the study of monolayers, including Langmuir–film and self-assembly monolayer [3]. Various experimental methods and related technologies such as ellipsometry imaging [4], nonlinear optical method [5, 6], *etc.*, have been proposed. Consequently, the physical and chemical properties of monolayers at the air–water interface have been clarified in terms of monolayer structure. Using X–ray diffraction [7], Brewster angle microscopy (BAM) [8, 9], sum-frequency generation (SFG) [10], and other state-of-the-art instruments, the structure of monolayer has been clarified, and the textures of monolayers have been visualized. For example, we can find an extensive study on Langmuir monolayers in a review article by Kaganer *et al.* [7], in which mainly the ordering structure of fatty acid monolayer is summarized with experimental data.

However, the physical properties of specific two-dimensional systems have not yet been discussed thoroughly from the viewpoint of dielectric physics. That is, the knowledge on monolayers based on dielectric physics related to spontaneous polarization, linear and nonlinear polarizations, chiral phase separation, piezoelectric effect, flexo-electric effect, *etc.*, is still not sufficient for the use in the fields of nano-electronics, soft material electronics, low-dimensional physics, interfacial physics, bio-physics, *etc.*

In order to explore the dielectric property of monolayers on the water surface in association with the two-dimensional behaviors, it is necessary to develop freshly both experimental and theoretical methods. Experimentally, measurements on surface potential, Maxwell displacement current (MDC) are useful to study the spontaneous polarization. Optical measurements using ellipsometer and Brewster Angle Reflectivity (BAR) instrument are useful to study the dielectric property of monolayers in terms of linear polarization in the optical frequency region, and BAM is useful to visualize monolayer structures macroscopically. Optical second-harmonic generation (SHG) [11–13] measurement is very helpful for probing nonlinear polarization of monolayers, which stems from the symmetrically broken structure of monolayers. It is instructive here to note that the generation of surface potential, MDC and SHG from monolayers happens in the presence of spontaneous and nonlinear polarizations in monolayers, in association with the polar non-centrosymmetric structure of monolayers, which is not observed in films with centrosymmetric structure. That is, the specific dielectric property of monolayers is due to the non-centrosymmetric structure of the films with respect to the surface, and these structures can be explored by surface potential, MDC, and SHG measurements [13].

With surface potential method, amphiphilic monolayers at the air–water interface, *e.g.*, monolayers of fatty acids and bipolar lipids, have been investigated. We can find typical examples in classical monolayer studies by Adam, Danielli and Harding [14], Schulman and Rideal [15], Macdonald and Barlow [16], and other researchers. In these studies, monolayers were treated very simply as parallel-plate capacitors of continuum medium with certain dielectric constant, where only classical experimental techniques were used.

After that monolayer studies have advanced according to the development of experimental techniques and theoretical analyses, where the information on the monolayer structure has been well adopted based on the advanced experimental techniques and analyses. Typical ones are studies using a two-layer and three-layer model, where the contribution of the permanent dipoles of hydrophilic and hydrophobic tails of constituent molecules was taken into consideration, and the relationship between the monolayer structure and the dielectric polarization of monolayers was argued as reviewed in [17]. Nevertheless, the statistical distribution of the constituent polar molecules was not considered.

Similarly, using the state-of-art SHG measurement, many studies on monolayers have been carried out by Shen [18], Heinz [19], Eisenthal [20], and others, in terms of the alignment of molecules on the water surface and liquid–solid interface. Under the assumption that the constituent molecules possess a single component of the molecular hyperpolarizability of second order susceptibility along the long axis of the constituent molecule, detailed information on the orientation and the electronic transitions in molecules in monolayers has been clarified step by step. Nevertheless, surface potential measurement, Maxwell displacement current measurement, BAR measurement, SHG measurement, *etc.*, have not been carried out based on common theoretical concept. It will be convenient and useful if we can discuss the spontaneous polarization leading to the generation of surface potential and MDC, linear polarization leading to BAM images, optical reflection, *etc.*, and nonlinear polarization leading to SHG, on the basis of common theoretical background.

For this purpose, monolayers at the air–water interface should be treated as quasi two-dimensional (2D) systems using a simple molecular model, where their states are characterized by the geometrical configuration and the statistical orientational distribution of the constituent polar molecules. With consideration of the crystal symmetry of organic monolayers, many theoretical studies have been done to express the physical quantities such as dielectric constant, linear and nonlinear polarization.

However, owing to the complexity in the mathematical treatment of nonlinear polarization, they are still not complete. For example, many theoretical studies on SHG generated from monolayers have been done, and the second order susceptibility (SOS) tensor of the monolayers [21, 22] have been obtained. Generalized formula was presented by Andrews *et al.* [23–25] and by Popov *et al.*, and documented in a book [26]. The formula is rather complex, but had been employed to analyze the experimental results for monolayers on the water surface. However, understanding of the physics of monolayer SOS tensor in association with the molecular SOS tensor of the constituent chiral and achiral molecules would be more useful, *e.g.*, for the study using spectroscopic tool such as SHG in coupling with MDC [27].

Furthermore, the analysis employed must be easily applicable to discuss the nanometric dielectric properties of the films. Some theoretical and experimental studies on monolayer systems for the MDC–SHG spectroscopy have been carried out. Also, the experiments by coupling with BAM and MDC measurements, and others have been developed. This chapter focuses on the interface dielectric polarization phenomena in monolayers. Results for a rod-like liquid crystal molecule of *4′–n–octyl–4–cyanobiphenyl* (8CB) are displayed as examples.

5.2 Order parameters and polarization of monolayer films

As mentioned in Sec. 5.1, the state of floating monolayers composed of rod-like polar molecules can be expressed by using two kinds of order parameters, planar order parameter [7] and orientation order

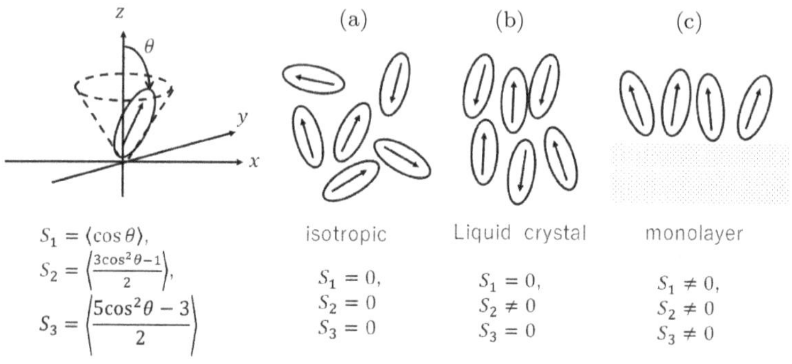

Figure 5.1. Symmetry breaking at surface and interface, represented using orientation order parameter S_n $(n = 1, 2, 3, \ldots)$. (a) Isotropic orientation in bulk of ordinary material. All orientation order parameters are zero. (b) Liquid crystals are specified by non-zero S_2. (c) Monolayer at the surface and interface where orientation ordering due to symmetry breaking is represented by non-zero S_1, S_2, and S_3.

parameter [13]. The former one expresses the molecular configuration and describes the positional distribution of the heads of molecules on the water surface. The latter one expresses the orientation distribution of rod-like polar molecules on the water surface. The orientation order parameter is important to specify the dielectric property of the monolayers. It is instructive to note that material systems composed of rod molecules can be classified using orientation order parameters S_n $(n = 1, 2, 3, \ldots)$, defined as $S_n = \langle P_n(\cos\theta)\rangle$ $(n = 1, 2,$ and $3)$ by using the thermodynamic average $\langle \ldots \rangle$ of the Legendre polynomials, $P_n(\cos\theta)$ $(n = 1, 2,$ and $3)$, of the orientation angle θ of molecules from the z–direction (see Fig. 5.1). Figure 5.1(a) shows the isotropic bulk materials, where rod molecules are randomly distributed in space. Consequently, we can easily find that

$$S_1 = \langle P_1(\cos\theta)\rangle = \langle\cos\theta\rangle = 0$$

$$S_2 = \langle P_2(\cos\theta)\rangle = \left\langle \frac{3\cos^2\theta - 1}{2} \right\rangle = 0$$

and

$$S_3 = \langle P_3(\cos\theta)\rangle = \left\langle \frac{5\cos^3\theta - 3\cos\theta}{2} \right\rangle = 0.$$

Figure 5.1(b) shows bulk liquid crystals (LCs) in the nematic phase, where the orientation direction of rod molecules is spatially restricted and molecules are totally distributed up and down directions, but there is no distinction between up and down molecular directions. Thereby, $S_1 = 0$, $S_3 = 0$, but $S_2 \neq 0$. Figure 5.1(c) shows the monolayer composed of rod molecules, where all molecules randomly orient in space, only in the upper direction. Thereby $S_1 \neq 0$, $S_2 \neq 0$, and $S_3 \neq 0$. From Fig. 5.1, it is found that the specific physical properties of nematic LCs can be specified by using S_2. Similarly, the physical properties of monolayers can be described using a set of non-zero order parameters, *e.g.*, S_1, S_2, and S_3. Obviously, among these non-zero parameters, S_1 and S_3 are the specific parameters that are connected to the nature related to symmetry breaking property, which is induced by the restriction of orientation. In other words, the specific physico-chemical properties of monolayers at the interface can be expressed using non-zero $S_1 \neq 0$ and $S_3 \neq 0$.

In this section, the polarization of a monolayer with C_∞–symmetry is explored using the orientation order parameters, S_1, S_2 and S_3. The specific properties of organic monolayers such as spontaneous polarization and nonlinear polarization can be interpreted using these three parameters, but without using higher order parameters S_n ($n \geq 4$). Therefore, in the development of SHG–MDC–BAR spectroscopy, the determination of these orientation order parameters is very important, and the establishment of theory of dielectric property of monolayers can be accomplished by using these three order parameters [28]. It is instructive here to note that the linear polarization is related to S_2, and this parameter is common to all materials. For example, the reflectivity of monolayers is expressed as a function of S_2, in the same way as that of bulk materials. Among nonlinear optical methods, SHG is expressed as a function of S_1 and S_3. MDC is a function of S_1.

5.2.1 *Monolayers with C_∞–symmetry*

We here consider a simple case, where each rod-like molecule has a permanent dipole moment μ along its long axis together with an anisotropic electronic polarizability, *i.e.*, electronic polarizability α_p

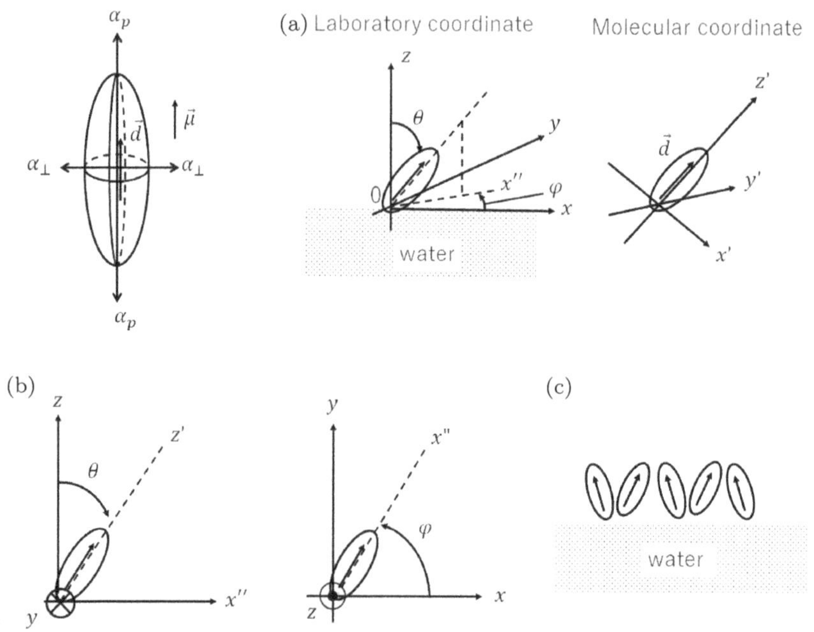

Figure 5.2. Rod-like molecule with a permanent dipole μ along its long direction. (a) Laboratory coordinate system (x, y, z) and molecular coordinate system (x', y', z') that specify orientation of the molecule. (b) Projection of molecular coordinate on x''–z plane and x–y plane. (c) C_∞–symmetry orientational distribution of rod-like molecules.

parallel to the molecular long axis, and $\alpha_\perp$ perpendicular to the molecular long axis (see Fig. 5.2). Furthermore, the molecule possesses a second-order susceptibility $\alpha_M^{(2)}$. The water surface is flat, and constituent molecules of the monolayer distribute randomly on the surface. The laboratory coordinate system is chosen in such that the monolayer surface is parallel to the $x - y$ plane and the monolayer normal is pointing along the positive z–axis, as illustrated in Fig. 5.2(a). The molecule at the origin is facing the water surface, and tilts with an angle θ from the positive z–axis.

Figure 5.3(a) illustrates a Langmuir trough where monolayer on water surface is investigated by compression and expansion of monolayer by using the barriers equipped at the trough. The molecular area A, *i.e.*, an area divided by surface molecular density (see Sec. 2.4), decreases or increases in accordance with the monolayer compression or the monolayer expansion, respectively.

The rod-like molecule also orientates with an azimuthal angle φ from the x–axis (Fig. 5.2). The molecular coordinate system (x', y', z') is attached to the dipolar molecule at the origin with z'–axis pointing toward the molecular long axis $\vec{d}$ where $\vec{d}$ is an unit vector. The permanent dipole moment $\vec{\mu}$ coincides with the molecular long axis and can be written as $\vec{\mu} = \mu\vec{d}$. In the laboratory coordinate system (x, y, z), the direction $\vec{d}$ of the polar molecule is given by $(\sin\theta\cos\varphi, \sin\theta\sin\varphi, \cos\theta)$. As the interaction between the rod-like molecules and the water surface is considered to be independent of the azimuthal angle φ, it is easy to expect that the average molecular direction, called director $\vec{m}$, is normal to the water surface, *i.e.*, in the direction $\vec{n} = (0, 0, 1)$ in the laboratory coordinate system (x, y, z). This means that the monolayer has C_∞–symmetry.

As mentioned earlier, the universal definition $S_n \equiv \langle P_n(\cos\theta)\rangle$ can be adopted as the orientation order parameters to describe the collective alignment properties of the molecules in monolayers [29–31], in the way as

$$S_n \equiv \langle P_n(\cos\theta)\rangle = \frac{\int_0^{2\pi}\int_0^{\pi} P_n(\cos\theta) f(\theta,\varphi)\sin\theta d\theta d\varphi}{\int_0^{2\pi}\int_0^{\pi} f(\theta,\varphi)\sin\theta d\theta d\varphi},$$

using the Boltzmann distribution function $f(\theta, \varphi)$ that describes the statistical spatial distribution of molecules, and it is proportional to $\exp(-W/k_BT)$. Here W is the interaction energy and it is a function of θ and φ. The energy W includes the interaction energy W_{in} between constituent molecules, the interaction energy W_{ext} between external field and dipoles, *etc.* k_B is Boltzmann constant and T is the absolute temperature.

For C_∞–symmetry molecular system on the water surface, $f(\theta, \varphi)$ and W are independent of φ and they are merely a function of θ. The order parameter

$$S_2 = \langle P_2(\cos\theta)\rangle = \left\langle \frac{3\cos^2\theta - 1}{2} \right\rangle$$

introduced by Tsvetkov [32, 33] is convenient for describing bulk properties, whereas

$$S_1 = \langle P_1(\cos\theta)\rangle = \langle \cos\theta\rangle$$

and

$$S_3 = \langle P_3(\cos\theta)\rangle = \left\langle \frac{5\cos^3\theta - 3\cos\theta}{2} \right\rangle$$

specify the state of the monolayer induced by breaking the symmetry from centrosymmetric bulk to non-centrosymmetric interface.

When molecules are subjected to an external electric field $\vec{E}$, the dielectric polarization given by the sum of the electronic polarization and the orientational polarization is induced [34]. The dielectric polarization of monolayers $\vec{P}$ is expressed by

$$\vec{P} = N_s\langle \vec{m}_p\rangle = N_s\langle\vec{\mu}\rangle + N_s\langle \overleftrightarrow{\alpha}^{(1)}\rangle \cdot \vec{E} + N_s\langle \overleftrightarrow{\alpha}^{(2)}\rangle : \vec{E}\vec{E}, \quad (5.1)$$

where the first term is due to dipolar spontaneous polarization and the other terms are due to the electronic polarization. N_s is the surface density of the molecules, $\overleftrightarrow{\alpha}^{(1)}$ is the linear electronic polarizability, and $\overleftrightarrow{\alpha}^{(2)}$ is the second-order molecular polarizability. With $\langle\vec{\mu}\rangle$, $\langle\overleftrightarrow{\alpha}^{(1)}\rangle$ and $\langle\overleftrightarrow{\alpha}^{(2)}\rangle$, the dielectric polarization is derived as a function of the orientation order parameters S_1, S_2, and S_3.

Under the external field $\vec{E} = \vec{E}_{ext}$ applied to the monolayer, the interaction energy W_{ext} is given by

$$W_{ext} = -\int \vec{E}_{ext} \cdot d\vec{m}_p \approx -\mu\vec{d} \cdot \vec{E}_{ext}, \quad (5.2)$$

$$\text{with} \quad \vec{m}_p \approx \vec{\mu} + \overleftrightarrow{\alpha}^{(1)} \cdot \vec{E}_{ext} + \overleftrightarrow{\alpha}^{(2)} : \vec{E}_{ext}\vec{E}_{ext}. \quad (5.2')$$

Noteworthy that the part of interaction energy between induced dipoles $\overleftrightarrow{\alpha}^{(1)} \cdot \vec{E}_{ext}$ and the external field, the interaction energy of higher order term $\overleftrightarrow{\alpha}^{(2)} : \vec{E}_{ext}\vec{E}_{ext}$, *etc.*, are neglected. The first and the third terms in Eq. (5.2′) are related to the generation of MDC and SHG, respectively, whereas the second term concerns with general optical phenomena related to linear polarization, *e.g.*, optical reflection.

It is noteworthy that the extended rod-like molecule model is available to investigate more general cases [35, 36]. The extended model means that a rod-like molecule is on the surface of cylinder and possesses a permanent dipole in a direction deviated away from the molecular long-axis at a distance l_0 perpendicular to the molecular long-axis. Here the direction of the cylinder is defined by its centerline. The impact of contribution of the component of dipole moment parallel and perpendicular with respect to molecular long axis to interaction energy W_{in} can be analyzed, with consideration of electrostatic image potential and dipole–dipole interaction between molecules. The chiral phase separation in monolayer is analyzed by merely assuming a model that the direction of dipole is tilting from the long-axis of rod-like molecule [36]. Basically, the dipole–dipole interaction in these systems is described using a set of orientation order parameters, and specific physical phenomena of interfacial monolayer is analyzed using these parameters.

5.2.2 *Spontaneous polarization and orientational polarization*

Suppose that the orientation distribution of the constituent rod-like polar molecules in the monolayer film obeys Boltzmann distribution, *i.e.*, $f(\theta, \varphi)$ is the Boltzmann distribution function proportional to $\exp(-W/k_B T)$, the dielectric polarization due to the orientational polarization $\vec{P}_d$, given by the first term of Eq. (5.2′), is obtained with Onsager approximation [37] as

$$\vec{P}_d = N_s \langle \vec{m}_p \rangle = N_s \mu S_1 \vec{n} + N_s \frac{\mu^2}{3k_B T} \overleftrightarrow{S}_{ori} \cdot \vec{E} \tag{5.3}$$

Here $\overleftrightarrow{S}_{ori}$ is defined by

$$\overleftrightarrow{S}_{ori} = \begin{pmatrix} 1 - S_2 & 0 & 0 \\ 0 & 1 - S_2 & 0 \\ 0 & 0 & 1 - 3S_1^2 + 2S_2 \end{pmatrix}$$

with the first and second order parameters

$$S_1 = \langle P_1(\cos\theta) \rangle = \langle \cos\theta \rangle,$$

and

$$S_2 = \langle P_2(\cos\theta)\rangle = \left\langle \frac{3\cos^2\theta - 1}{2} \right\rangle,$$

when $\vec{E} = 0 \cdot \overleftrightarrow{S}_{ori}$ describes the influence of the orientation distribution of polar molecules in the monolayer films. It is clear that $\overleftrightarrow{S}_{ori}$ depends on the monolayer structure characterized by the two orientation parameters S_1 and S_2.

In Eq. (5.3), the first term is the spontaneous polarization and the second term is the orientational dipolar polarization induced under external electric field. From Eq. (5.3), it is found that the contribution of orientational dipolar polarization is given by

$$N_s \frac{\mu^2}{3k_B T} E,$$

if $S_1 = 0$ and $S_2 = 0$, *i.e.*, $\overleftrightarrow{S}_{ori} = \overleftrightarrow{I}$. This is nothing but the Debye–Langevin equation which expresses the orientational polarization of permanent dipoles in the isotropic bulk states [34, 37]. In other words, the monolayer has specific property originating from non-zero orientation order parameters, S_1 and S_2, owing to the symmetry breaking and spatial restriction.

In the MDC measurement, MDC generated from monolayers flows in a closed-circuit ($\vec{E} = 0$) due to the change of the non-zero S_1 whilst the monolayers are compressed and expanded, as will be described in Sec. 5.3. Consequently, only the contribution of the first term of Eq. (5.3) is recorded as an MDC. As a result, the orientation order parameter S_1 is determined using the recorded MDC.

5.2.3 *Electronic polarization and nonlinear polarization*

The main contribution to the electronic polarization is induced dipoles, which are characterized by the electronic polarizability of the constituent molecules of the monolayer. The second term of Eq. (5.3) describes the linear electronic polarization. In most cases, the electronic polarizability is not a scalar, but a tensor as a result of the dielectric anisotropy originating from the specific molecular shape.

The linear parts of the electronic molecular polarizability of rod-like molecules are α_p and $\alpha_\perp$. Thus, in the (x', y', z') molecular frame, an anisotropic electronic polarizability $\alpha^{(1)}_{M\,i',j'}$ $(i', j' = (x', y', z'))$ is written as

$$\overleftrightarrow{\alpha}^{(1)}_M = \begin{pmatrix} \alpha_\perp & 0 & 0 \\ 0 & \alpha_\perp & 0 \\ 0 & 0 & \alpha_p \end{pmatrix} \tag{5.4}$$

For C_∞–symmetry monolayer comprising of molecules expressed by Eq. (5.4), under the assumption that the interaction energy of monolayer with C_∞–symmetry is specified only as a function of angle θ defined from the normal direction to the surface, we obtain

$$\langle \overleftrightarrow{\alpha}^{(1)} \rangle = \bar{\alpha}^{(1)} \overleftrightarrow{I} - \frac{1}{3} \Delta\alpha^{(1)} \Gamma S_2 \tag{5.5}$$

with $\Delta\alpha^{(1)} = \alpha_p - \alpha_\perp$ and $\bar{\alpha}^{(1)} = (\alpha_p + 2\alpha_\perp)/3$. Here $\overleftrightarrow{I}$ is 3×3 unit tensor and Γ is given by

$$\Gamma = \begin{pmatrix} 1 & 0 & 0 \\ 0 & 1 & 0 \\ 0 & 0 & -2 \end{pmatrix}, \quad \overleftrightarrow{I} = \begin{pmatrix} 1 & 0 & 0 \\ 0 & 1 & 0 \\ 0 & 0 & 1 \end{pmatrix}. \tag{5.6}$$

As shown in Eq. (5.5), the anisotropic electronic polarization associates only with the second orientation order parameter S_2, representing the centrosymmetric structure of materials, but not with order parameter S_1 or S_3, representing the non-centrosymmetric structure of materials. This result is reasonable because induced dipole, not permanent dipole, contributes to the electronic polarization induced by linear polarization, which is proportional to the external electric field. With the components of the anisotropic electronic polarizability expressed by Eq. (5.5), the linear electronic polarization $\vec{P}_e$ is given. The linear electronic polarization of monolayer may be explored by a technique based on optical ellipsometry method, with which the anisotropic dielectric constant of monolayer can be determined together with the monolayer thickness. With the ellipsometry method and other optical techniques, orientation order parameter S_2 can be estimated. Noteworthy that the dipolar polarization given by the second term of Eq. (5.3) is linear polarization and it is proportional to the electric field, but this polarization does not appear in the optical frequency region.

On the other hand, higher order polarization arises from the nonlinear electronic polarizability $\langle\overleftrightarrow{\alpha}^{(2)}\rangle$, there are additional nonlinear electronic polarizability produced by the local-field molecular interaction, and the nonlinear orientational induced dipole moment as a result of the classical electrostatic interaction between the dipole moment and the external field, but they are generally very small. On the other hand, in case of SHG, the nonlinear polarizabilities originate from the quantum interaction between electrons in molecules and the external electric field, and they are quite large at the specific material-dependent angular frequency [21, 22]. The main contribution to the SHG is characterized by the second-order nonlinear electronic polarizability $\langle\overleftrightarrow{\alpha}^{(2)}\rangle$ at an angular frequency ω of the external electric field. This polarization is expressed as

$$\vec{P}^{(2)} = \chi^{(2)} : \vec{E}\vec{E} = N_s\langle\overleftrightarrow{\alpha}^{(2)}\rangle : \vec{E}\vec{E}. \tag{5.7}$$

The second-order susceptibility (SOS)

$$\chi^{(2)} \equiv [\chi_{ijk}]$$

of the monolayer is related to the components of the nonlinear molecular electronic polarizability $\alpha_{i'j'k'}$ by $\chi_{ijk} = N_s\langle T_{ijk}^{i'j'k'}\rangle\alpha_{i'j'k'}$, where $T_{ijk}^{i'j'k'} = R_i^{i'}R_j^{j'}R_k^{k'}$ with the usual Euler rotation matrix

$$R(\phi,\theta,\psi) = [R_i^{i'}]$$

describes the coordinate transformation between the molecular system (x', y', z') and the laboratory system (x, y, z) [38], that is,

$$R_i^{i'} = \begin{pmatrix} \cos\phi\cos\theta\cos\psi - \sin\phi\sin\psi & -\cos\phi\cos\theta\sin\psi - \sin\phi\cos\psi & \cos\phi\sin\theta \\ \sin\phi\cos\theta\cos\psi + \cos\phi\sin\psi & -\sin\phi\cos\theta\sin\psi + \cos\phi\cos\psi & \sin\phi\sin\theta \\ -\sin\theta\cos\psi & \sin\theta\sin\psi & \cos\theta \end{pmatrix}$$

Equation (5.7) takes no account of the local field correction factors. However, the underlying physics does not change even when this factor is introduced on the right hand side of Eq. (5.7).

With SHG, one can deduce $\chi^{(2)}$, but has to have $T_{ijk}^{i'j'k'}$ to obtain SOS from $\alpha^{(2)}$. The macroscopic SOS tensor components χ_{ijk} of a monolayer with C_∞–symmetry are immediately obtained as a function of the components of the molecular nonlinear electronic polarizability $\alpha_{i'j'k'}$ and the orientation order parameters $S_n \equiv \langle P_n(\cos\theta)\rangle$. Nonlinear polarization $\vec{P}^{(2)} = \vec{P}^N$ is found to be a sum of the polarization $\vec{P}^N_{ch}$ associated with the chirality of the monolayer and the polarization $\vec{P}^N_{ach}$ associated with the non-chirality of the monolayer (achiral). The two polarizations in vector form are [27, 39–44]:

$$\begin{aligned}\vec{P}^N_{ch} &= \frac{1}{2}s_{14}[(\vec{E}\cdot\vec{n})(\vec{F}\times\vec{n}) + (\vec{F}\cdot\vec{n})(\vec{E}\times\vec{n})]\\ &\quad + \frac{1}{2}a_{14}(\vec{E}\times\vec{F}) + \frac{1}{2}(a_{36}-a_{14})\vec{n}\cdot(\vec{E}\times\vec{F})\vec{n}\end{aligned} \tag{5.8}$$

$$\begin{aligned}\vec{P}^N_{ach} &= (s_{33}-s_{15}-s_{31})(\vec{n}\cdot\vec{E})(\vec{n}\cdot\vec{F})\vec{n} + s_{31}(\vec{E}\cdot\vec{F})\vec{n}\\ &\quad + \frac{1}{2}s_{15}[(\vec{n}\cdot\vec{F})\vec{E} + (\vec{n}\cdot\vec{E})\vec{F}] + \frac{1}{2}a_{15}(\vec{E}\times\vec{F})\times\vec{n}\end{aligned} \tag{5.9}$$

with the 7 independent nonzero elements given by

$$\begin{aligned}
s_{14} &= \frac{N_s}{2}S_2\,(\sigma_{14}-\sigma_{25}) = \chi_{123}+\chi_{132}\\
s_{15} &= \frac{N_s}{5}(S_1-S_3)(2\sigma_{33}-\sigma_{32}-\sigma_{31})\\
&\quad + \frac{N_s}{10}(3S_1+2S_3)(\sigma_{24}+\sigma_{15}) = \chi_{131}+\chi_{113}\\
s_{31} &= \frac{N_s}{10}(S_1-S_3)(2\sigma_{33}-\sigma_{24}-\sigma_{15})\\
&\quad + \frac{N_s}{10}(4S_1+S_3)(\sigma_{32}+\sigma_{31}) = \chi_{311}\\
s_{33} &= \frac{N_s}{5}(S_1-S_3)(\sigma_{32}+\sigma_{31}+\sigma_{24}+\sigma_{15})\\
&\quad + \frac{N_s}{5}(3S_1+2S_3)\sigma_{33} = \chi_{333}
\end{aligned}$$

$$a_{14} = \frac{N_s}{3}(\lambda_{14}+\lambda_{25}+\lambda_{36}) + \frac{N_s}{6}S_2(\lambda_{14}+\lambda_{25}-2\lambda_{36}) = \chi_{123}-\chi_{132}$$

$$a_{15} = \frac{N_s}{2}S_1(\lambda_{15}-\lambda_{24}) = \chi_{131}-\chi_{113}$$

$$a_{36} = \frac{N_s}{3}(\lambda_{14}+\lambda_{23}+\lambda_{36}) + \frac{N_s}{3}S_2(2\lambda_{36}-\lambda_{14}-\lambda_{25}) = \chi_{312}-\chi_{321}. \tag{5.10}$$

where the two 3×6 matrices, σ_{ij} and λ_{ij} are defined by the components of the nonlinear molecular polarizability $\alpha_{i'j'k'}$ in the same way as the conventional contracted notation of s_{ij} and a_{ij} defined from $\chi^{(2)}$, *i.e.*, $\sigma_{11} = \alpha_{111}$, $\sigma_{14} = \alpha_{123} + \alpha_{132}$, $\lambda_{14} = \alpha_{123} - \alpha_{132}$ *etc.* $\vec{E}$ and $\vec{F}$ represent the two kinds of external fields. In Eq. (5.8) and Eq. (5.9), these are used.

In the SHG measurement, the electric fields $\vec{E}$ and $\vec{F}$ are the same and they are coming from the same optical source, but in the SFG measurement (sum frequency measurement) *etc.* [10–12, 21, 22], these are coming from two independent electric field sources $\vec{E} \neq \vec{F}$. In this section, we confine our discussion to the case of SHG, but we can carry the discussion on SFG *etc.*, in the same way, using Eq. (5.8) and Eq. (5.9). In the following analysis, thereby $\vec{E} = \vec{F}$ is assumed.

From the nonlinear polarization point of view (see from Eq. (5.8) to Eq. (5.10)), the orientation order parameters S_1 and S_3 are related to the nonlinear polarization $\vec{P}^N_{ach}$ associated with the nonchirality of the monolayers, whereas S_2 is related to the nonlinear polarization $\vec{P}^N_{ch}$ associated with the chirality of the monolayers.

In case of monolayers with C_∞–symmetry, where the constituent molecules are achiral (non-chiral), *i.e.*, $s_{14} = a_{14} = a_{36} = 0$ in Eq. (5.8) and Eq. (5.9), the nonlinear electric polarization in SHG ($\vec{E} = \vec{F}$) reduces to

$$\vec{P}^N = (s_{33} - s_{31} - s_{15})(\vec{n}\cdot\vec{E})^2\vec{n} + s_{31}(\vec{E}\cdot\vec{E})\vec{n} + s_{15}(\vec{n}\cdot\vec{E})\vec{E}. \tag{5.11}$$

The nonlinear dielectric polarization is a function of the order parameters S_1 and S_3 together with the three elements related to the nonchirality, s_{31}, s_{33}, and s_{15}.

Similarly, in case of monolayers with C_∞–symmetry, where the constituent molecules are chiral, *i.e.*, s_{14}, a_{14}, $a_{36} \neq 0$ in Eq. (5.8) and Eq. (5.9), the nonlinear electric polarization in SHG ($\vec{E} = \vec{F}$) is given by

$$\vec{P}^N = s_{14}(\vec{E}\cdot\vec{n})\cdot(\vec{E}\times\vec{n}) + (s_{33} - s_{31} - s_{15})(\vec{n}\cdot\vec{E})^2\vec{n}$$
$$+ s_{31}(\vec{E}\cdot\vec{E})\vec{n} + s_{15}(\vec{n}\cdot\vec{E})\vec{E}. \tag{5.12}$$

The first term on the right hand side of Eq. (5.12) as a function of s_{14} with S_2 (see Eq. (5.10)) represents the chiral effect. As mentioned earlier, the quantum interaction between electrons in molecules and the external electric field makes a main contribution to SHG, and thus the nonlinear dielectric polarization given by Eq. (5.11) and Eq. (5.12) becomes the main contribution to the SHG.

5.3 SHG–BAR–MDC measurement

5.3.1 *Method*

Figure 5.3(a) shows a schematic diagram of the MDC measurement coupled with SHG measurement. Electrode 1 is suspended in air, and is placed parallel to the water surface. Electrode 2 is immersed in the water. These two electrodes 1 and 2 are connected to each other through an electrometer. The induced charge on Electrode 1 changes in accordance with the orientational motion of the molecules on the water surface and the change of the surface density of the molecules. Thus, MDC flows through the closed circuit by monolayer compression using two barriers crossing water surface (see Fig. 5.3(a)). The induced charge Q_1 on Electrode 1 coming from the spontaneous polarization $\vec{P} = N_s\mu S_1\vec{n}$ is given as (see also Sec. 2.4, Fig. 2.12 and Eq. (2.9)).

$$Q_1 = -\vec{P}\cdot\vec{z}\frac{B}{L} - C\phi_s = -P_z\frac{B}{L} - C\phi_s. \tag{5.13}$$

Here the contribution of the polarization normal to the water surface P_z is $\vec{P}\cdot\vec{z}$ ($\vec{z}$: unit vector normal to the water surface, and it is the same with $\vec{n} = (0,0,1)$) and B is the working area of Electrode 1, C is the capacitance between Electrode 1 and the water surface, L is

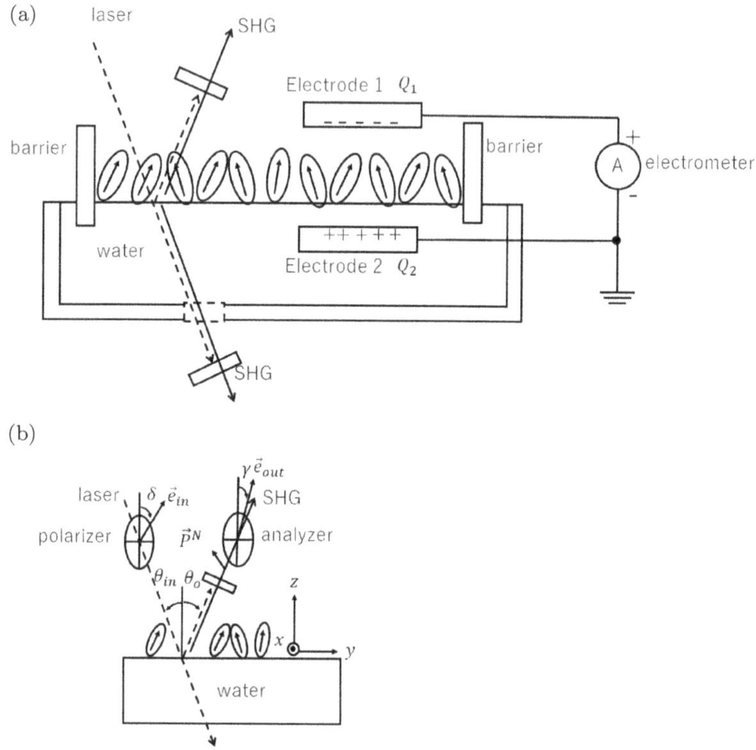

Figure 5.3. (a) MDC–SHG measurement system for determining orientation order parameters S_1 and S_3. (b) Optical arrangement of SHG measurement. θ_{in} and θ_o represent the incident and the output angles, respectively. The angles δ and γ are the polarized angles of the incident light and the output SH light, respectively. $\vec{e}_{in}$ is an unit vector pointing in vibrating electric field of incidence laser beam and $\vec{e}_{out}$ is that of SH light. Laboratory coordinate system (x, y, z) is set as that z-axis points in normal to water surface and y-axis is in the incident plane of laser beam.

the distance between Electrode 1 and the water surface, and ϕ_s is the surface potential of the water. In the MDC measurement, Q_1 changes by monolayer compression, and current $I = -dQ_1/dt$ flows through the short circuit. In other words, MDC flows due to the change of P_z. In the low surface pressure region, the density of molecule is very low and the molecules are lying on the water surface on account of the electrostatic Coulomb interaction between polar molecules and the water. This means that $P_z = 0$, *i.e.*, the orientation order parameter is zero, $S_1 = 0$. In case of monolayer with C_∞–symmetry, by using Eq. (5.13) the order parameter S_1 is determined as a function of the molecular area under the assumption that S_1 is zero in the region of low molecular densities and the contribution to the change of the surface potential of water ϕ_s is negligibly small [45, 46]. Furthermore, it is clear from Eq. (5.13) that phase transition can be detected easily, since it accompanies an abrupt change of the polarization P_z at the transition point.

It is instructive to note that in the surface potential measurement, compensation voltage V_{comp} is applied to electrode 1 such that $Q_1 + CV_{comp} = 0$. As a result, Q_1 does not change even when electrode 1 vibrates up and down mechanically at $V = V_{comp}$. That is, no current flows through the external circuit. Here the compensation voltage V_{comp} is obviously given by

$$V_{comp} = P_z \frac{B}{CL} + \phi_s = \frac{P_z}{\varepsilon_0} + \phi_s. \tag{5.14}$$

That is V_{comp} depends on the monolayer polarization P_z. Consequently, P_z can be determined by measuring the surface potential V_{comp}, whereas the change of P_z with time is difficult to detect. In comparison with the measurement of surface potential, the MDC measurement is effective in the detection of phase transition.

On the other hand, for the SHG measurement, second-harmonic (SH) light is generated from the monolayer by laser irradiation. The experimental arrangement is illustrated in Fig. 5.3(b), where θ_{in} and θ_o represent the incident and the output angles, respectively. The angles δ and γ are the polarized angles of the incident light and the output SH light, respectively. $\vec{e}_{in}$ is an unit vector pointing in vibrating electric field of incidence laser beam and $\vec{e}_{out}$ is that of SH light. The input light consists of s- and the p-polarized waves and is expressed as

$$\vec{E}_{in} = S_\omega \vec{s}_{in} + P_\omega \vec{p}_{in} \tag{5.15}$$

where $\vec{s}_{in}$ and $\vec{p}_{in}$ are unit vectors for the s- and the p-polarized direction of the waves, and S_ω and P_ω are the amplitude of the s- and the p-polarized waves, respectively.

With the incident plane in the $y - z$ plane (see Fig. 5.3(b)), the elements of $\vec{s}_{in}$ and $\vec{p}_{in}$ are given by $(-1, 0, 0)$ and $(0, \cos\theta_{in}, \sin\theta_{in})$, respectively. With δ as the angle of the polarizer for the input light (see Fig. 5.3(b)), S_ω and P_ω are given by $S_\omega = E_{in}\sin\delta$ and $P_\omega = E_{in}\cos\delta$, respectively. Similarly, the output light $\vec{P}^N$ is the sum of the s- and the p-polarized wave. With these, the output light intensity $I_{2\omega}$ is proportional to the square of $\vec{e}_{out}(2\omega)\cdot\vec{P}^N$. Noteworthy that the SHG is due to the second order nonlinear electronic polarization induced by the quantum interaction between the electrons and external electric field. With proper choice of the angles δ and γ under the assumption that $\alpha^{(2)}_{3'3'3'}$ is the main contributor, the orientation order parameters, S_1 and S_3, of monolayers with C_∞–symmetry can be determined, experimentally, since $\vec{e}_{out}(2\omega)\cdot\vec{P}^N$ is a function of S_1 and S_3 [43, 47]. Furthermore, the discussion in Sec. 5.2 shows that the SHG intensity depends on the polarization and the direction of the electric field of the output signals. Thus, we may conclude that phase transitions such as the transition from planar to orientational alignment, chiral–achiral transition, and others can be detected by the SHG measurement. Furthermore, we can examine the chirality of monolayers by taking into account the contribution of the first term of Eq. (5.12) related to chiral term s_{14} (see also Eq. (5.8), Eq. (5.9), and Eq. (5.10)).

5.3.2 *BAR–MDC measurement*

Brewster Angle Microscopy (BAM) is used to directly visualize the texture of monolayers that floats on the water surface. At the same time, probing reflectometry at Brewster's angle (BAR) is possible and has been established as another means of investigating monolayer films [28]. When p-polarized light is incident to a Fresnel interface at Brewster's angle, Θ_B, from the normal direction to the interface, all of the reflected light vanishes. If the interface is altered, *i.e.*, by applying a monolayer to the water surface, the reflected light will no longer vanish and can be collected. From this reflected light, direct information about the monolayer can be obtained.

Noteworthy that the reflection concerns with linear polarization of the monolayer in the optical frequency region, and it is related to the optical phenomena given by the second term of Eq. (5.2′) with linear electronic polarizability given by Eq. (5.4). Reflected light is due to the interaction with electric field and induced polarization, *i.e.*, $\vec{e}_{out}(\omega)\cdot\vec{P}_e$. As we already discussed above, the electronic polarization $\vec{P}_e$ is merely a function of S_2. We can say that S_2 can be determined using BAR. Consequently, we can investigate the change of S_2 during monolayer compression and expansion, in the way like in the MDC and SHG measurements.

5.4 Experimental system

The MDC, SHG and BAR experiments can be carried out with the experimental setup [28, 43] shown schematically in Fig. 5.3 and Fig. 5.4. Briefly, the experimental setup consists of a Langmuir–trough equipped with a two-electrode arrangement for the MDC measurement, and the optical measurement arrangement with a Q–switched Nd^{3+}:YAG laser (wavelength 0.532 μm), pulse duration $<$7 ns, fundamental pulse rate $\leq$15 Hz) for the SHG measurement (see Fig. 5.3). The rectangular Langmuir–trough (600 mm $\times$ 150 mm in length and width, 10 mm in depth) is made of polytetrafluoroethylene (PTFE) and filled with pure water (resistivity 17 MΩcm). A transparent silica glass plate is attached to the bottom of the Langmuir–trough for the SHG measurement.

For MDC measurement, transparent glass slide coated with Indium–Tin Oxide (ITO) is used as Electrode 1 (electrode area, 45.4 cm^2). It is placed in the air and is parallel to the water surface at a distance of 1 mm. Electrode 2 is a spiral gold-wire (1 mm in diameter and 500 mm in length) and is immersed in the water. Electrodes 1 and 2 are connected to each other through an electrometer with negligible internal electric resistance. For SHG measurement, the laser beam is irradiated onto the monolayer with an energy of about 6 mJ/pulse at a repetition rate of 2 Hz. The size of the laser spot is about 56 mm^2. After spreading a 0.001 mol/l chloroform solution of molecules, *e.g.*, *4′–n–octyl–4–cyanobiphenyl*

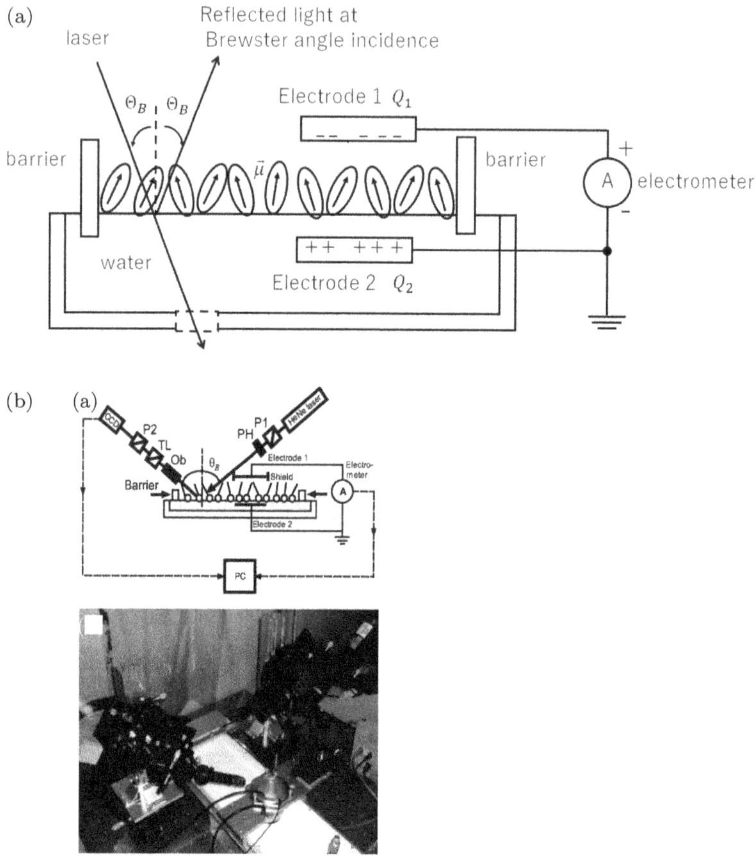

Figure 5.4. (a) MDC–BAR measurement system for determining orientation order parameters S_1 and S_2. (b) Experimental setup of BAR–SHG system. Input and output optical paths are set at the Brewster's angle Θ_B at air–water interface. P1 and P2 are linear polarizer. Objective lens (Ob), tube lens (TL), and CCD camera form a microscope imaging system. PH is a pin-hole used for optical adjustment *etc.* Reprinted from R. Wagner, T. Yamamoto, T. Manaka, M. Iwamoto, Rev. Sci. Instrum., Vol. 76, 083902, 2005, with the permission of AIP Publishing [28].

(8CB, see Fig. 5.5), onto the water surface of the Langmuir–trough to form a floating monolayer, the MDC–SHG measurement can be carried out with monolayer compression at a speed of 10 mm/min by moving the barrier. The surface pressure of the monolayer was also measured during the monolayer compression by a Wilhelmy plate.

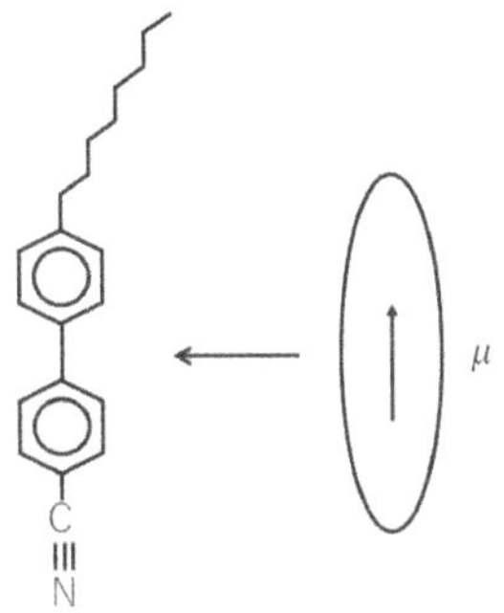

Figure 5.5. 4′–*n*–octyl–4–cyanobiphenyl molecule (8CB). 8CB molecule is rod-like shape and modelled as a rod with a dipole μ along the long-axis of the rod.

For the BAR experiment, the experimental set up is similar (Fig. 5.4), but a 15 mW He–Ne laser at a wavelength of 633 nm, which is *p*-polarized by a Glan–Thompson polarizer (P1) is used and it is incident onto the water surface at an angle of $\Theta_B = 53.1°$ of Brewster's angle for an air–water surface. Basically, the experimental setup for BAR is the same with the SHG system illustrated in Fig. 5.3, except that the incident angle θ_{in} (and θ_o) is chosen at 53.1°, to satisfy the Brewster angle for an air–water interface. Noteworthy, by recording the image of reflections of monolayers, we can visualize the texture of monolayers on the water surface. The image is nothing but the BAM images. Experimentally, a programable CCD camera is utilized for the visualization.

5.5 Probing orientation order parameters

5.5.1 *BAR measurement and determination of S_2*

Figure 5.6 shows the general surface pressure–area isotherm of 4′–*n–octyl–4–cyanobiphenyl* (8CB) monolayers. 8CB molecules are well established as rod shaped. Initially, in the very dilute region with an area per molecule greater than that of 80 Å^2, the monolayer is described as a 2D gas. The molecules are randomly oriented and generally laying parallel to the water surface. In this phase, the orientation order parameter S_2 is zero as a result of the non-establishment of the director due to the randomness of the molecules. Here, the

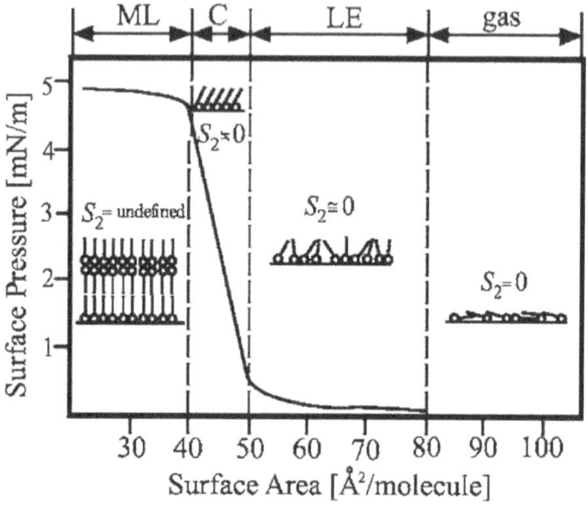

Figure 5.6. Surface pressure–area (π–A) isotherm of 8CB monolayer on water surface. Initially, the monolayer is a 2D gas and upon compression enters into the liquid expanded (LE) phase, where the molecules gradually standup. Compressing further into the condensed (C) phase, the molecules become ordered and the director tilt angle is defined. Compressing past this region results in the formation of a multilayer (ML) film. Reprinted from R. Wagner, T. Yamamoto, T. Manaka, M. Iwamoto, Rev. Sci. Instrum., Vol. 76, 083902, 2005, with the permission of AIP Publishing [28].

director means principal direction of non-zero order parameter S_2. As the monolayer is compressed below 80 Å^2, the molecules begin to gradually standup relative to the water surface as seen in the liquid expanded (LE) phase in Fig. 5.6, but are still randomly oriented. Thus, the director is still not defined in this region, and the order parameter S_2 is still thought to be zero. But with that said, there is the possibility that a small amount of order is established if the molecules are not entirely random. This region is still slightly unclear in terms of orientation order. Compressing the monolayer below 50 Å^2 results in a steep rise of the surface pressure as seen in the condensed (C) phase in Fig. 5.6. In this phase the area between the molecules has decreased sufficiently to restrict their motion, thus establishing a director and orientational order. Compressing the monolayer past 40 Å^2, limits the area per molecule such that the molecules are forced to form a multilayer (ML) film.

To determine the orientation order parameters S_1, S_2 and S_3, individually, MDC, SHG and BAR are available. For the determination of S_2, BAR is utilized. In BAR experiment, it is only possible to define S_2 in the condensed phase (C) due to the establishment of the director in this region. In the case of monolayers comprised of rod-shaped polar molecules, the reflectivity, r, can be derived using the Berreman's 4×4 matrix formalism [48, 49]. Since the irradiated area on the monolayer is large enough ($\sim$0.3 mm^2), the square of reflectivity, proportional to the intensity, averaged in terms of tilt azimuth can be, in a similar way to [50], expressed as [28, 51]

$$|r|^2 \propto \frac{3}{8}A^2 + AB + B^2, \tag{5.16}$$

where A and B are given as

$$\begin{aligned} A &= k\Delta z \frac{2n_1 \cos\Theta_B \cos^2\Theta_t}{(n_2 \cos\Theta_B + n_1 \cos\Theta_t)^2} \frac{\varepsilon_\perp (\varepsilon_p - \varepsilon_p) \sin^2\vartheta}{\varepsilon_\perp \sin^2\vartheta + \varepsilon_p \cos^2\vartheta} \\ B &= k\Delta z \frac{2n_1 \cos\Theta_B \cos^2\Theta_t}{(n_2 \cos\Theta_B + n_1 \cos\Theta_t)^2} \\ &\quad \times \frac{\begin{array}{l}\{ \left[n_1^2 \left(1 + \tan^2\Theta_B\right) - \varepsilon_\perp\right] (\varepsilon_p - \varepsilon_\perp) \cos^2\vartheta \\ \quad + \left(n_1^2 - \varepsilon_\perp\right)\left(\varepsilon_\perp - n_1^2 \tan^2\Theta_B\right) \}\end{array}}{\varepsilon_\perp \sin^2\vartheta + \varepsilon_p \cos^2\vartheta}. \end{aligned} \tag{5.17}$$

The constant k is the magnitude of the wave vector in a vacuum, k_x is the parallel component of the wave vector for the monolayer, n_1 is the refractive index of air, n_2 is the refractive index of water, Θ_t is the transmitted angle of the light through the water, Δz is the monolayer thickness, ϑ is the director tilt angle, and $\varepsilon_\perp$ and ε_p are a dielectric constant of the monolayer in the normal and parallel directions to the director of the monolayer surface, respectively, and are expressed as follows:

$$\begin{aligned} \varepsilon_\perp &= \frac{1}{3}\Delta\xi S_2 + \bar{\varepsilon} \\ \varepsilon_p &= \frac{2}{3}\Delta\xi S_2 + \bar{\varepsilon} \end{aligned} \tag{5.18}$$

where $\Delta\xi$ is the proportional constant of S_2 to the anisotropic term of the dielectric coefficient, and $\bar{\varepsilon}$ is the dielectric constant when $S_2 = 0$. It can be seen from the above equations that the reflectivity of the monolayer is a function of both the orientation order parameter S_2 and the director tilt angle ϑ. From the discussion above, it is found that S_2 can be determined using BAR, and we can investigate the change of S_2 during monolayer compression and expansion, in the way same as in the MDC and SHG measurements.

Figure 5.7 displays typical results collected by the BAR–MDC system for an 8CB monolayer. From top to bottom, the Brewster angle reflectance, the orientation order parameter S_2, Maxwell's displacement current, the first orientation order parameter S_1, and the $\pi - A$ isotherm. In prior to a monolayer spreading on the water surface, all of the signals were immeasurably small. The graphs shown in Fig. 5.7 are decomposed into four regions, the gas, LE, condensed, and multilayer phases (see Fig. 5.6) correspond to regions 1, 2, 3, and 4, respectively. Since the tilt angle can be defined only in region 3, it was determined from SHG measurements for a monolayer comprised of 8CB molecules that the director tilt angle is approximately $\vartheta \sim 57$ degrees [43]. Enderle *et al.* found a similar result for an 8CB monolayer in this region [52]. The constants $\Delta\xi$ and $\bar{\varepsilon}$ in Eq. (5.18) were extrapolated from literature [53].

In the S_2 plot, the gas and LE phases were found to have values of $S_2 = 0$ and $S_2 = 0.73$ after the computer calculation described in Appendix E. As the molecules in these phases are randomly oriented, and in the gas phase are generally laying parallel to the water surface, it is reasonable to assume that the order parameter is zero $S_2 = 0$ in these two regions as there is no orientational order. As the monolayer is compressed further into the condensed phase, the $\pi - A$ isotherm increases steeply, indicating the area between the molecules has decreased sufficiently to cause stronger intermolecular interactions. Thus, a director can be defined and orientational order is established. The plot of S_2 in this region revealed a linearly increasing S_2 beginning at $S_2 = 0.73$ and ending at $S_2 = 0.78$. Thus, in this region it was assumed that the value of S_2 would be that of the linearly increasing values. At this point it should be made clear

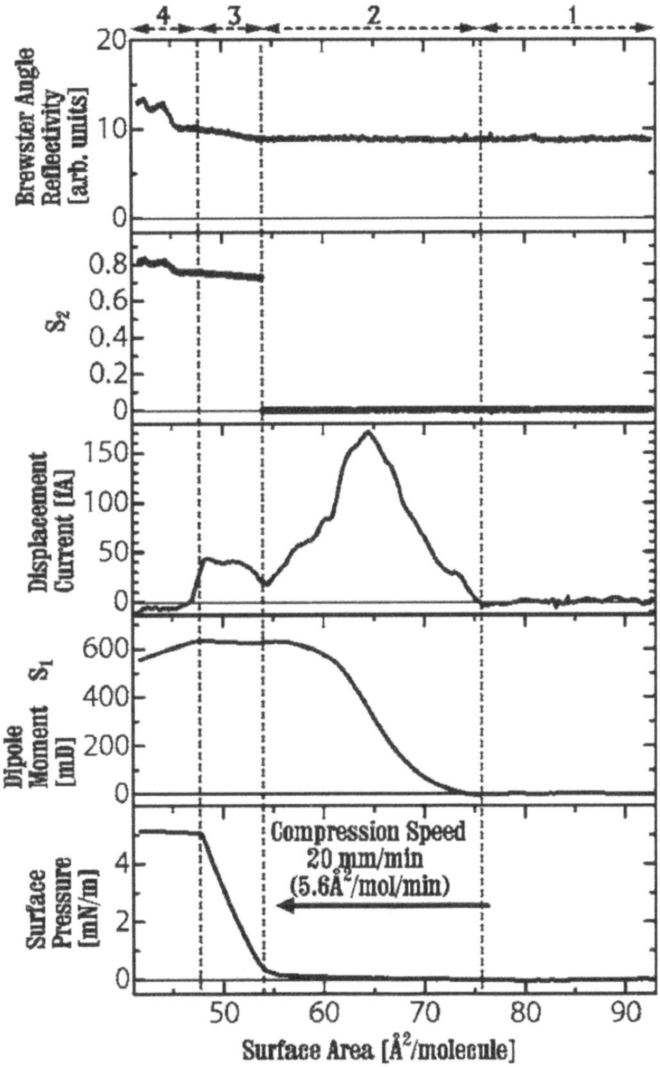

Figure 5.7. BAR–MDC measurement of 8CB monolayer on water surface. A typical example of BAR and MDC measurements of an 8CB monolayer under compression. From top to bottom, BAR, the order parameter S_2, MDC, dipole moment proportional to S_1, and the $\pi - A$ isotherm. The numbers 1–4 described on the top of the figure represent monolayer phases illustrated in Fig. 5.6 (1: gas, 2: liquid expanded (LE), 3: condensed (C), 4: multilayer (ML) phase. Reprinted from R. Wagner, T. Yamamoto, T. Manaka, M. Iwamoto, Rev. Sci. Instrum., Vol. 76, 083902, 2005, with the permission of AIP Publishing [28].

that there is an instantaneous jump from the S_2 value in the LE phase to the S_2 value in the condensed phase.

In region 1 of Fig. 5.7, at a molecular area greater than 80 Å^2, it can be noted that while there is no displacement current generated in this region there is a non-zero reflectance $|r|^2 \neq 0$. In this region both S_1 and S_2 are approximately zero, *i.e.*, there is neither polar symmetry breaking nor nematic orientational order. If we move into region 2, at a molecular area between 55–80 Å^2, we can see that while the MDC signal gradually increases, the BAR signal remains constant. In this region, the S_1 and S_2 graphs begin to deviate from each other. While nematic order is still not observed, polar symmetry is beginning to break. In region 3, between 47–55 Å^2, a gradual increase in the reflectivity occurs, and the surface pressure begins to increase steeply, but the MDC signal decays and becomes steady. These results indicate a sharp increase in S_2, a decrease of the order parameter S_1. That is, nematic order is established whereas polar symmetry breaking stops increasing. In region 4, in a molecular area less than 47 Å^2, it has been found from BAM images that multilayer domains form on the water surface (see Fig. 5.8). The MDC decreases to zero as the technique is insensitive for multilayer domains. The reflectivity on the other hand shows a large increase in intensity, as well as fluctuations. This rise corresponds to the domains forming and entering

Figure 5.8. BAM image of the 8CB monolayer in (a) gas/LE phase, (b) condensed phase, and (c) multilayer phase. Reprinted from R. Wagner, T. Yamamoto, T. Manaka, M. Iwamoto, Rev. Sci. Instrum., Vol. 76, 083902, 2005, with the permission of AIP Publishing [28].

into the laser spot area. As described in this section, BAR–MDC measurement system is useful for the determination of S_1 and S_2 of interface monolayer. MDC well reflects S_1 while BAR and BAM utilizes electronic polarization due to S_2. Noteworthy that chiral phase separation in monolayer is identified by using BAM [51, 54].

5.5.2 *MDC–SHG measurement and determination of S_1 and S_3*

The determination of order parameters S_1 and S_3 can be done by using the MDC and SHG measurements. As an example of monolayers, rod-like 8CB molecules are used. This molecule is achiral and the second-harmonic polarization P^N is well described by Eq. (5.11). The molecule has amphiphilic property. The alkyl group is hydrophobic, whereas the cyano group is hydrophilic. Both alkyl and cyano groups have permanent dipole moment along the molecular long-axis. The presence of permanent dipole moments leads to the generation of MDC, whereas the second order nonlinear electronic polarizability of the biphenyl group with alkyl and cyano groups leads to SHG. As the molecule is rod-like, the main contribution to the SHG is coming from the nonlinear polarization component of the molecule, $\alpha^{(2)}_{3'3'3'}$. It should be noted that the contribution of the permanent dipole moment of the hydrophilic part is diminished owing to the screening effect of the water with a high dielectric constant ($\approx$78) when the hydrophilic part is protruding into the water. As a result, although the dipole moment of the methyl group is very small in comparison with that of the nitrile, the long alkyl chain is a main contributor of the generation of MDC. This situation is the same for fatty acid monolayers [2, 13].

Figures 5.9(a) and 5.9(b) show typical examples of the MDC–SHG measurements for 8CB monolayers, corresponding to the case of p-output and s-output, respectively, where the laser irradiated with a wavelength of 0.532 μm [55]. The incident and the output lights are polarized (p- or s-polarization) waves, and they are indicated in the figures as p–p, s–p, s–p, and s–s (input polarization–output polarization). From the bottom to the top in the figures, the dependences

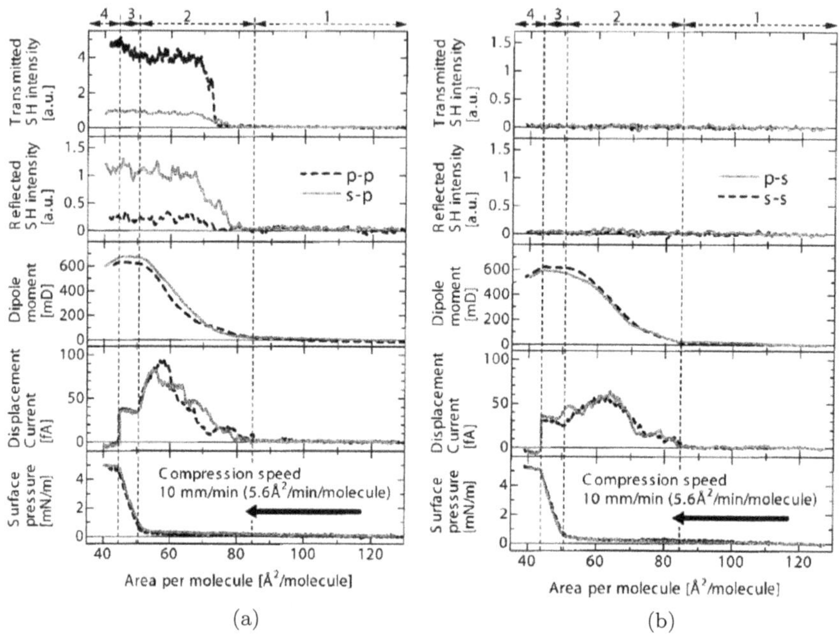

Figure 5.9. Example of MDC–SHG measurement of 8CB monolayer on water surface, during a course of compression. (a) p-polarized SHG. (b) s-polarized SHG. Reprinted from A. Tojima, T. Manaka, M. Iwamoto, O.-Y. Zhong-can, J. Chem. Phys., Vol. 118, 5640, 2003, with the permission of AIP Publishing [42].

of the surface pressure–area, the MDC–area, the vertical component of the dipole moment–area, the reflected SH intensity–area, and the transmitted SH intensity–area are plotted. The MDC and the SHG signals are generated steadily, and they are generated in accordance with the change of the molecular area during a course of monolayer compression (from right to left in the figure). The isotherms can be divided into four regions. In region 1, the surface pressure is immeasurably low and the MDC is almost zero. Similarly, the SH intensity is also very low. These suggest that the 8CB molecules are lying randomly on the water surface. In other words, in region 1, the monolayer is in an isotropic state, *i.e.*, P_z $(=N_s\mu S_1) = 0$, and the SOS components in $\vec{P}^N$ given by Eq. (5.11) are zero, *i.e.*, $\vec{P}^N = 0$. In region 2, the surface pressure is still immeasurably small, but the MDC begins to flow by monolayer compression, possibly due to the standing up of the 8CB molecules on the water surface. As the MDC is generated from the change of the spontaneous polarization P_z, the MDC current in region 2 reflects the change of the orientation order parameter S_1 (see Eq. (5.13)). Integration of the MDC with respect to the molecular area gives the vertical component of the dipole moment of 8CB molecules, μS_1. The dipole moment–area isotherm in Fig. 5.9 represents the relationship between μS_1 and the molecular area A ($\equiv 1/N_s$). It is found that, by monolayer compression, S_1 increases gradually and finally saturates at the end of region 2.

SHG is observed for p-polarized light output (p–p and s–p), but not for s-polarized light output (p–s and s–s). These can be explained by Eq. (5.11). Briefly, for incident s-polarized wave, the external electric field $\vec{E}$ is given by $\vec{E}_{in} = S_\omega \vec{s}_{in}$, with $\vec{s}_{in} = (-1, 0, 0)$ in the laboratory system (see Eq. (5.15)). With $\vec{s}_{in} \cdot \vec{n} = 0$, Eq. (5.11) leads to $\vec{P}^N = s_{31} S_\omega^2 \vec{n}$. Thus the s- and the p-polarized radiations are proportional to $\vec{P}^N \cdot \vec{s}_{out} = s_{31} S_\omega^2 \vec{n} \cdot \vec{s}_{out} = 0$ and $\vec{P}^N \cdot \vec{p}_{out} = s_{31} S_\omega^2 \vec{n} \cdot \vec{p}_{out} \neq 0$, respectively. Similarly, for the p-polarized wave irradiation, the external electric field $\vec{E}$ is given by $\vec{E}_{in} = P_\omega \vec{p}_{in}$ with $\vec{p}_{in} = (0, \cos\theta_{in}, \sin\theta_{in})$ in the laboratory system (see Eq. (5.15)). Substitution of $\vec{E}_{in}$ into Eq. (5.11) gives $\vec{P}^N = (s_{33} - s_{31} - s_{15}) P_\omega^2 (\vec{n} \cdot \vec{p}_{in})^2 \vec{n} + s_{31} P_\omega^2 (\vec{p}_{in} \cdot \vec{p}_{in}) \vec{n} + s_{15} P_\omega^2 (\vec{n} \cdot \vec{p}_{in}) \vec{p}_{in}$. Thus, it is expected that the s- and the p-polarized radiation are proportional to $\vec{P}^N \cdot \vec{s}_{out} = 0$ and $\vec{P}^N \cdot \vec{p}_{out} \neq 0$, respectively, since $\vec{n} \cdot \vec{s}_{out} = 0$, $\vec{n} \cdot \vec{p}_{out} = \vec{n} \cdot \vec{p}_{in} \neq 0$, and $\vec{p}_{out} \cdot \vec{p}_{in} \neq 0$ and $\vec{p}_{in} \cdot \vec{p}_{in} = 1$. These results agree very well with the observation of SHG–area isotherm in region 2, where the output p-polarized wave is detected but no s-polarized wave is observed. Based on these results, it is confirmed that the 8CB molecules stand up on the water surface by monolayer compression, and the non-zero nonlinear second order polarization expressed by Eq. (5.11) is induced by the laser irradiation, and the 8CB monolayer has $C_{\infty v}$–symmetry in region 2. It is interesting here to study the possibility of the formation of C_s–symmetry monolayer in region 2. The feature of the C_s–symmetry can be seen when monolayers are irradiated by s-polarized wave. The substitution of $\vec{E}_{in} = S_\omega \vec{s}_{in}$ into Eq. (5.11) gives $\vec{P}^N = (s_{33} - s_{31} - s_{15}) S_\omega^2 (\vec{m} \cdot \vec{s}_{in})^2 \vec{m} + s_{31} S_\omega^2 (\vec{s}_{in} \cdot \vec{s}_{in}) \vec{m} + s_{15} S_\omega^2 (\vec{m} \cdot \vec{s}_{in}) \vec{s}_{in}$ where $\vec{m}$ represents the director, *i.e.*, tilting direction of constituent molecules in monolayers. Thus it is expected that both s- and p- polarized SH waves are radiated from the monolayer, because the radiated s- and p-polarized outputs are proportional to $\vec{P}^N \cdot \vec{s}_{out}$ and $\vec{P}^N \cdot \vec{p}_{out}$, respectively, and both of them are non-zero. For more detail, $\vec{m} \cdot \vec{s}_{out}$ and $\vec{m} \cdot \vec{s}_{in}$ are calculated to the order of $\cos \gamma_d$, where γ_d is the angle between the director $\vec{m}$ and $\vec{s}_{in}$ (or $\vec{s}_{out}$) ($\vec{m}$ and x-axis). This means that $\vec{P}^N \cdot \vec{s}_{out}$ is not zero, although it becomes zero only when the condition $\vec{m} = \vec{n}$ is satisfied (in case of $C_{\infty v}$–symmetry). As mentioned above, it is expected that with incident s-polarized wave irradiated on C_s–symmetric monolayer, both p- and s-polarized SH waves will appear. This is the most probable prediction for monolayers with C_s–symmetry. But the situation changes if the domains with C_s–symmetry are small in comparing with the spot size of the laser beam and the director $\vec{m}$ of the domains is randomly distributed on the water surface. In such a case, the radiated s-polarized wave is the vector sum of the radiated waves from all domains that are irradiated by the laser beam. Obviously, the average value of $\cos \gamma_d$ over all the radiated domains is zero if azimuth distribution of $\vec{m}$ is random. In other words, the polarization of monolayers that are composed of small domains with C_s–symmetry looks like a monolayer with $C_{\infty v}$–symmetry. In the experiment shown in Figs. 5.9(a) and 5.9(b), this seems not quite possible since the generation of MDC is not observed in region 1 by monolayer compression. If the domains with C_s–symmetry make significant contribution to the generation of MDC, on account of the condensation of domains with the spontaneous polarization [13, 45], MDC should flow even in region 1. The generation of MDC due to condensation is inversely proportional to the molecular area, but such MDCs are not seen in Fig. 5.9. However, the fact that the $s - p$ SH intensity is greater than the $p - p$ SH in reflected SHG intensity, indicates the existence of condensation of small but local domains with C_s–symmetry [42]. Based on the above discussion, we may conclude that the 8CB monolayer on the water surface has $C_{\infty v}$–symmetry in region 2, though there is a possibility of the condensation of small domains with C_s–symmetry.

In region 3, the surface pressure increases rapidly but MDC flows steadily by further compression. SHG signals are also generated in steadily, where the transmitted and the reflected lights are generated in steady, and the intensity of these transmitted and reflected signals increase somewhat by compression, on account of the condensation of the molecules. These results indicate that, in this region, the orientational order is nearly saturated and that only the density of the molecules increases by monolayer compression. As the MDC and the SHGs are proportional to the spontaneous polarization, *i.e.*, the 1st term of Eq. (5.2′), and the nonlinear polarization, *i.e.*, Eq. (5.11), respectively, they are only proportional to the density of the molecules when the orientational order does not change under monolayer compression. As a result, under monolayer compression, only very small increases of MDC and of SHG are seen in region 3. Obviously, these increases are not drastic in comparison with the increase of the surface pressure. At the end of region 3, MDC decreases abruptly by further compression, whereas the SHG is generated in a way similar to that in region 3. These results suggest that, during monolayer compression, the transition from one layer to an interdigitated *three*-layer is induced in region 4 [56]. In such three-layer structure, the two layers near the water surface form a bulk smectic A state with centro-symmetry and contributes nothing to the spontaneous polarization P_z. That is, MDC is zero and SHG is unchanged. Briefly, with the formation of the *three*-layer structure, the polarization P_z does not change. Thus, MDC is not generated by further compression, because the MDC should flow on account of the change of the polarization P_z. Furthermore, the unchangeable nonlinear polarization given by Eq. (5.11) indicates that the density of the molecules on the top monolayer does not change during this transition. As a result, the intensity of SHG does not change with further compression.

The orientation order parameters are estimated from the SHG–MDC measurement under appropriate polarized angles δ and γ [43]. Figure 5.10 shows an example of the determination of the order parameters S_1, and S_3 for 8CB monolayers with $C_{\infty v}$–symmetry.

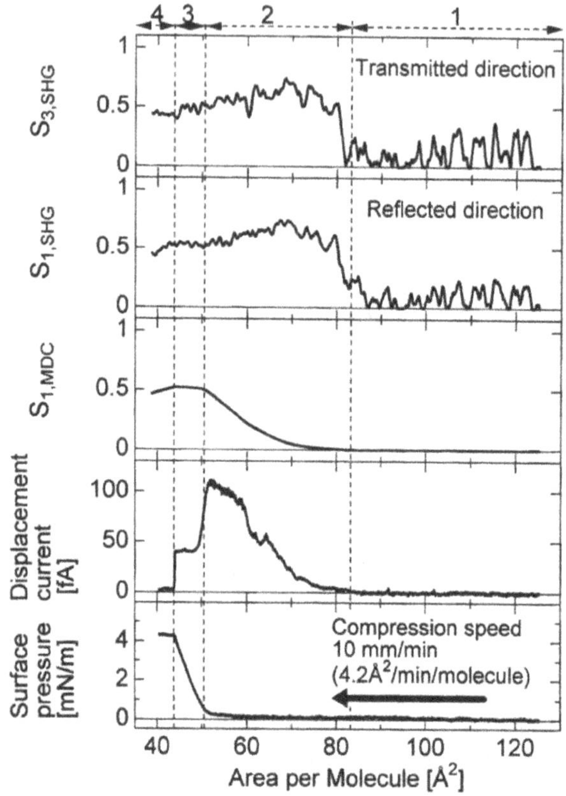

Figure 5.10. Example of MDC–SHG measurement for determining orientation order parameter S_1 and S_3. "Study of orientational order parameters of 4–octyl–4′–n–cyanobiphenyl langmuir films by Maxwell displacement current measurement coupled with optical second harmonic generation measurement", A. Tojima, R. Hiyoshi, H. Fujimaki, H. Ootake, T. Manaka, M. Iwamoto, Molecular Crystals and Liquid Crystals, Vol. 412, 189/[1799], 2004, reprinted by permission of Taylor & Francis Ltd, (https://www.tandfonline.com/) [55].

In region 1, $S_1$and S_3 of SHG are very small although fluctuation coming from the formation of domains and the high sensitivity of the detection system is observed. In region 2, S_1 and S_3 increase gradually and show the maximum. In region 3, they are nearly constant about 0.5. By contrast, S_1 of MDC is nearly zero in region 1, and increases monotonously in region 2. S_1 is about 0.5 in region 3. From these results, it is estimated that molecules lying on a water surface (region 1) stand up by monolayer compression, and with an average tilting angle of about 60° in region 2. As mentioned above, with the MDC–SHG measurement, surface polarization phenomena can be detected and the phase transition and the orientation orders of the monolayers can be explored in terms of S_1 and S_3.

5.6 Piezoelectric effect and flexo-electric effect of monolayer

As mentioned in Sec. 5.1 and Sec. 5.2, the spontaneous polarization and nonlinear dielectric polarization of organic monolayers are due to their non-centrosymmetric structure (Fig. 5.1). The spontaneous polarization changes by monolayer compression and thus leads to the generation of MDC. This fact indicates that monolayers possess piezoelectric property as illustrated in Fig. 5.11(a). As described earlier, the spontaneous polarization of monolayers are given by P_z $(=N_s\mu S_1)$, and MDC is generated due to either the change of surface density of molecules N_s or the change of orientation order parameter S_1 by monolayer compression, assuming that these changes happen homogeneously over the entire region of monolayer covering area. That is, director $\vec{m}$ $(=\vec{n})$ is the same over the entire region of monolayer covering area.

By the way, as is well-known, many kinds of textures appear in liquid crystals and monolayers. This means that the spatial variation of director $\vec{m}$ distribution $(\vec{m} \neq \vec{n})$ may happen in bulk LCs

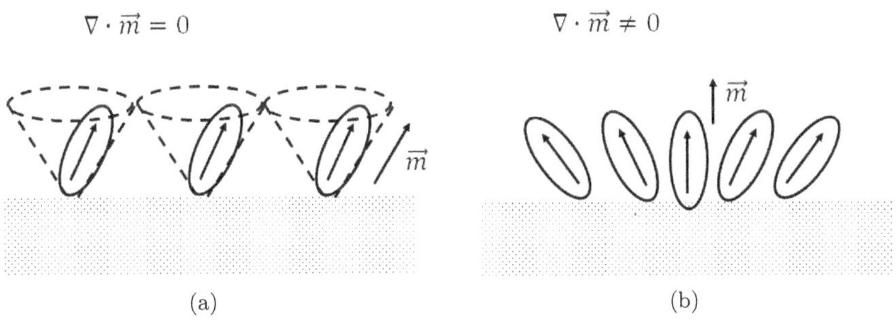

Figure 5.11. (a) Piezo electric effect and (b) flexo electric effect of monolayer with splay deformation.

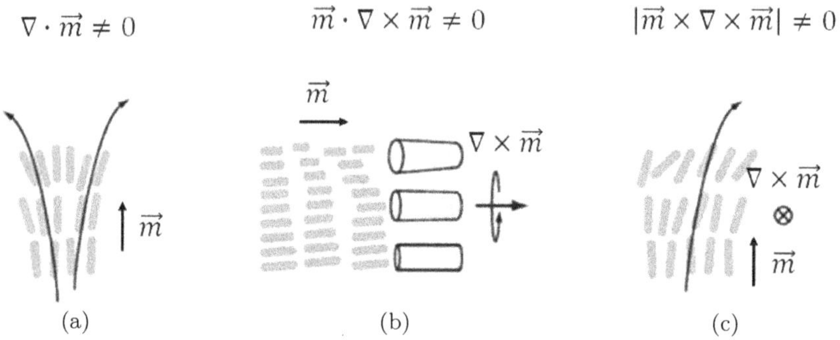

Figure 5.12. The three basic orientational elastic deformations of bulk liquid crystal. (a) Splay, (b) twist, and (c) bend deformation. $\vec{m}$ is liquid crystal director.

and monolayers. Of course, to give such spatial distribution, some additional elastic deformation energy is required. According to the Oseen–Zocher–Frank theory of the curvature elasticity based on the continuum theory [29], the elastic deformation energy F for nematic LCs where directors are distributing in a way as shown in Fig. 5.12 is given by the sum of splay, bend and twist energies. The energy is called Frank energy F and expressed as [29]

$$F = \int \left[\frac{1}{2} k_{11} (\nabla \cdot \vec{m})^2 + \frac{1}{2} k_{22} (\vec{m} \cdot \nabla \times \vec{m})^2 + \frac{1}{2} k_{33} (\vec{m} \times \nabla \times \vec{m})^2 \right] dV \tag{5.19}$$

where k_{11}, k_{22}, and k_{33}, is Frank constant for splay, twist, and bend deformation, respectively. V is volume. Though such deformations do not accompany the orientational order change and the molecular density change locally, but they have the possibility to give rise to the electric flux density changes in a manner similar to the piezoelectric effect, and lead to the generation of polarization in LCs. This is so-called flexo-electric effect proposed by Meyer [57, 58]. For nematic LCs, bend and splay deformations are origins of flexo-electric effect. On the other hand, for monolayers, due to the symmetry breaking, all constituent molecules orient toward upper direction (Fig. 5.11). For this reason, monolayers in liquid crystalline phases have the possibility to have the spontaneous splay deformation, though this is not probable for nematic bulk LCs. Furthermore, all molecules are confined to the flat surface in monolayers. This fact is quite different from bulk LCs. Thus, the flexo-electric effect due to bend deformation will not be allowed.

Using the MDC and SHG measurement, shear-induced tilt azimuthal orientation of amphiphilic *4-heptyloxy-4′-cyanobiphenyl*(7OCB) monolayers at the air–water interface has been successfully detected in a course of monolayer compression [59, 60], and the experimental results were found to be explained by a model based on the $C_\infty \to C_{2v}$ transition in the flow of a two-dimensional hexatic lattice structure coupled with the shear flow in monolayers, where we could show that the Ericksen–Leslie theory for LCs was applicable to this analysis [29].

Using SHG microscope, polarization enhancement along the lobe of lipid monolayer domains has been visualized during a course of monolayer compression [61, 62]. The results show that dendrite-shaped lobes of domains develop during monolayer compression, and SHG image reflects polarization distribution inside the domains (Fig. 5.13). Interestingly, the enhancement of SHG intensity localized

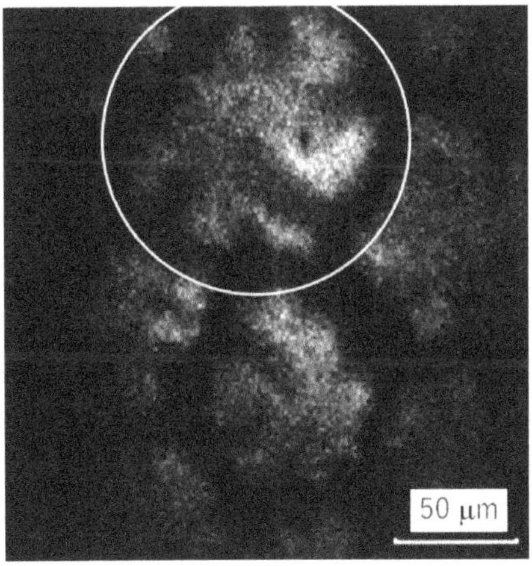

Figure 5.13. Flexoelectric effect of chiral lipid monolayer. The polarization enhancement, indicated by circle, in the bend-lobe of monolayer domain is visualized by using SHG microscope [61].

at the bend-lobes. The input–output polarization condition is p–p, and this SHG enhancement suggests that the director tilt ϑ inside the domain is spatially changing in the inner part of the domain. The spatial change of director tilt angle evidenced that the flexo-electric polarization is produced in the condition $\nabla \cdot \vec{m} \neq 0$ with Frank energy F stored in interfacial monolayer. It is instructive here to note that, due to symmetry breaking of monolayer, the first orientation order parameter is non-zero $S_1 \neq 0$ for monolayers. Consequently, not only Frank energy in Eq. (5.19) but also electrostatic energy contribution governs free energy of domain. The details are described in Ref. [63] and in a book [64].

References

1. Langmuir, I. (1917). The constitution and fundamental properties of solids and liquids. II. Liquids, *J. Am. Chem. Soc.*, 39, 1848–1906 https://doi.org/10.1021/ja02254a006; Japanese translation by Fukuda, K. (1975) *Kagaku no Genten No.7 Kaimen Kagaku (Originality of Chemistry Series, No. 7, Interface Science)*, Sec. 2, The Chemical Society of Japan ed. (Tokyo Daigaku Shuppan Kai, Tokyo).
2. Gaines, JR., G. L. (1966). *Insoluble Monolayers at Liquid–Gas Interfaces* (Interscience, New York) Chap. 3.
3. Ulman, A. (1995). *Characterization of Organic Thin Films* (Butterworth–Heinemann, Boston London Oxford Singapore Sydney Toronto Welington, Manning, Connecticut).
4. Miranda, A. and De Beule, P. A. A. (2016). Microscopic thin film optical anisotropy imaging at the solid–liquid interface, *Rev. Sci. Instrum.*, 87, 043701, https://doi.org/10.1063/1.4947258.
5. Stiehm, T., Schneider, R., Kern, J., Niehues, I., de Vasconcellos, S. M. and Bratchitsch, R. (2019). Supercontinuum second harmonic generation spectroscopy of atomically thin semiconductors, *Rev. Sci. Instrum.*, 90, 083102, https://doi.org/10.1063/1.5100593.
6. Franz, J., van Zadel, M.-J. and Weidner, T. (2017). A trough for improved SFG spectroscopy of lipid monolayers, *Rev. Sci. Instrum.*, 88, 053106, https://doi.org/10.1063/1.4982050.
7. Kaganer, V. M., Möhwald, H. and Dutta, P. (1999). *Mod. Rev. Phys.*, 71, 779–819, https://doi.org/10.1103/RevModPhys.71.779.
8. Hénon, S. and Meunier, J. (1991). Microscope at the Brewster angle: Direct observation of first-order phase transition in monolayers, *Rev. Sci. Instrum.*, 62, 936, https://doi.org/10.1063/1.1142032.
9. Hönig, D. and Möbius, D. (1991). Direct visualization of monolayers at the air–water interface by Brewster angle microscopy, *J. Phys. Chem.*, 95, 4590–4592, https://doi.org/10.1021/j100165a003.

10. Miranda, P. B. and Shen, Y. R. (1999). Liquid interfaces: A study by sum-frequency vibrational spectroscopy, *J. Phys. Chem. B*, 103, 3292–3307, https://doi.org/10.1021/jp9843757.
11. Sung, W., Kim, D. and Shen, Y. R. (2013). Sum-frequency vibrational spectroscopic studies of Langmuir monolayers, *Curr. Appl. Phys.*, 13, 619–632, https://doi.org/10.1016/j.cap.2012.12.002.
12. Tian, C. S. and Shen, Y. R. (2014). Recent progress on sum-frequency spectroscopy, *Surf. Sci. Rep.*, 69, 105–131, https://doi.org/10.1016/j.surfrep.2014.05.001.
13. Iwamoto, M. and Wu, C.-X. (2001). *The Physical Properties of Organic Monolayers* (World Scientific, Singapore New Jersey London Hong Kong), https://doi.org/10.1142/4575.
14. Adam, N. K., Danielli, J. F. and Harding, J. B. (1934). The structure of surface films XXI—Surface potentials of dibasic esters, alcohols, aldoximes, and ketones, *Proc. Roy. Soc. A*, 147, 491–499, https://doi.org/10.1098/rspa.1934.0231.
15. Schulman, J. H. and Rideal, E. K. (1931). On the surface potentials of unimolecular films of long chain fatty acids. — Part I. Experimental method, *Proc. Roy. Soc. A*, 813, 259–270, https://doi.org/10.1098/rspa.1931.0002.
16. Macdonald, J. R. and Barlow, Jr., C. A. (1963). Work function change on monolayer adsorption, *J. Chem. Phys.*, 39, 412–422, https://doi.org/10.1063/1.1734263.
17. Taylor, D. M. (2000). Developments in the theoretical modelling and experimental measurement of the surface potential of condensed monolayers, *Adv. Colloid Interface Sci.*, 87, 183–203, https://doi.org/10.1016/S0001-8686(99)00044-5.
18. Guyot-Sionnest, P., Hsiung, H. and Shen, Y. R. (1986). Surface polar ordering in liquid crystal observed by optical second-harmonic generation, *Phys. Rev. Lett.*, 57, 2963–2966, https://doi.org/10.1103/PhysRevLett.57.2963.
19. Chen, C. K., Heinz, T. F., Ricard, D. and Shen, Y. R. (1981). Detection of molecular monolayers by optical second-harmonic generation, *Phys. Rev. Lett.*, 46, 1010–1012, https://doi.org/10.1103/PhysRevLett.46.1010.
20. Yan, E. C. Y., Liu, Y. and Eisenthal, K. B. (1998). *J. Phys. Chem. B*, 102, 6331–6336, https://doi.org/10.1021/jp981335u.
21. Shen, Y. R. (1984). *The Principles of Nonlinear Optics* (John Wiley & Sons, New York).
22. Boyd, R. W. (2008). *Nonlinear Optics*, 3rd ed. (Academic, London).
23. Andrews, D. L. and Thirunamachandran, T. (1977). On three-dimensional rotational averages, *J. Chem. Phys.*, 67, 5026–5033, https://doi.org/10.1063/1.434725.
24. Andrews, D. L. and Harlow, M. J. (1984). Phased and Boltzmann-weighted rotational averages, *Phys. Rev. A*, 29, 2796–2806, https://doi.org/10.1103/PhysRevA.29.2796.
25. Andrews, D. L. a nd Hands, I. D. (1996). Second harmonic generation in partially ordered media and at interfaces: Analysis of dynamical and orientational factors, *Chem. Phys.*, 213, 277–294, https://doi.org/10.1016/S0301-0104(96)00236-4.

26. Popov, S. V., Svirko, Y. P. and Zheludev, N. I. (1995). *Susceptibility Tensors for Nonlinear Optics* (IOP); English tarnslation by Kisin V. I. (2017). *Susceptibility Tensors for Nonlinear Optics* (ebook, Taylor and Francis, New York), https://doi.org/10.1201/9780203737699.
27. Iwamoto, M., Wu, C.-X. and Zhong-can, O.-Y. (2000). Second-order susceptibility tensor of a monolayer at the liquid–air interface: SHG spectroscopy by compression, *Chem. Phys. Lett.*, 325, 545–551, https://doi.org/10.1016/S0009-2614(00)00658-8.
28. Wagner, R., Yamamoto, T., Manaka, T. and Iwamoto, M. (2005). Determination of the complete dielectric polarization of Langmuir monolayers, *Rev. Sci. Instrum.*, 76, 083902. https://doi.org/10.1063/1.1988207.
29. Chandrasekhar, S. (1992). *Liquid Crystals*, 2nd ed. (Cambridge University, London), https://doi.org/10.1017/CBO9780511622496.
30. de Gennes, P. G. (1991). *The Physics of Liquid Crystals* (Clarendon, Oxford).
31. Luckhurst, G. R. and Gray, G. W. (1979). *The Molecular Physics of Liquid Crystals* (Academic, London).
32. Saupe, A. (1964). Kernresonanzen in kristallinen Flüssigkeiten und in kristallinflüssigen lössungen, *Z. Natureforsch.*, 19a, 161–171, https://doi.org/10.1515/zna-1964-0201.
33. Tsvetkov, V. (1942). *Acta. Physicochim. USSR*, 16, 132.
34. Kittel, C. (1974). *Introduction to Solid State Physics* (Wiley, New York).
35. Iwamoto M. and Wu, C.-X. (2001). *The Physical Properties of Organic Monolayers* (World Scientific, Singapore New Jersey London Hong Kong), https://doi.org/10.1142/4575 Chap. 3.
36. Yamamoto, T., Manaka, T. and Iwamoto, M. (2007). *J. Chem. Phys.* 126, 125106, https://doi.org/10.1063/1.2709644.
37. Frohlich, H. (1949). *Theory of Dielectrics* (Oxford at the Clarendon Press).
38. Rose, M. E. (1957). *Elementary Theory of Angular Momentum* (Wiley, New York).
39. Tojima, A., Manaka, T. and Iwamoto, M. (2001). Orientational order study of monolayers at the air–water interface by Maxwell–displacement current and optical second harmonic generation, *J. Chem. Phys.*, 115, 9010–9017, https://doi.org/10.1063/1.1413738.
40. Wang, C. C. (1969). Second-harmonic generation of light at the boundary of an isotropic medium, *Phys. Rev.* 178, 1457–1461, https://doi.org/10.1103/PhysRev.178.1457.
41. Giordmaine, J. A. (1965). Nonlinear optical properties of liquids, *Phys. Rev.*, 138, A1599–A1606, https://doi.org/10.1103/PhysRev.138.A1599.
42. Tojima, A., Manaka, T., Iwamoto, M. and Zhong-can O.-Y. (2003). Detection of phase transition of monolayers at the air–water interface by compression using Maxwell displacement current and optical second harmonic generation, *J. Chem. Phys.*, 118, 5640–5649, https://doi.org/10.1063/1.1556093.
43. Tojima, A., Manaka, T. and Iwamoto, M. (2003). Instrument equipped with Maxwell displacement current and optical second-harmonic generation measurement system, *Rev. Sci. Instrum.*, 74, 2828–2835, https://doi.org/10.1063/1.1568536.

44. Fujimaki, H., Manaka, T., Ohtake, H., Tojima, A. and Iwamoto, M. (2003). Second-harmonic generation and Maxwell displacement current spectroscopy of chiral organic monolayers at the air–water interface, *J. Chem. Phys.*, 119, 7427–7434, https://doi.org/10.1063/1.1605943.
45. Iwamoto, M., Kubota, T. and Muhamad, M. R. (1995). Detection of phase transitions in liquid crystals on a water surface by a Maxwell displacement current-measuring technique, *J. Chem. Phys.*, 102, 9368–9374, https://doi.org/10.1063/1.468804.
46. Majima, Y., Naruse, H. and Iwamoto, M. (1992). Reversible displacement current generation due to photochromism in a spread monolayer: Influence of molecular orientation, *Jpn. J. Appl. Phys.*, 31, 864–867, https://doi.org/10.1143/JJAP.31.864.
47. Tojima, A., Matsuo, Y., Hiyoshi, R., Manaka, T., Majima, Y. and Iwamoto, M. (2001). Orientational order study of 4–alkyl–4′–cyanobiphenyl Langmuir films by Maxwell displacement current and optical second harmonic generation measurements, *Thin Solid Films*, 393, 86–91, https://doi.org/10.1016/S0040-6090(01)01103-8.
48. Berreman, D. W. (1972). Optics in stratified and anisotropic media: 4×4–Matrix formulation, *J. Opt. Soc. Am.*, 62, 502–510, https://doi.org/10.1364/JOSA.62.000502.
49. Puebla, C. (1989). Non-linear optics in liquid crystalline phases A 4×4 formalism, *Liq. Cryst.*, 5, 1319–1322, https://doi.org/10.1080/02678298908026437.
50. Zhang, T. G., Zhang, C. H. and Wong, G. K. (1990). Determination of molecular orientation in molecular monolayers by second-harmonic generation, *J. Opt. Soc. Am. B*, 7, 902–907, https://doi.org/10.1364/JOSAB.7.000902.
51. Aida, T., Yamamoto, T., Ou-Yang, W., Manaka, T. and Iwamoto, M. (2008). Study of domain shapes and orientational structure of phospholipid monolayers using Maxwell displacement current and Brewster angle microscopy, *Jpn. J. Appl. Phys.*, 47, 411–415, https://doi.org/10.1143/JJAP.47.411.
52. Enderle, Th., Meixner, A. J. and Zschokke-Gränacher, I. (1994). Phase transitions in mixed Langmuir films of 8CB and stearic acid, *J. Chem. Phys.*, 101, 4365–4372, https://doi.org/10.1063/1.467487.
53. Chirtoc, I., Chirtoc, M., Glorieux, C. and Thoen, J. (2004). Determination of the order parameter and its critical exponent for nCB ($n = 5 - 8$) liquid crystals from refractive index data, *Liq. Cryst.*, 31, 229–240, https://doi.org/10.1080/02678290310001642540.
54. Ou-yang, W., Yamamoto, T., Aida, T., Manaka, T. and Iwamoto, M. (2008). Domain observation and molecular orientation of chiral dipalmitoylphosphatidylcholine monolayer by Brewster angle microscopy and Maxwell displacement current, *Thin Solid Films*, 516, 2649–2651, https://doi.org/10.1016/j.tsf.2007.04.077.
55. Tojima, A., Hiyoshi, R., Fujimaki, H., Ootake, H., Manaka, T. and Iwamoto, M. (2004). Study of orientational order parameters of 4–octyl–4′–n–cyanobiphenyl Langmuir films by Maxwell displacement current

measurement coupled with optical second harmonic generation measurement, *Mol. Cryst. Liq. Cryst.*, 412:1, 189–196, https://doi.org/10.1080/15421400490439815.

56. Xue, J., Jung, C. S. and Kim, M. W. (1992). Phase transitions of liquid-crystal films on an air–water interface, *Phys. Rev. Lett.*, 69, 474–477, https://doi.org/10.1103/PhysRevLett.69.474.
57. Meyer, R. B. (1969). Piezoelectric effects in liquid crystals, *Phys. Rev. Lett.*, 22, 918–921, https://doi.org/10.1103/PhysRevLett.22.918.
58. Derzhanski, A. I. and Petrov, A. G. (1979). Flexoelectricity in nematic liquid crystals, *Acta Phys. Polonica (A)*, 55, 747–767.
59. Iwamoto, M., Tojima, A., Manaka, T. and Zhong-can, O.-Y. (2003). Compression-shear-induced tilt azimuthal orientation of amphiphilic monolayers at the air–water interface: A $C_\infty \rightarrow C_{2v}$ transition in the flow of a two-dimensional hexatic structure, *Phys. Rev. E*, 67, 041711, https://doi.org/10.1103/PhysRevE.67.041711.
60. Yamamoto, T., Taguchi, D., Manaka, T. and Iwamoto, M. (2005). Detection of flexoelectric effect from 4–heptyloxy–4′–cyanobiphenyl monolayers at an air–water interface by Maxwell displacement current and optical second harmonic generation, *J. Chem. Phys.*, 122, 164703, https://doi.org/10.1063/1.1884597.
61. Matsuoka, Y., Taguchi, D., Manaka, T. and Iwamoto, M. (2014). Visualizing polarization structure of lipid Langmuir monolayer by surface second-harmonic generation technique, *Thin Solid Films*, 554, 8–12, https://doi.org/10.1016/j.tsf.2013.04.099.
62. Fukunaga, K., Taguchi, D., Manaka, T. and Iwamoto, M. (2014). Observation of polar-structure in domains of phospholipid monolayers by optical second-harmonic generation microscopy, *15th International conference on organized molecular films (ICOMF15–LB15)*, S7:LLO5, International Convention Center, Jeju, Korea, 10 July 2014.
63. Yamamoto, T., Taguchi, D., Weis, M., Manaka, T. and Iwamoto, M. (2008). Electrostatic Maxwell stress model of the shapes of condensed phase domains in monolayers at the air–water interface, *J. Chem. Phys.*, 128, 204706, https://doi.org/10.1063/1.2920474.
64. Iwamoto, M., Yamamoto T. and OuYang, Z.-C. (in preparation) *Electrical and Geometrical Properties of Organic Monolayers* (World Scientific, Singapore).

Chapter 6

Optical Second-Harmonic Generation

This chapter describes the basis of physics on optical second-harmonic generation (SHG), which is available for directly probing nonlinear polarization originated from the presence of electrons and alignment of dipoles. The optical second-harmonic generation (SHG) is the quantum effect generated due to the coupling of electromagnetic waves with electrons in molecules. From the viewpoint of *Analysis and Application of organic materials for organic electronics*, two kinds of optical second-harmonic generation mechanisms are of particular interest. One is the SHG generated from organic materials under no d.c. electric field, and the other is the SHG generated in the presence of d.c. electric field. In this book, we show that the former one is available for characterizing material structures in a manner as described in Chap. 5, while the latter one is available for probing electron transfer in electronic materials. For both cases, understanding the physics on nonlinear polarization induced in organic materials is key. This chapter is devoted to this understanding for the use of SHG.

6.1 MDC and SHG for probing charge displacement and dipole alignment

Figures 6.1 and 6.2 illustrate the concept of probing charge displacement and dipolar alignment by means of MDC and SHG, respectively. As described in Chap. 2 and Chap. 3, electronic charges, *e.g.*, electrons, holes, and dipoles are main origins of dielectric polarization of organic materials.

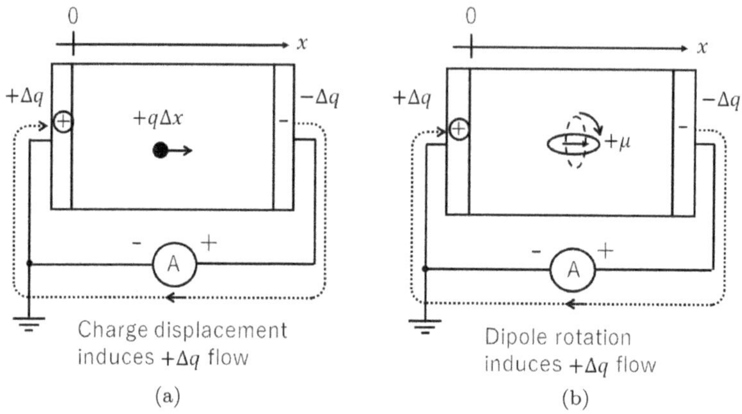

Figure 6.1. Concept of MDC measurement. (a) Displacement of electronic charge and (b) change of dipolar orientation. Charge behaviors in (a) and (b) are origins of MDC, where the change of induced charge "on electrodes" is indirectly detected using an ammeter. By contrast, the origins of SHG are nonlinear polarization induced "in material", and these are directly detected with non-contact in the SHG measurement by using laser.

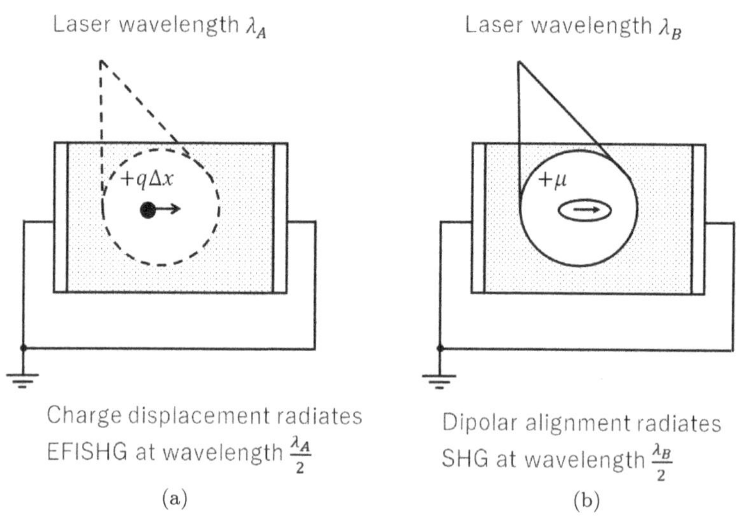

Figure 6.2. Concept of SHG measurement. (a) At laser wavelength λ_A, charge displacement is visualized at EFISHG wavelength $\lambda_A/2$. (b) At laser wavelength λ_B, dipolar alignment is visualized by detecting SHG at wavelength $\lambda_B/2$. Charge behaviors in (a) and (b) are visualized by means of SHG, where the induced nonlinear polarization originating from charge displacement and dipolar alignment "in material" is detected. This point is contrast to MDC that detects induced charge "on electrode".

Their dynamical behaviors in a material can be traced by detecting the change of induced charge on electrodes by using an ammeter in electrical measurements. Time-of-Flight (TOF) measurement is used for probing electron transit across a film. Thermally Stimulated Current (TSC) measurement is helpful for probing rotational motion of dipoles in film. These electrical measurements are typical ones among MDC measurements.

Main origins of the dielectric polarization are the positional displacement of charge $qd\vec{r}$ and the rotation of permanent dipole $\vec{m}$, and these are serving as the source of MDC. Here q is elementary charge of charged particles, *e.g.*, electrons and holes, and $d\vec{r}$ is the displacement of charge. MDC is generated during the charge transfer and dipole rotation, *i.e.*, during the polarization change with respect to time. Main dielectric polarization induced in organic materials $\vec{P}_0$ can be defined microscopically as the sum of the contribution of $qd\vec{r}$ and $\vec{m}$, and MDC is generated due to the change of polarization $\vec{P}_0$ with time.

MDC is short circuit current and it is generally probed by using a sample sandwiched between parallel electrode plates. The change of polarization of the sample in the direction normal to the surface of parallel electrodes is measured indirectly through the detection of the change of induced charges Δq on the electrodes in a closed circuit (see Chap. 1 and Chap. 2).

Define the normal direction to electrode surface to be $\vec{x}$, we can derive the MDC i_{MDC} generated in a material as follows:

$$i_{MDC} = \frac{d}{dt}\vec{P}_0 \cdot \vec{x} \quad \text{with } \vec{P}_0 = \sum_i q\vec{r}_i + \sum_j \vec{m}_j \tag{6.1}$$

Here subscripts i $(= 1, 2, 3, \ldots, N-1, N)$ and j $(= 1, 2, 3, \ldots, N'-1, N')$ denote i-th electron and j-th dipole in N electrons and N' dipoles in the bulk of materials, respectively. In a closed-circuit, MDC i_{MDC} flowing in a material is identical with the current flowing in an external circuit. Consequently, an ammeter installed in the external circuit can measure MDC i_{MDC} flowing in a material. Noteworthy that the two terms of Eq. (6.1) become nonzero when the sum of the charge displacement toward the x–direction is non-zero and the sum of the dipolar rotation toward the x–direction is non-zero, respectively. The non-zero situation of the first term is related to the electrostatic polarization phenomena induced in a film, due to short-range and long-range displacement of excess charges in the bulk. Meanwhile, the second term concerns with the molecular ordering and disordering of dipoles. MDC measurement can detect the transitional change of dielectric polarization induced by these two fundamental electronic processes in organic materials.

This situation brings out two practical aspects of MDC measurement. One aspect is that MDC provides a conventional way to probe charge displacement and dipole orientational ordering (see Fig. 6.1). Electronic materials and devices can generate MDC due to charge displacement in them. Polar molecules in materials and in devices contribute to generate MDC during ordering and disordering of the dipoles. The other aspect is that we must carry MDC measurement without information on the origins of MDC, particularly, in the case of new materials. That is, we have no idea whether the origin of MDC is electronic charge displacement or dipole rotation, before we carry the MDC measurement. It takes a lot of time to identify the microscopic origin of MDC by the trace of MDC itself. The basic reason is as suggested from Fig. 6.1. In Fig. 6.1(a), the charge q displaces Δx in the space between two electrodes. Meanwhile charge $\Delta q = q\Delta x/d$ flows through the ammeter, as described in Sec. 2.1. Here, d is distance between two electrodes. In Fig. 6.1(b), a dipole μ $(= q\Delta x)$ rotates in the space between two electrodes. During a course of the dipolar rotation, MDC flows with accompanying effective charge flow $\Delta q = \mu/d$ as described in Sec. 2.2. Consequently, solely using the fact that Δq flows through an ammeter by means of MDC measurement, it is not easy to identify the microscopic origin, *i.e.*, charge q or dipole μ. We need lots of experiments to identify the origins, on keeping in mind the characteristic features of MDC due to electronic charges and dipoles (see Chap. 2 and Chap. 3).

On the other hand, optical SHG measurement directly probes dielectric polarization phenomena related to electronic charges and alignment of dipoles (see Fig. 6.2), where nonlinear polarization $\vec{P}_{NL}$

is probed by detecting SHG intensity I_{SHG}. The relationship between I_{SHG} and $\vec{P}_{NL}$ is given as

$$I_{SHG} \propto \left|\vec{P}_{NL}\right|^2 \quad \text{with}$$
$$\vec{P}_{NL} = \sum_i \gamma(qd\vec{r}_i)\vec{E}_0\vec{E}_\omega\vec{E}_\omega + \sum_j \beta(\vec{m}_j)\vec{E}_\omega\vec{E}_\omega \qquad (6.2)$$

Here $\gamma(qd\vec{r})$ concerns with the displacement of charges $qd\vec{r}$ (see Fig. 6.2(a)), and $\beta(\vec{m})$ concerns with the alignment of dipoles $\vec{m}$ (see Fig. 6.2(b)). Note that the first and second terms of Eq. (6.2) correspond to the first and second terms of MDC described in Eq. (6.1), respectively.

In SHG measurement, the second-order molecular polarizability $\beta(\vec{m})$ makes a main contribution to probe dipolar orientation. On the other hand, in the probing of displacement of electronic charges $qd\vec{r}$, charges produce electrostatic field $\vec{E}_0$, and third-order molecular polarizability $\gamma(qd\vec{r})$ makes a significant contribution to SHG. That is, we can measure the charge displacement and dipolar alignment by measuring the SHG light intensity.

This situation is the same with the MDC measurement. MDC allows one to measure charge displacement and dipole rotation by using an ammeter. SHG allows one to measure these two contributions by using optical detector. However, SHG differs from MDC in that SHG is resonantly enhanced at a specific laser wavelength. This fact allows one to identify the contribution of charges and dipoles, selectively, as illustrated in Fig. 6.2. At a wavelength λ_A, $\gamma(qd\vec{r})$ resonates, and charge displacement is selectively probed (Fig. 6.2(a)). On the other hand, at a wavelength λ_B, $\beta(\vec{m})$ dominates, and dipolar alignment is selectively visualized (Fig. 6.2(b)). Figure 6.3 summarizes an idea of measurement of charge displacement and dipolar orientation by MDC and SHG, in terms of the dielectric dispersion property of materials. MDC probes charge displacement and dipole orientation by measuring the sum of electrostatic polarization $\vec{P}_0$ in Eq. (6.1) at d.c. frequency. SHG selectively probes dipole orientation by using laser with angular frequency ω_1 (wavelength λ_1), and charge displacement with different angular frequency ω_2 (wavelength λ_2).

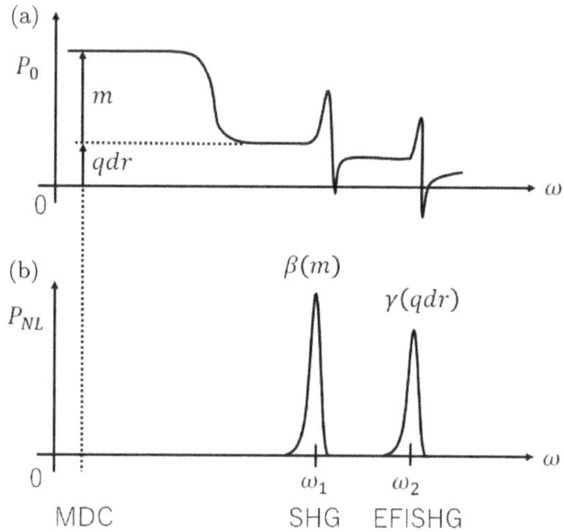

Figure 6.3. Idea of MDC and SHG in dielectric dispersion relation. (a) Dispersion of electrostatic polarization P_0. (b) Dispersion of nonlinear polarization P_{NL}.

In SHG measurement, the different angular frequency dependence on the resonant condition for $\beta(\vec{m})$ and $\gamma(qd\vec{r})$ allows one to selectively probe dipolar orientation and charge displacement. This difference comes from the structure of molecule in energy space. Figure 6.4 displays wave functions of electron, ψ_1, ψ_2, and ψ_3, of a molecule that are responsible for $\beta(\vec{m})$ and $\gamma(qd\vec{r})$ of the molecule. Here, ψ_1 is a wave function of ground state. Highest occupied molecular orbital (HOMO) is a candidate for ψ_1. The wave functions ψ_2 and ψ_3 represent molecular orbitals of excited states. Lowest unoccupied molecular orbital (LUMO) and the orbital next to LUMO (LUMO + 1), respectively, are candidates for ψ_2 and ψ_3. Figure 6.4 illustrates the shape and energy levels of these wave functions, *i.e.*, E_1 (wave function ψ_1), E_2 (ψ_2), and E_3 (ψ_3). We assume here that the symmetry of ψ_2 is different from the ψ_1, with respective to the spatial origin in x–direction, as illustrated in Fig. 6.4(a), whereas ψ_3 has the same symmetry with ψ_1 as illustrated in Fig. 6.4(b). Optical transition between ψ_1 and ψ_2 in Fig. 6.4(a) contributes to $\beta(\vec{m})$, while the transition between ψ_1 and ψ_3 in Fig. 6.4(b) represents optical process for $\gamma(qd\vec{r})$.

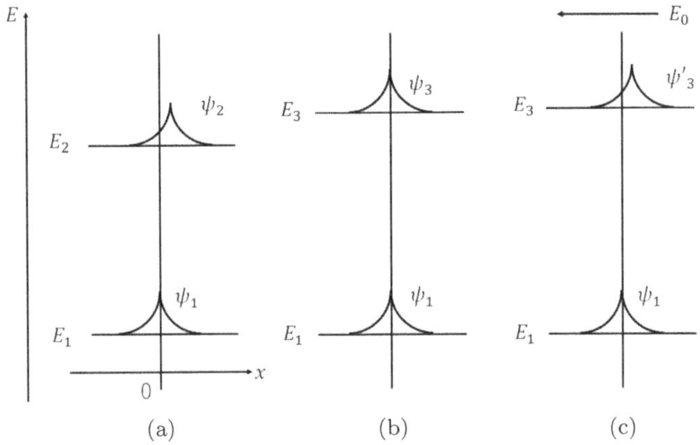

Figure 6.4. Wave functions of molecular orbital concerning SHG processes. (a) Molecular orbitals with wave function ψ_1 and ψ_2 for β, (b) ψ_1 and ψ_3 for γ in zero external electric field. (c) ψ_1 and ${\psi'}_3$ for γ in the presence of non-zero electrostatic field $E_0 \neq 0$.

Firstly, we discuss the optical second-harmonic generation (SHG) due to $\beta(\overrightarrow{m})$ (Fig. 6.4(a)). Prior to laser irradiation, electrons occupy ψ_1 and they are in stationary state with energy E_1. Under laser irradiation, photon energy $\hbar\omega_1$ is transferred to the electron in ψ_1 whilst photons are traveling through the absorption cross-section of the electron in molecule. The probability of this energy transfer is p (<1). After the energy transfer, energy of the electron changes from E_1 (<0) to $E_1 + \hbar\omega_1$. This state is no longer stationary state. Under pulse laser irradiation, where photon density is very high, another photon transfers energy $\hbar\omega_1$ to the electron at probability p. After that, the electron energy is $E_1 + 2\hbar\omega_1$. After the two-photon energy transfer, the angular frequency of excited electron is $\omega_e = \frac{E_1}{\hbar} + 2\omega_1$, and has the same angular frequency with the angular frequency of excited state ψ_2, $\omega_e = \frac{E_2}{\hbar} = \frac{E_1}{\hbar} + 2\omega_1$. In other words, optical transition fulfils the energy condition $E_2 - E_1 = 2\hbar\omega_1$, and excited electron is in stationary state in ψ_2. Afterwards, the electron in ψ_2 undergoes transition to ψ_1, on accompanying with SHG emission of photon energy $\hbar \cdot 2\omega_1$. The probability of transition from ψ_2 to ψ_1 is proportional to transition dipole moment $\langle\psi_1 \,|\, q\vec{r} | \psi_2\rangle$, in the manner same as that of linear polarization. In total, SHG intensity is proportional to the square of $p^2|\langle\psi_1\,|\,q\vec{r}\,|\,\psi_2\rangle|$, *i.e.*, $p^4\,|\langle\psi_1|q\vec{r}|\psi_2\rangle|^2$. Assuming that the absorption cross-section is very small and does not overlap with cross-section of other molecules, the probability p increases linearly in proportion to vibrating electric field of laser $\vec{E}_\omega$. As a result, SHG intensity increases in proportion to the square of laser intensity I_{Laser}, $I_{SHG} \propto p^4 \propto |\vec{E}_\omega|^4 \propto I_{Laser}^2$. This is the situation expressed by Eq. (6.2), $I_{SHG} \propto |\vec{E}_\omega|^4 \propto I_{Laser}^2$. On the other hand, $\langle\psi_1|q\vec{r}|\psi_2\rangle$ is non-zero, where ψ_1 and ψ_2 has different symmetry. This is the origin of SHG expressed by β of Eq. (6.2). SHG from non-zero $\beta(\vec{m})$ is used to visualize aligned dipole $\vec{m}$. This SHG vibrates at an angular frequency $2\omega_1$ corresponding to the energy gap $E_2 - E_1 = \hbar \cdot 2\omega_1$. The dipolar alignment is visualized at the SHG wavelength $\lambda_1 = 2\pi c/2\omega_1$ (c: velocity of laser light) and laser wavelength $2\pi c/\omega_1$.

Secondly, we focus on the SHG emission process $\gamma(qd\vec{r})$ concerning with ψ_3 and ψ_1, where ψ_1 and ψ_3 have the same symmetry. Consequently $\langle\psi_1|q\vec{r}|\psi_3\rangle$ is zero. SHG intensity is in proportion to $p^4\,|\langle\psi_1|q\vec{r}|\psi_3\rangle|^2$, but it is obviously zero because of $\langle\psi_1|q\vec{r}|\psi_3\rangle = 0$. In the presence of excess charges in materials, non-zero electrostatic field $\vec{E}_0$ is generating from these charges, according to Gauss law of electromagnetic theory. With non-zero electrostatic field $\vec{E}_0 \neq 0$, electron cloud of a molecule deviates as shown in Fig. 6.4(c) from its original distribution shown in Fig. 6.4(b). In the presence of the electric field $\vec{E}_0$, ψ_3 changes the symmetry from ψ_1-like to the symmetry of ψ_2-like (see the similarity of Figs. 6.4(a) and 6.4(c)). Consequently, $\langle\psi_1|q\vec{r}|\psi_3\rangle$ is non-zero in the presence of electrostatic field $\vec{E}_0$, and SHG can be activated. This SHG process is expressed by $\gamma\,(qd\vec{r})$ in Eq. (6.2), and is called the electric-field-induced optical second-harmonic generation (EFISHG) process. Note that there is an electric-field-induced SHG caused by dipolar polarization, but in this book, EFISHG indicates the SHG due to electron displacement and not the SHG due to dipolar polarization. The resonant angular frequency $2\omega_2$ for γ is governed by the energy gap $E_3 - E_1 = \hbar{\cdot}2\omega_2$. This means that EFISHG is resonantly enhanced at an SHG wavelength $\lambda_2 = 2\pi c/2\omega_2$ (laser wavelength $2\pi c/\omega_2$), in response to electrostatic field $\vec{E}_0$.

It is very important to point out that ψ_2 and ψ_3 are different wave functions of the same molecule, leading to different specific wavelength for SHG and EFISHG, *i.e.*, $\lambda_1 \neq \lambda_2$. For example, ψ_2 and ψ_3 are wave functions of LUMO and LUMO+1, respectively. Accordingly, by using SHG at wavelength λ_1 (laser wavelength $2\lambda_1$), dipole alignment is visualized. By using EFISHG at wavelength λ_2 (laser wavelength $2\lambda_2$), charge displacement is visualized. After identifying the specific wavelength for dipolar alignment λ_1 and charge displacement λ_2 in SHG spectrum, one can probe charge displacement and dipolar alignment, individually.

Similarly, it is obvious that SHG measurement has material selectivity. Different materials have different electronic energy structures, leading to different SHG spectra for β and γ. The SHG wavelength λ_2 for probing charge displacement in material A, *i.e.*, $\lambda_2 = \lambda_a$ is different from λ_2 for material B, *i.e.*, $\lambda_2 = \lambda_b$. This fact allows one to selectively probe charges moving in a layer of material A among multiple layers, *e.g.*, material A on material B, by choosing SHG wavelength at λ_a ($\neq \lambda_b$). In this way, charge motion in selected layer of multilayer device [1], charge motion in hole transport layer of organic light-emitting diodes (OLED) [2], charge motion in donor-phase of bulk heterojunction (BHJ) solar cells [3], and so on, are selectively probed (see Chap. 7). Motion of holes and electrons injected in an active layer of organic field-effect transistors (OFET) is also selectively visualized (Chap. 8).

Noteworthy, the abovementioned situation concerning the SHG is different in the MDC measurement. MDC merely probes $\vec{P}_0$ by electrostatic instruments at d.c. frequency. It is very hard to discuss the location of charges in multiple layer system solely from the trace of MDC. As shown in Fig. 6.3, both MDC and SHG are helpful for probing charge displacement and dipolar alignment, but SHG can probe them selectively, one-by-one, on the basis of dielectric dispersion in nonlinear polarization. Making use well of nonlinear optical spectra of materials and devices, one can selectively probe dipolar alignment and charge displacement by choosing laser wavelengths appropriately. This allows us to study the impact of dipolar alignment and the charge displacement in materials one-by-one by means of the optical SHG measurement, though it is difficult to distinguish them by means of the electrical MDC measurement.

6.2 Optical second-harmonic generation (SHG)

Optical second-harmonic generation (SHG) probes dipole alignment and electronic charges in materials and devices. The wavelength λ of incident light changes to half wavelength, *i.e.*, $\lambda/2$, in response to dielectric polarization in materials. The generated light with the half wavelength $\lambda/2$ is SHG. Figure 6.5 illustrates a string vibration model imitating an induction of second-harmonic wave. Fundamental wavelength is $2L$ for a string with length L. For generating a sound, we make displacement of string. Figure 6.5(a) gives an example where the string is pulled at the position nearby one end of string. This displacement is in a form of sine wave function with wavelength $4L$, *i.e.*, double wavelength of the fundamental tone. This form is not a steady state for a given string. After releasing the string, a sound is radiated from the string vibrating with a wavelength of $2L$ (Fig. 6.5(b)). That is, external force is given in a form of wavelength $4L$, but vibration is produced at halved wavelength $2L$. Note that SHG is radiated through optical transition, and the analogy of a string vibration applies to this transition, not to steady state.

SHG is a quantum mechanical process due to the coupling of electromagnetic waves with electrons in materials. In response to electronic charge displacement and dipole alignment in materials and devices, the color of incident laser changes to one octave higher. In the similar way to the generation of great tone by music instruments

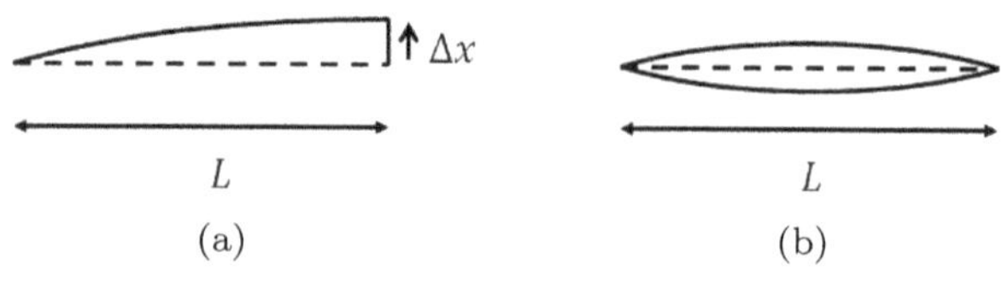

Figure 6.5. String vibration model imitating induction of SHG. (a) Induction of a string at double wavelength $4L$ of the fundamental tone. (b) Steady state of a string vibrating with wavelength $2L$ at the fundamental tone.

which harmonizes with material nature and air condition resonating inside, optical second-harmonics is resonantly enhanced reflecting the dielectric nature of molecules and electric field at molecule. For example, invisible infrared light (wavelength 1064 nm) is accompanied by generation of green light (wavelength 532 nm) in the presence of d.c. electric field in polyimide insulating layer. A green laser beam (wavelength 532 nm) induces ultra-violet light (wavelength 266 nm) due to dipolar alignment in the polyimide molecules.

Microscopically, the origins of SHG are modelled using displacement $d\vec{r}$ of charge q and orientational alignment of dipole $\overrightarrow{m}$. Charge q and dipole $\overrightarrow{m}$ are associated with electrostatic field $\vec{E}_0$ diverging from these charged particles according to the Gauss law of the electro-magnetism theory, and they are given as

$$q = 4\pi\varepsilon r^2(\vec{E}_0 \cdot \hat{r}) \tag{6.3a}$$

and

$$\overrightarrow{m} = 2\pi\varepsilon r^3(\vec{E}_0 \cdot \hat{r})\hat{r} - 4\pi\varepsilon r^3(\vec{E}_0 \cdot \hat{\theta})\hat{\theta}, \tag{6.3b}$$

with $\hat{r} \cdot \hat{\theta} = 0$ (see also Appendix A).

Here $\hat{r}$ and $\hat{\theta}$ represent a radial unit vector and an angular unit vector, respectively.

Equation (6.3) inspires us that there are two ways for probing charges and dipoles by using SHG. One is to directly probe charge q and dipole $\overrightarrow{m}$. This simple idea arises by looking at the left-hand-side term of Eq. (6.3). The other one is to probe the electrostatic field $\vec{E}_0$ diverging from charge q and dipole $\overrightarrow{m}$. This idea comes out from the right-hand-side term of Eq. (6.3). As it will be discussed in this chapter, SHG provides a way for measuring charge q and dipole $\overrightarrow{m}$, and electric-field-induced optical second-harmonic generation (EFISHG) gives a way to probe the electric field $\vec{E}_0$ diverging from charge q and dipole $\overrightarrow{m}$. In the followings, generation of SHG and EFISHG is briefly discussed. More precise description is found in books [4, 5], if necessary.

Dielectric polarization $\vec{P}$ is comprised of many kinds of polarizations related to displacement of charge q and orientational ordering of dipole $\vec{m}$. It is given as

$$\vec{P} = \vec{P}_0 + \vec{P}_l + \vec{P}_{NL}. \tag{6.4}$$

Here, $\vec{P}_0$ is spontaneous polarization induced in materials in the absence of external electric field, where electrostatic charge displacement and permanent dipole alignment are main contributors. $\vec{P}_0$ is detectable by means of MDC as described in Eq. (6.1) and discussed in Sec. 2.4. $\vec{P}_l$ is linear polarization induced under the non-zero external electric field, and it is proportion to this field. Electronic polarization and dipolar polarization mainly govern $\vec{P}_l$, where charge displacement and permanent dipole alignment induced by the external electric field are main contributors [6–8]. $\vec{P}_{NL}$ is nonlinear polarization, and this polarization is higher order terms with respect to external electric field. SHG and EFISHG are the nonlinear polarization. The vibrating electric field of laser beam serves as an external electric field $\vec{E}_\omega$ to induce nonlinear polarization $\vec{P}_{NL}$.

Noteworthy that spontaneous polarization $\vec{P}_0$ is defined as the result of alignment of permanent dipoles in the absence of external electric field. The electrostatic charge effect caused by displacement of electrons is also included here in the definition of spontaneous polarization $\vec{P}_0$. For example, in TSC measurement described in Chap. 3, electronic trapped charge displacement is frozen at low temperature, and it is included in $\vec{P}_0$. In organic solar cells, excitons are created under light illumination. Afterwards they are separated into holes and electrons, resulting in electromotive force. Electromotive force is also induced by triboelectric generator with an impact of the rubbing on materials. These electromotive forces are included in $\vec{P}_0$. They are discussed in Chap. 7 and Chap. 9. By focusing on electrostatic polarization $\vec{P}_0$, MDC probes charge displacement and dipole alignment. By focusing on nonlinear polarization $\vec{P}_{NL}$, SHG probes charge displacement and dipole alignment.

In SHG measurement, probe light propagates through materials to be visualized. The probe light is electro-magnetic wave and it accompanies vibrating electric field $\vec{E}_\omega$ at an angular frequency ω. This vibrating field acts on an electron in molecule to displace it. As a result, $\vec{P}$ with multiple frequency, ω, 2ω, *etc.*, is induced in materials. The vibration of polarization $\vec{P}$ at angular frequency ω causes electromagnetic wave that radiates at angular frequency ω. The vibration at a doubled frequency 2ω is a source of SHG and EFISHG at 2ω. Pulsed lasers are used as a probe light in SHG measurement. The electric field $\vec{E}_\omega$ of pulsed laser is very strong. SHG system equipped with Q-switched Nd:YAG laser is available, where, for example, pulse duration $a = 4\,\text{ns}$, repetition rate $b = 10\,\text{Hz}$, average power $c = 10\,\text{mW}$, and focus area $d = 1\,\text{mm}^2$. The electric field $\vec{E}_\omega$ of the laser is calculated from the amplitude of Poynting vector $\vec{E} \times \vec{H}$ as [9]

$$I_\omega = |\vec{E}_\omega \times \vec{H}| = 2c_0\varepsilon_0 \frac{|\vec{E}_\omega|^2}{2}. \tag{6.5}$$

In air, by using the speed of light $c_0 \approx 3.0 \times 10^8\,\text{m/s}$, and the dielectric permittivity of vacuum $\varepsilon_0 \approx 8.9 \times 10^{-12}\,\text{F/m}$, and peak power of pulsed laser $I_\omega = c/(abd)$, electric field of the laser is evaluated as $\left|\vec{E}_\omega\right| = 9.7 \times 10^4\,\text{V/cm}$. In this calculation, relative dielectric constant of air is assumed unity, because it is always nearly 1. In another SHG system, a femto-second laser, *e.g.*, pulse duration $a = 80\,\text{fs}$, repetition rate $b = 1\,\text{kHz}$, average power $c = 10\,\text{mW}$, and laser beam radius $d = 1\,\text{mm}$, is utilized. The laser intensity is $I_\omega = c/(\pi abd^2)$ and electric field is estimated as $\left|\vec{E}_\omega\right| = 1.2\,\text{MV/cm}$.

It is instructive here to examine the impact of the laser radiation on the molecule in organic electronic devices, *i.e.*, the situation of electrons in a molecule under this high electric field. For example, as typical molecules used in organic devices, we may consider pentacene, phthalocyanine, and other conjugated molecules. Assuming one positive charge with electricity $q = +1.6 \times 10^{-19}\,\text{C}$, *i.e.*, hole, is on a pentacene molecule, it is estimated that the electrostatic field 14 MV/cm is formed at the location where the neighbor pentacene molecule exists, *e.g.*, 1 nm away from the charged pentacene molecule. The magnitude of the electric field of pulse laser is comparable to this field. The strong electric field of pulsed laser promotes electrons from a ground state of a molecule to excited energy states.

Nonlinear polarization $\vec{P}_{NL}$ with doubled frequency 2ω, induced by incidence probe laser with angular frequency ω, is given in a tensor form as

$$\vec{P}_{NL} = \varepsilon_0 \overleftrightarrow{\chi}^{(2)}(-2\omega;\omega,\omega) : \vec{E}_\omega \vec{E}_\omega + \varepsilon_0 \overleftrightarrow{\chi}^{(3)}(-2\omega;0,\omega,\omega) \vdots \vec{E}_0 \vec{E}_\omega \vec{E}_\omega + \cdots \quad (6.6)$$

Here, ε_0 is vacuum permittivity. $\overleftrightarrow{\chi}^{(2)}(-2\omega;\omega,\omega)$ is the second order nonlinear susceptibility, which serves to convert the electric field of probe laser at angular frequency ω to the second-harmonic polarization at angular frequency 2ω. This leads to the generation of SHG. $\overleftrightarrow{\chi}^{(3)}(-2\omega;0,\omega,\omega)$ is the third order nonlinear susceptibility, which serves to convert the electric field of probe laser to nonlinear polarization at angular frequency 2ω in the presence of d.c. electric field $\vec{E}_0$, resulting in electric-field-induced optical second-harmonic generation (EFISHG). The detail of the expression of $\overleftrightarrow{\chi}^{(2)}$ and $\overleftrightarrow{\chi}^{(3)}$ is described in books [4, 5]. The tensor calculation of the right-hand-side term of Eq. (6.6) is carried out by means of matrix transformation in a manner described in the book [10].

EFISHG due to non-zero $\chi^{(3)}$ allows one to measure the impact of electrostatic charge displacement in materials, because displaced charges form electrostatic field $\vec{E}_0$ in materials. This measurement is based on an idea coming out from right-hand-side terms of Eq. (6.3a). On the other hand, SHG due to $\chi^{(2)}$ allows one to evaluate dipolar alignment. This evaluation is related to $\vec{m}$ on the left-hand-side of Eq. (6.3b). In this book, the use of SHG and EFISHG, respectively, for probing dipolar alignment and charge displacement, is described with these ideas. It is instructive here to note that both SHG and EFISHG are available in other ways to probe electronic charge and dipole. SHG is capable of probing electrostatic charge, on focusing q of left-hand-side of Eq. (6.3a). EFISHG provides a way to evaluate dipole alignment by visualizing electrostatic field $\vec{E}_0$ produced by dipoles by looking at right-hand-side of Eq. (6.3b).

6.2.1 *Microscopic nonlinear polarizabilities* $\boldsymbol{\beta}$, $\boldsymbol{\gamma}$, *and macroscopic nonlinear susceptibilities* $\boldsymbol{\chi^{(2)}}$, $\boldsymbol{\chi^{(3)}}$

In nonlinear polarization process of materials, electrons in a molecule response nonlinearly to vibrating electric field $\vec{E}_\omega$ of incident probe laser. Nonlinear polarization $\vec{p}_{NL}$ of a molecule is given as

$$\vec{p}_{NL} = \overleftrightarrow{\beta} : \vec{E}_\omega \vec{E}_\omega + \overleftrightarrow{\gamma} \vdots \vec{E}_0 \vec{E}_\omega \vec{E}_\omega + \cdots \tag{6.7}$$

$\overleftrightarrow{\beta}$ is the second-order nonlinear polarizability, $\overleftrightarrow{\gamma}$ is the third-order nonlinear polarizability. They are the third rank tensor $\overleftrightarrow{\beta} = \beta_{ijk}$, and the fourth rank tensor $\overleftrightarrow{\gamma} = \gamma_{ijkl}$, respectively. β_{ijk} converts the product of j-th component of electric field vector $\vec{E}_\omega$ and k-th component of $\vec{E}_\omega$ to i-th component of polarization $\vec{p}_{NL}$. Here the subscripts i, j, k = 1, 2, 3 denote the x, y, z direction of 3-dimensional (3D) coordinate system. For example, β_{123} converts the product of y–component of $\vec{E}_\omega$ of probe laser and z–component of $\vec{E}_\omega$ to the x–component of $\vec{p}_{NL}$, β_{111} converts the x–component of $\vec{E}_\omega$ to the x–component of $\vec{p}_{NL}$.

These polarizabilities, $\overleftrightarrow{\beta}$ and $\overleftrightarrow{\gamma}$, concern with the potentiality of displacement of electrons in a molecule at the double-frequency of probe laser. The energy structure of molecule directly impacts on the displacement of electrons, and this impact appears in the SHG spectrum. That is, on scanning laser wavelength λ in the region from λ_1 to λ_2 ($\lambda_1 > \lambda > \lambda_2$), SHG spectrum due to non-zero $\overleftrightarrow{\beta}$ is obtained by measuring light intensity from aligned dipoles at halved wavelength $\lambda/2$. Meanwhile, by scanning laser wavelength λ in the region from λ_1 to λ_2 and measuring light intensity at halved wavelength $\lambda/2$, in response to electrostatic field $\vec{E}_0$, EFISHG spectrum due to non-zero $\overleftrightarrow{\gamma}$ is obtained.

The energy structure depends on molecular structure, where optical transition is allowed only at specific frequencies. SHG ($\overleftrightarrow{\beta}$) and EFISHG ($\overleftrightarrow{\gamma}$) are resonantly enhanced at different angular frequencies. We therefore can identify microscopic origins of electrostatic phenomena, *e.g.*, electric field from charge displacement or dipolar alignment, by means of the SHG measurement.

In material system where the density of N molecules responses to the vibrating electric field $\vec{E}_\omega$ in an identical manner, nonlinear polarization $\vec{P}_{NL}$ is given as

$$\begin{aligned}\vec{P}_{NL} &= N\vec{p}_{NL} \\ &= N\overleftrightarrow{\beta} : \vec{E}_\omega\vec{E}_\omega + N\overleftrightarrow{\gamma} \vdots \vec{E}_0\vec{E}_\omega\vec{E}_\omega.\end{aligned} \tag{6.8}$$

As illustrated in Fig. 6.3, $\overleftrightarrow{\beta}$ and $\overleftrightarrow{\gamma}$ contribute to generate different SHG spectra, depending on a quantum mechanical wave function that governs $\overleftrightarrow{\beta}$ and $\overleftrightarrow{\gamma}$. Assuming that the second order polarizability $\overleftrightarrow{\beta}$ dominates at angular frequency ω_1 and the third order polarizability $\overleftrightarrow{\gamma}$ is minor at the same angular frequency, nonlinear polarization $\vec{P}_{NL}$ in Eq. (6.8) at angular frequency ω_1 is

$$\vec{P}_{NL} = N\overleftrightarrow{\beta} : \vec{E}_\omega\vec{E}_\omega. \tag{6.8a}$$

On the other hand, at angular frequency ω_2, the third order molecular polarizability $\overleftrightarrow{\gamma}$ is enhanced whereas the second order polarizability $\overleftrightarrow{\beta}$ is not enhanced. At this frequency, Eq. (6.8) is

$$\vec{P}_{NL} = N\overleftrightarrow{\gamma} \vdots \vec{E}_0\vec{E}_\omega\vec{E}_\omega. \tag{6.8b}$$

In SHG experiment, by using laser beam at $\omega = \omega_1$, we can selectively probe dipole orientation, where non-zero $\overleftrightarrow{\beta}$ is a main contributor. On the other hand, by setting the laser wavelength at $\omega = \omega_2$, electronic charge displacement is selectively probed by utilizing the fact that $\overleftrightarrow{\gamma}$ is enhanced at this angular frequency, but $\overleftrightarrow{\beta}$ is not.

In material system where molecular orientation is distributed, nonlinear polarization $\vec{P}_{NL}$ is given as

$$\begin{aligned}\vec{P}_{NL} &= N\langle\vec{p}_{NL}\rangle \\ &= N\langle\overleftrightarrow{\beta}\rangle : \vec{E}_\omega\vec{E}_\omega + N\langle\overleftrightarrow{\gamma}\rangle \vdots \vec{E}_0\vec{E}_\omega\vec{E}_\omega + \cdots\end{aligned} \tag{6.9}$$

Here $\langle\cdots\rangle$ represents the statistical average of molecular orientation in space. Looking at Eq. (6.6) and Eq. (6.9), relationship between macroscopic susceptibility $\chi^{(2)}_{ijk}$ and microscopic molecular polarizability β_{ijk} is derived as

$$\chi^{(2)}_{ijk} = \frac{N\langle\beta_{ijk}\rangle}{\varepsilon_0}, \beta_{ijk} = a_{ii'}a_{jj'}a_{kk'}\beta_{i'j'k'}. \tag{6.10a}$$

Similarly, relationship between macroscopic susceptibility $\chi^{(3)}_{ijkl}$ and microscopic molecular polarizability γ_{ijkl} is given by

$$\chi^{(3)}_{ijkl} = \frac{N\langle\gamma_{ijkl}\rangle}{\varepsilon_0}, \gamma_{ijkl} = a_{ii'}a_{jj'}a_{kk'}a_{ll'}\gamma_{i'j'k'l'}. \tag{6.10b}$$

In Eq. (6.10), denominator ε_0 on right-hand-side term indicates that susceptibility is defined in reference to dielectric permittivity of vacuum. Subscripts i, j, k, l and i', j', k', l' denote cartesian axis 1,2,3 in reference to laboratory and molecular coordinates, respectively. $a_{11'}, a_{12'}, \ldots a_{32'}, a_{33'}$ are transform matrix elements from molecular coordinates to laboratory coordinates [11, 12]. β_{ijk} and $\beta_{i'j'k'}$ have $3^3 = 27$ components for 3 subscripts changing from 1 to 3. γ_{ijkl} and $\gamma_{i'j'k'l'}$ have $3^4 = 81$ components for 4 subscripts changing from 1 to 3. The product of transform matrix element $a_{ii'}a_{jj'}a_{kk'}$ has $3^6 = 729$ components for 6 subscripts changing from 1 to 3. $a_{ii'}a_{jj'}a_{kk'}a_{ll'}$ have $3^8 = 6561$ components for 8 subscripts changing from 1 to 3. The derivation is very lengthy, but they can be immediately calculated by using a computer. In most cases, only a few of them are non-zero among 27 members of $\chi^{(2)}_{ijk}$ and 81 members of $\chi^{(3)}_{ijkl}$ [13] and other members are zero, depending on the symmetry of a material system. This situation is the same as that of other tensors used for describing other physical quantities. For example, electronic polarizability α_{ij} is second-rank tensor and has 9 members, α_{11}, α_{12}, α_{13}, α_{21}, α_{22}, α_{23}, α_{31}, α_{32}, α_{33}. These 9 members represent physical quantity of polarization p_i in terms of electric field components E_j as $p_i = \alpha_{ij}E_j$. Noteworthy that this is the most general case, and actual case is different. For example, electronic polarizability of spherical helium atom is isotropic. Accordingly, only one parameter α $(= \alpha_{11} = \alpha_{22} = \alpha_{33})$ is sufficient to represent electronic polarization of helium atom. Other elements are zero $\alpha_{12} = \alpha_{13} = \alpha_{21} = \alpha_{23} = \alpha_{31} = \alpha_{32} = 0$. In the same way, we can reduce the number of parameters, depending on the symmetry of molecules, *i.e.*, geometry and shape of molecules.

6.2.2 *Nonlinear polarizabilities β and γ in terms of perturbation method*

For calculating the second and third order polarizability $\overleftrightarrow{\beta}$ and $\overleftrightarrow{\gamma}$ on the basis of quantum mechanics, for example, perturbation method is available [14]. Using the result of perturbation method, we can evaluate SHG spectrum coming from nonlinear polarizabilities $\overleftrightarrow{\beta}$ and $\overleftrightarrow{\gamma}$, and thus can determine proper probe laser wavelength to be used for SHG and EFISHG measurement [15]. For this evaluation, semi-empirical calculation, *ab-initio* calculation, and others are available [16–19]. Use of perturbation method means that our calculation assumes that unperturbed stationary states are known *a priori,* for example, by semi-empirical quantum chemical calculation, and the motion of electron is described on the basis of these unperturbed stationary states. This point is different from an idea described in Sec. 6.1 where an electron motion is considered with an idea of non-stationary states, but provides a useful way for evaluation of $\overleftrightarrow{\beta}$ and $\overleftrightarrow{\gamma}$.

In the following, calculation of $\beta = \beta_{111}$ and $\gamma = \gamma_{1111}$ is shown. At first, β is derived. Probe laser beam is an electromagnetic wave with vibrating electric field $\overrightarrow{E}$ and magnetic field $\overrightarrow{B}$ with angular frequency ω. Here, we assume that main force working a charge in molecule to move is electric Coulomb force, and magnetic force is discarded. The electric field E_ω along the x–axis is alternately changing at an angular frequency ω, and it is expressed as

$$\begin{aligned} E_\omega &= E\cos\omega t \\ &= E\frac{e^{+j\omega t} + e^{-j\omega t}}{2}. \end{aligned} \tag{6.11}$$

In response to this vibrating sinusoidal electric field, an electron in molecule moves a distance of x at an angular frequency ω by the Coulomb force $f = qE_\omega$. That is, the motion is perturbed by the potential energy V' of this Coulomb force. The potential $V = V_0$ that represents the stationary state of a molecule in zero external electric field changes to $V = V_0 + V'$ in the presence of vibrating sinusoidal electric field. The potential V' is given by

$$V' = fx = qE_\omega x. \tag{6.12}$$

Before laser irradiation, highest occupied molecular orbital (HOMO) is occupied with electrons, and the electrons are in stationary state ψ. Under perturbation V' of Eq. (6.12), the motion of the electron is deformed. Accordingly, the wave-function ψ changes to the wave-function ψ' of the perturbed state. The expectation value of polarization p' of this perturbed state is

$$p' = \langle \psi' | ex | \psi' \rangle. \tag{6.13}$$

The perturbation calculation results in ψ' as

$$\begin{aligned} | \psi' \rangle &= |0\rangle + \lambda b_0 \, |0\rangle \, e^{-j\omega_0 t} + \lambda b_1 \, |1\rangle \, e^{-j\omega_1 t} + \lambda b_2 \, |2\rangle \, e^{-j\omega_2 t} \\ &\quad + \lambda^2 c_0 \, |0\rangle \, e^{-j\omega_0 t} + \lambda^2 c_1 \, |1\rangle \, e^{-j\omega_1 t} + \lambda^2 c_2 \, |2\rangle \, e^{-j\omega_2 t} \end{aligned} \tag{6.14}$$

where $|0\rangle$ is HOMO, $|1\rangle$ is LUMO+1, and $|2\rangle$ is LUMO, as illustrated in Fig. 6.6(a). λ represents an order of the perturbation, b_n and c_n are the 1st and 2nd order perturbation coefficients of $|n\rangle$ $(n = 0, 1, 2)$, $\omega_n = E_n/\hbar$ (E_n: energy level of $|n\rangle$, $\hbar$: Dirac's constant).

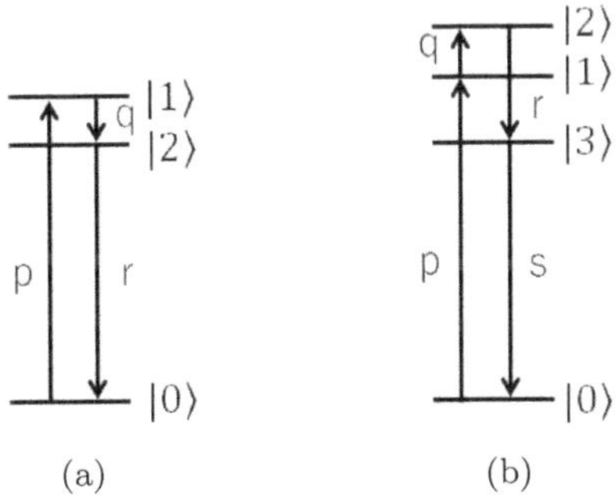

Figure 6.6. Energy diagram of SHG based on perturbation method. (a) SHG process for β. (b) EFISHG process for γ.

The polarization p' in Eq. (6.13) is calculated as

$$\begin{aligned} p' = \langle \psi' | ex | \psi' \rangle &= \langle 0 | ex | 0 \rangle + \cdots + \{ b_0 \langle 0 | ex | 0 \rangle + \cdots \} E \\ &+ \{ c_0 \langle 0 | ex | 0 \rangle + \cdots \} EE + \{ b_0^* b_0 \langle 0 | ex | 0 \rangle + \cdots \} EEE \\ &+ \{ c_0^* c \langle 0 | ex | 0 \rangle + \cdots \} EEEE. \end{aligned} \tag{6.15}$$

The right-hand-side of Eq. (6.15) is comprised of a number of terms. Among them, the terms vibrating at 2ω concerns with SHG. From Eq. (6.15), nonlinear polarization p_{NL} vibrating at doubled frequency 2ω is derived

$$\begin{aligned} p_{NL} &= \left[\frac{\mu_{01}^r \mu_{12}^q \mu_{20}^p}{4\hbar^2} \frac{1}{(\omega_{10} - \omega_q - \omega_p)(\omega_{20} - \omega_p)} \right. \\ &\quad \left. + \frac{\mu_{02}^r \mu_{21}^q \mu_{10}^p}{4\hbar^2} \frac{1}{(\omega_{20} - \omega_q - \omega_p)(\omega_{10} - \omega_p)} \right] E^2 e^{i(-2\omega)t} \\ &= \beta E^2 e^{i(-2\omega)t}. \end{aligned} \tag{6.16}$$

The first term represents transition from $|0\rangle$ to $|2\rangle$ to $|1\rangle$ to $|0\rangle$. The second term represents transition from $|0\rangle$ to $|1\rangle$ to $|2\rangle$ to $|0\rangle$. Assuming that SHG wavelength is corresponding to the energy gap between $|0\rangle$ and $|2\rangle$ in experiment, the second term is accepted as an expression for β. In detail, the idea for taking terms with $\langle 0|ex|2\rangle$ is based on the quantum mechanical transition of SHG. Figure 6.6(a) illustrates three states $|0\rangle$, $|1\rangle$, and $|2\rangle$, which appear in the terms of Eq. (6.14). $|0\rangle$ is HOMO at, for example, $E_0 = -8.0\,\text{eV}$. $|1\rangle$ is LUMO+1 at $E_1 = -5.5\,\text{eV}$. $|2\rangle$ is LUMO $E_2 = -6.0\,\text{eV}$. These states are represented by unperturbed wave functions. We are setting SHG wavelength at 620 nm corresponding to the energy gap between HOMO and LUMO $E_2 - E_0 = 2.0\,\text{eV}$, meanwhile laser wavelength is set at 1240 nm (photon energy 1.0 eV), that is, double of SHG wavelength. Prior to laser irradiation, an electron is in ground state $|0\rangle$. In the presence of electric field of laser beam, an electron motion is deformed from $|0\rangle$. As a result, this electron has a possibility to be promoted to an excited state $|1\rangle$. For excitation from $|0\rangle$ to $|1\rangle$, photon energy $\hbar\omega$ and $\hbar\omega'$ are transferred to the electron with energy condition $E_{10} = \hbar\omega + \hbar\omega'$ where E_{10} is energy gap between

$|0\rangle$ and $|1\rangle$, $\hbar\omega$ is photon energy of laser (1.0 eV), and $\hbar\omega'$ (1.5 eV) is photon energy from vacuum. Accordingly, electron at E_0 is promoted to LUMO + 1 at $E_1 = -5.5$ eV. The electron in $|1\rangle$ transits to $|2\rangle$ in the presence of second order perturbation. During the course of second perturbation, another energy $\hbar\omega - \hbar\omega'$ is transferred to the electron. $\hbar\omega$ (1.0 eV) is photon energy of laser and $-\hbar\omega'$ (-1.5 eV) is photon energy going back to vacuum. In total, two-photon energy quanta $2\hbar\omega$ is transferred from laser to an electron in molecule, and excited to $|2\rangle$. In vacuum, a set of creation and annihilation gives no total effect $+\hbar\omega' - \hbar\omega' = 0$. Afterwards, electron at LUMO $|2\rangle$ transits to the ground state $|0\rangle$ accompanying with SHG emission of photon energy $\hbar \cdot 2\omega = E_2 - E_0$ (2.0 eV). Assuming that transition from LUMO to HOMO, *i.e.*, $|2\rangle$ to $|0\rangle$, dominates, the transitions of SHG process terminates with $\langle 0|ex|2\rangle$:

$$\begin{aligned} p_{NL} &= \frac{\mu_{02}^r \mu_{21}^q \mu_{10}^p}{4\hbar^2} \frac{1}{(\omega_{20} - \omega_q - \omega_p)(\omega_{10} - \omega_p)} E^2 e^{i(-2\omega)t} \\ &= \beta E^2 e^{i(-2\omega)t}, \text{ with } \beta = \frac{\mu_{02}^r \mu_{21}^q \mu_{10}^p}{4\hbar^2} \frac{1}{(\omega_{20} - \omega_q - \omega_p)(\omega_{10} - \omega_p)}. \end{aligned} \tag{6.17}$$

In SHG measurement, ω_p $(= \omega_q = \omega)$ is angular frequency ω of laser beam. For this SHG, second order molecular polarizability β is

$$\beta = \frac{\mu_{02}^r \mu_{21}^q \mu_{10}^p}{4\hbar^2} \frac{1}{(\omega_{20} - 2\omega)(\omega_{10} - \omega)}. \tag{6.17a}$$

Equation (6.17) shows that second order polarizability β is resonantly enhanced at an angular frequency, on satisfying the relation $2\omega = \omega_{20}$. For example, using Nd:YAG laser (wavelength 1064 nm) as a probe laser, angular frequency of laser is calculated as ω $(=\omega_q = \omega_p) = 1.8 \times 10^{15}$ rad/s $(2.9 \times 10^{14}$ Hz) and frequency of p_{NL} is $2\omega = 3.6 \times 10^{15}$ rad/s $(5.7 \times 10^{14}$ Hz). By using Nd:YAG laser with 1064 nm wavelength, SHG at 532 nm is induced. The photon energy of SHG is 2.3 eV. A molecule with HOMO–LUMO energy gap of 2.3 eV radiates SHG due to resonantly enhanced β by using Nd:YAG laser.

The third order molecular polarizability γ is calculated by 3rd order perturbation as a sum of terms such as

$$\begin{aligned}\gamma = &\frac{\mu_{03}^{s}\mu_{32}^{r}\mu_{21}^{q}\mu_{10}^{p}}{8\hbar^3}\frac{1}{(\omega_{30}-\omega_r-\omega_q-\omega_p)(\omega_{20}-\omega_q-\omega_p)(\omega_{10}-\omega_p)}\\ &+\frac{\mu_{01}^{s}\mu_{12}^{r}\mu_{21}^{q}\mu_{10}^{p}}{8\hbar^3}\frac{1}{(\omega_{10}-\omega_r-\omega_q-\omega_p)(\omega_{20}-\omega_q-\omega_p)(\omega_{10}-\omega_p)}\\ &+\frac{\mu_{02}^{s}\mu_{23}^{r}\mu_{31}^{q}\mu_{10}^{p}}{8\hbar^3}\frac{1}{(\omega_{20}-\omega_r-\omega_q-\omega_p)(\omega_{30}-\omega_q-\omega_p)(\omega_{10}-\omega_p)}\\ &+\frac{\mu_{02}^{s}\mu_{21}^{r}\mu_{13}^{q}\mu_{30}^{p}}{8\hbar^3}\frac{1}{(\omega_{20}-\omega_r-\omega_q-\omega_p)(\omega_{10}-\omega_q-\omega_p)(\omega_{30}-\omega_p)}+\cdots\end{aligned} \tag{6.18}$$

In Eq. (6.18), the first term terminates with the transition from $|3\rangle$ to $|0\rangle$, *i.e.*, μ_{03}^{s}, and is considered for EFISHG in the following. The optical transition corresponding to this process is illustrated in Fig. 6.6(b). With assuming that E_p is electrostatic field, the angular frequency ω_p is zero, *i.e.*, $\omega_p = 0$. By taking the first term of Eq. (6.18), the third order molecular polarizability γ is expressed as

$$\gamma = \frac{\mu_{03}^{s}\mu_{32}^{r}\mu_{21}^{q}\mu_{10}^{p}}{8\hbar^3}\frac{1}{(\omega_{30}-\omega_r-\omega_q)(\omega_{20}-\omega_q)\omega_{10}} \tag{6.19}$$

In SHG measurement, ω_r $(= \omega_q = \omega)$ is angular frequency ω of laser beam. For this SHG, third order molecular polarizability γ is

$$\gamma = \frac{\mu_{03}^{s}\mu_{32}^{r}\mu_{21}^{q}\mu_{10}^{p}}{8\hbar^3}\frac{1}{(\omega_{30}-2\omega)(\omega_{20}-\omega)\omega_{10}} \tag{6.19a}$$

As expressed by Eq. (6.17) and Eq. (6.19), the second and third nonlinear polarizability β of SHG and γ of EFISHG are dependent on the laser angular frequency ω. These spectra are specific to molecules and optical transition dipoles concerning with the nonlinear optical process.

6.2.3 *Resonant enhancement in nonlinear polarizabilities β and γ*

From view point of dielectric physics, wavelength dependence of β and γ is discussed in relation to dipoles $\vec{m}$ and electronic charge displacement $qd\vec{x}$. Accordingly, one finds selectivity of SHG measurement for charge displacement $qd\vec{x}$ by γ and dipolar orientation $\vec{m}$ by β. In linear optical process, optical transition is resonantly enhanced at angular frequency of incidence light of $\omega = \omega_{10}$, where ω_{10} satisfies $\hbar\omega_{10} = E_{10}$. E_{10} is an energy difference between ground state $|0\rangle$ and excited state $|1\rangle$. An electron in ground state $|0\rangle$ transits to $|1\rangle$, meanwhile photon energy $\hbar\omega$ transfers to the electron. Afterwards, an excited electron in $|1\rangle$ transits to ground state $|0\rangle$, and a photon with energy quanta $\hbar\omega_{10}$ is emitted. That is, the linear polarization generated by optical transition is in proportion to transition dipole moment $\mu_{10} = \langle 1|ex|0\rangle$ and $\mu_{01} = \langle 0|ex|1\rangle$. This is the so-called one-photon process.

The situation of resonant enhancement in SHG is similar, but in nonlinear polarization process, an involved electron goes through multiple energy states. Figures 6.7(a–1) and (a–2) describe molecular energy levels, where $|1\rangle$ is symmetric (S) to $|3\rangle$, while $|2\rangle$ is asymmetric (A) to $|3\rangle$. For SHG process that is governed by the second order polarizability β, transition is illustrated in Fig. 6.7(a–1), and an electron undergoes multiple transitions from $|0\rangle$ to $|1\rangle$ to $|3\rangle$ to $|0\rangle$. As Eq. (6.17) shows, polarization of this SHG process is proportional to $\mu_{03}\mu_{31}\mu_{10}$. In EFISHG process where the third order nonlinear polarizability γ dominates, an electron transits from $|0\rangle$ to $|1\rangle$ to $|2\rangle$ to $|3\rangle$ to $|0\rangle$ (Fig. 6.7(a–2)). As Eq. (6.19) shows, polarization of this SHG process is proportional to $\mu_{03}\mu_{32}\mu_{21}\mu_{10}$. In β process, transition dipole is zero $\mu_{31} = 0$ for transition from $|1\rangle$ to $|3\rangle$. As a result, this SHG process is forbidden, and β is minor. In γ process, transition dipole is from $\mu_{32} \neq 0$ and $\mu_{21} \neq 0$. As a result, this SHG process is allowed, and γ is enhanced. For the set of energy levels illustrated in Fig. 6.7(a–1) and 6.7(a–2), γ dominates among β and γ. That is, SHG mainly reflects charge displacement qdx.

By contrast, Fig. 6.7(b–1) and (b–2) illustrate other set of energy levels, where $|1\rangle$ is asymmetric (A) to $|3\rangle$, while $|2\rangle$ is symmetric (S) to $|3\rangle$. In this case, $\mu_{31} \neq 0$ and β is non-zero. But $\mu_{32} = 0$, and γ is zero. For this set of energy levels, β is enhanced but γ is small. That is, SHG mainly probes dipolar alignment $\vec{m}$.

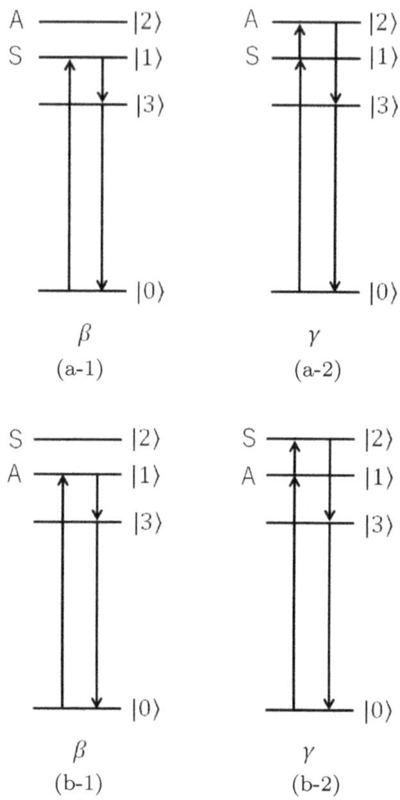

Figure 6.7. Energy diagram of SHG based on perturbation method. "S" and "A" labeled on left-hand-side of energy level denote, respectively, symmetric and asymmetric wave function in contrast to $|3\rangle$. $|1\rangle$ is symmetric while $|2\rangle$ is asymmetric, for (a–1) β and (a–2) γ. $|1\rangle$ is asymmetric while $|2\rangle$ is symmetric, for (b–1) β and (b–2) γ.

6.2.4 *Dielectric polarization and β and γ*

Molecular nonlinear polarizabilities β and γ are responsible for SHG and EFISHG, respectively. The dipolar alignment contributes to β. The electrostatic field induces electronic displacement in molecule, that governs γ. The nonlinear polarization of a molecule due to β of Eq. (6.17) is

$$p_{NL} = \beta E E e^{i(-2\omega)t} = \mu_{02}^{r} e^{i(-2\omega)t} \frac{1}{4} \frac{\mu_{21}^{q} E}{\hbar\omega_{20} - 2\hbar\omega} \frac{\mu_{10}^{p} E}{\hbar\omega_{10} - \hbar\omega} \tag{6.20}$$

That is, SHG of β is double-frequency vibration $\mu_{02}^{r}e^{i(-2\omega)t}$ that occurs with multiple probabilities

$$\frac{\mu_{21}^{q}E}{\hbar\omega_{20}-2\hbar\omega}$$

and

$$\frac{\mu_{10}^{p}E}{\hbar\omega_{10}-\hbar\omega}$$

The probability $\mu_{21}^{q}E/(\hbar\omega_{20}-2\hbar\omega)$ is the ratio of electrostatic potential energy $\mu_{21}^{q}E$ to an energy quantum $\hbar\omega_{20}-2\hbar\omega$ which is necessary to promote electrons from $|0\rangle$ to $|2\rangle$, *i.e.*, $\hbar\omega_{20}$, in addition to double of laser beam photon $2\hbar\omega$. The SHG process of Eq. (6.20) during a course of SHG spectrum measurement brings out a mechanism for resonant enhancement of β. We here discuss SHG spectrum measurement where the laser wavelength is scanned from long wavelength to short wavelength, *e.g.*, from 1200 nm to 800 nm. The optical absorption peak wavelength of material is assumed to be located at 450 nm, corresponding to HOMO–LUMO transition energy $\hbar\omega_{20} = 2.76\,\text{eV}$. Photon energy of $\lambda = 1200\,\text{nm}$ laser beam is $1240/\lambda$ (λ in nm) eV and is 1.03 eV. Double photon energy is 2.06 eV. The laser beam excites electrons in HOMO by 2.06 eV during a course of two photon absorption, and additionally $\Delta E = 0.70\,\text{eV}$ $(= 2.76\,\text{eV} - 2.06\,\text{eV})$ is needed to promote the electron to LUMO. The magnitude of electric field by femto-second laser is, for example, $E = 1.2 \times 10^{8}\,\text{V/m}$. The electrostatic potential stabilized on transition dipole moment $\mu_{21} = 10$ Debye (1 Debye $= 3.33 \times 10^{-30}\,\text{Cm}$) in this field is $\mu_{21}E = 4.0 \times 10^{-21}\,\text{J}$ $(= 0.02\,\text{eV})$. The energy ratio of electric polarization $\mu_{21}E$ to ΔE is $0.02\,\text{eV}/0.70\,\text{eV} = 3\,\%$, and not sufficient to enhance SHG. In course of sweeping the laser wavelength from 1200 nm to shorter wavelength, the wavelength arrives at 905 nm, laser photon energy is 1.37 eV, double photon excites electron in molecules by 2.74 eV, and electrostatic potential is stabilized 0.02 eV. The energy ratio of electric polarization $\mu_{21}E$ to ΔE is 1. That is, one electron in HOMO is promoted by energy quantum by $2.74\,\text{eV} + 0.02\,\text{eV} = 2.76\,\text{eV}$ reaching the transition energy for the HOMO–LUMO transition. Accordingly, by scanning the laser wavelength from 1200 nm to 800 nm, SHG due to β is zero at laser wavelength 1200 nm (SHG wavelength 600 nm), and resonantly enhanced from 905 nm (SHG wavelength 453 nm), and reaching nearly maximum at 900 nm (SHG wavelength 450 nm, corresponding to HOMO–LUMO transition), and again zero at 800 nm (SHG 400 nm). In the shorter wavelength region, β is zero. The vibration at doubled frequency is not allowed in this shorter region, where no stationary state is allowed.

In a similar way, the probability $\mu_{10}^{p}E/(\hbar\omega_{10} - \hbar\omega)$ is the ratio of electrostatic potential energy $\mu_{10}^{p}E$ to $\hbar\omega_{10} - \hbar\omega$. As illustrated in Fig. 6.6(a), $|1\rangle$ is LUMO + 1, and $\hbar\omega_{10} > \hbar\omega_{20}$. For example, assuming that LUMO + 1 is 1.34 eV above LUMO, $\hbar\omega_{10} = 4.10\,\text{eV}$, giving second optical absorption peak at wavelength of 300 nm. At the laser wavelength 900 nm ($\hbar\omega = 1.38\,\text{eV}$) satisfying the resonant condition $\hbar\omega_{20} - 2\hbar\omega = 0$, $\hbar\omega_{10} - \hbar\omega$ is 2.72 eV. $\mu_{10}^{p}E/(\hbar\omega_{10} - \hbar\omega)$ is in the order of 10^{-2}, and slowly increasing toward shorter wavelength. The contribution of this term to resonant enhancement of SHG is minor.

It is noteworthy to clarify the relationship between transition dipole and permanent dipole. In Eq. (6.17), μ_{10}^{p}, μ_{21}^{q}, and μ_{02}^{r} is a transition dipole:

$$\begin{aligned}\mu_{10}^{p} &= \langle 1|ex^{p}|0\rangle \\ \mu_{21}^{q} &= \langle 2|ex^{q}|1\rangle \\ \mu_{02}^{r} &= \langle 0|ex^{r}|2\rangle \end{aligned} \tag{6.21}$$

$|0\rangle$ is HOMO and $|2\rangle$ is LUMO. Following the standard way of linear combination of atomic orbitals, HOMO and LUMO are produced as a result of the linear combination of two orbitals $|a\rangle$ and $|b\rangle$,

$$\begin{aligned}|0\rangle &= |a\rangle + |b\rangle \\ |2\rangle &= |a\rangle - |b\rangle\,. \end{aligned} \tag{6.22}$$

Transition dipole $\mu_{02}^{r} = \langle 0|ex^{r}|2\rangle$ is calculated as

$$\begin{aligned}\langle 0|ex^{r}|2\rangle &= (\langle a\,| + \langle b|)\,|ex^{r}|\,(\,|a\rangle - |b\rangle) \\ &= \langle a|ex^{r}|a\rangle - \langle a|ex^{r}|b\rangle + \langle b|ex^{r}|a\rangle - \langle b|ex^{r}|b\rangle \\ &= \langle a|ex^{r}|a\rangle - \langle b|ex^{r}|b\rangle \end{aligned} \tag{6.23}$$

with assuming the relation $\langle a|ex^r|b\rangle = \langle b|ex^r|a\rangle$. $\overrightarrow{m}_a = \langle a|ex^r|a\rangle$ is permanent dipole of orbital $|a\rangle$ and $\overrightarrow{m}_b = \langle b|ex^r|b\rangle$ is permanent dipole of orbital $|b\rangle$, and permanent dipole of ground state is $\overrightarrow{m} = \overrightarrow{m}_a + \overrightarrow{m}_b$. In a simple case of $|\overrightarrow{m}_a| \gg |\overrightarrow{m}_b|$, the permanent dipole of ground state is approximately $\overrightarrow{m} = \overrightarrow{m}_a$. The transition dipole is $\mu_{02}^r = \langle 0|ex^r|2\rangle = \overrightarrow{m}_a$. The transition dipole μ_{02}^r is pointing in the same direction as permanent dipole $\overrightarrow{m}$. In other words, SHG due to $\beta = \beta(\overrightarrow{m})$ is enhanced in the direction of permanent dipole $\overrightarrow{m}$ as expressed in Eq. (6.2). On the other hand, in case $|\overrightarrow{m}_a| \sim |\overrightarrow{m}_b|$, $\langle 0|ex^r|2\rangle$ is normal to permanent dipole. SHG reflects orthogonal of permanent dipole. Accordingly, SHG due to β is capable of identifying alignment of dipole $\overrightarrow{m}$.

The second-harmonic polarization induced by the third nonlinear polarizability γ is

$$p_{NL} = \gamma E_0 EE = (\mu_{03}^s e^{i(-2\omega)t} \frac{1}{8} \frac{\mu_{32}^r E}{\hbar\omega_{30} - \hbar 2\omega} \frac{\mu_{21}^q E}{\hbar\omega_{20} - \hbar\omega}) \cdot \frac{\mu_{10}^p E_0}{\hbar\omega_{10}}. \tag{6.24}$$

The term in parenthesis, $\mu_{03}^s \ldots$ to $\mu_{21}^q E/(\hbar\omega_{20} - \hbar\omega)$, is similar to the second order polarizability β. The last part $\mu_{10}^p E_0/\hbar\omega_{10}$ is enhancement of SHG in the presence of electrostatic field E_0. That is, charge displacement $qd\vec{r}$ in the electrostatic field $E^p = E_0$ is created as part $\mu_{10}^p E_0/\hbar\omega_{10}$ of transition dipole $\langle 1|ex|0\rangle$. This is an electronic polarization process described by quantum mechanics. That is, in the absence of electrostatic field, ground state is $|0\rangle$. In the presence of electrostatic field E_0, the ground state is mixed with higher state $|1\rangle$, in the perturbation theory, and change its symmetry (see the change of wave function from Fig. 6.4(b) to 6.4(c)). The mixture of $|0\rangle$ and $|1\rangle$ causes static electronic displacement dx, and electronic polarization $qd\vec{x}$ is formed. In the presence of electrostatic field E_0, wave function change from $|0\rangle$ to

$$|\psi\rangle = |0\rangle + cE_0\,|1\rangle \tag{6.25}$$

Here, c is perturbation coefficient and given by $c = \langle 0|ex|1\rangle / \hbar\omega_{10}$. Charge central of $|\psi\rangle$ is the induced dipole m' in x-direction of perturbed state

$$\begin{aligned} m' &= \langle \psi|ex|\psi\rangle = (\langle 0| + c^* E_0 \langle 1|)ex(|0\rangle + cE_0 |1\rangle) \\ &= \langle 0|ex|0\rangle + 2cE_0 \langle 1|ex|0\rangle + |c|^2 E_0^2 \langle 1|ex|1\rangle. \end{aligned} \tag{6.26}$$

The first term is dipole of ground state $m = \langle 0|ex|0\rangle$. The last term is a small quantity in proportional to $|c|^2$, and is discarded. Consequently, charge displacement is given as

$$qdx = m' - m = 2cE_0 \langle 1|ex|0\rangle \tag{6.27}$$

Substituting perturbation coefficient $c = \langle 0|ex|1\rangle / \hbar\omega_{10}$

$$qdx = 2\frac{\langle 0|ex|1\rangle \langle 1|exE_0|0\rangle}{\hbar\omega_{10}} = 2\mu_{10}\frac{\mu_{10}E_0}{\hbar\omega_{10}} \tag{6.28}$$

This means that charge displacement qdx in electrostatic field E_0 is a part of dipole μ_{10} by the ratio of $\mu_{10}E_0$ to $\hbar\omega_{10}$. That is, charge displacement is proportional to the ratio of electrostatic potential energy stabilized in electric field E_0 and energy quanta that is needed to promote an electron in $|0\rangle$ to $|1\rangle$. The ratio $\mu_{10}E_0/\hbar\omega_{10}$ governs the third order nonlinear polarizability γ, as described in Eq. (6.19). Accordingly, the EFISHG of third order molecular polarizability γ reflects electronic charge displacement qdx in molecule.

It is instructive here to mention electronic polarization and dipolar polarization from viewpoint of energy space. We assume electrostatic field E_0 is pointing in the x-direction. For electronic polarization,

$$qdx = 2\mu_{10}\frac{\mu_{10}E_0}{\hbar\omega_{10}}. \tag{6.29}$$

For dipolar polarization

$$\vec{m} \cdot \vec{x} = \mu\frac{\mu E_0}{3kT}. \tag{6.30}$$

In the linear approximation, electronic charge displacement is due to displacement from $|0\rangle$ to $|1\rangle$ where transition energy is $\hbar\omega_{10}$.

Under electrostatic potential $\mu_{10}E_0$, effectively the electronic polarization is a part of μ_{10} by the energy ratio $\mu_{10}E_0/\hbar\omega_{10}$. Dipolar polarization is due to rotation of permanent dipole μ. The x-component of a single independent electric dipole changes from zero to μ by the rotation toward electric field direction, on keeping its magnitude μ. In thermodynamic equilibrium, many dipoles exchange their directions. In classical mechanics, motion of a single body in potential energy has the same average potential energy and kinetic energy. In thermal equilibrium, average kinetic energy is $kT/2$ and potential energy is stabilized by $kT/2$. By using thermal energy $3kT$, dipole is allowed to change its orientation toward any direction in three-dimensional (3D) space. Under finite stabilization by electrostatic potential μE_0, the resulting dipole is a part of μ by $\mu E_0/3kT$.

In summary, electronic polarization is probed by γ for EFISHG. γ reflects electronic displacement. At the resonant angular frequency $2\omega = \omega_{30}$, optical second harmonic light is emitted as a result of vibration of transition dipole μ_{03} at double-frequency. The nonlinear polarization is enhanced in proportion to electrostatic field E_0. SHG probes dipolar orientation by β. At the resonant angular frequency $2\omega = \omega_{20}$, optical second-harmonic light is emitted due to the vibration of transition dipole μ_{02} at double angular frequency of laser beam. The two resonant angular frequency is different, *i.e.*, $\omega_{30} \neq \omega_{20}$, allowing one to selectively probe electronic charge displacement and dipolar orientation, one-by-one, by means of SHG measurement.

6.2.5 *Example of β and γ*

Nonlinear polarizabilities β and γ are responsible for nonlinear susceptibilities $\chi^{(2)}$ and $\chi^{(3)}$, respectively (see Eq. (6.10)). The second order polarizability β well probes dipole $\vec{m}$, while the third order polarizability γ is responsible for detecting charge displacement qdx. And wavelength dependence of $\chi^{(2)}$ and $\chi^{(3)}$ is originating from the wavelength dependence of β and γ, respectively.

As an example of calculating SHG spectrum of β and γ, semi-empirical calculation of β and γ of N–phenylmaleimide (cas registry number 941–69–5) is shown here. The molecular structure of

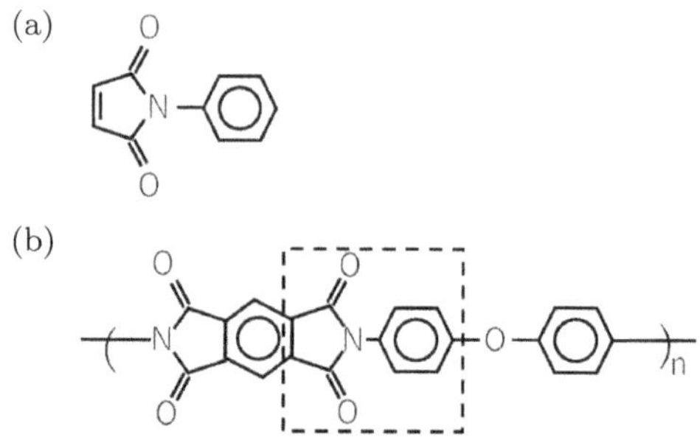

Figure 6.8. Molecular structure of (a) N–phenylmaleimide (b) and PMDA–ODA polyimide. A part in dotted square corresponds to N–phenylmaleimide.

N–phenylmaleimide is displayed in Fig. 6.8(a), whose energy levels have been reported in [20]. For comparison with experimental SHG spectrum, SHG spectrum is measured for polyimide. Figure 6.8(b) shows the chemical structure of polyimide. This polyimide is assumed corresponding to a polymer of N–phenylmaleimide, and used for experiment to compare with calculation. Semi-empirical quantum mechanical calculation software MOPAC and MOS–F is used to calculate the transition dipoles $\vec{\mu}_{ij} = \langle\phi_i|e\vec{r}|\phi_j\rangle$ of N–phenylmaleimide. The obtained transition dipole moments are used to evaluate multiplied transition dipole components for β and γ in Eq. (6.17) and Eq. (6.19). That is, in Eq. (6.17) and Eq. (6.19), the term $\mu_{02}\mu_{21}\mu_{10}$ and $\mu_{03}\mu_{32}\mu_{21}\mu_{10}$, respectively, are evaluated by using transition dipole $\mu_{01}, \mu_{02}, \ldots$ of N–phenylmaleimide. Note that the calculation for a single molecule gives molecular polarizability β and γ, while experimental SHG spectrum reflects second order and third order susceptibility $\chi^{(2)}$ and $\chi^{(3)}$. The measured SHG gives an average β and γ over molecules that comprise a film.

In the calculation, the molecular geometry is optimized by setting AM1 Hamiltonian. After that, energy states of the optimized molecule are calculated using INDO/S Hamiltonian. For calculation, 1s of hydrogen, 2s, 2p of carbon, 2s, 2p of nitrogen, and 2s, 2p of oxygen atomic orbitals are used. The resulting transition dipole moment is stored in output file by setting TMOM (transition dipole moment) as a keyword with INDO/S in input file.

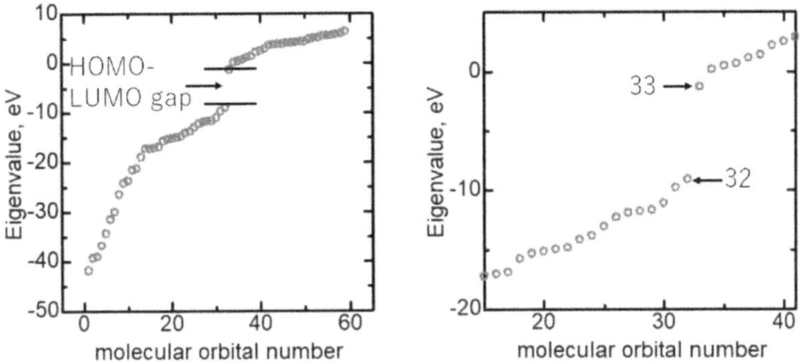

(a) (b)

Figure 6.9. (a) Energy level (Eigenvalue) of molecular orbital of N–phenylmaleimide. The molecular orbital number is assigned from the lower energy orbital to higher energy orbital. (b) Close look up around HOMO–LUMO gap. HOMO and LUMO is molecular number 32 and 33, respectively.

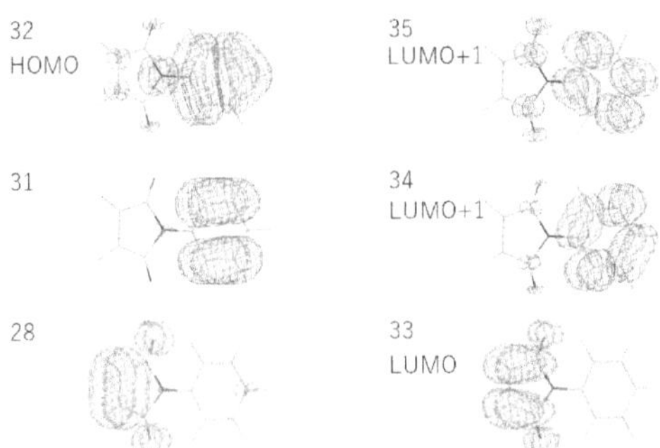

(a) (b)

Figure 6.10. π-electron orbitals of (a) HOMO and lower energy orbitals, (b) LUMO and higher orbitals. The numbers are molecular orbital numbers in the lateral axis in Fig. 6.9.

Figure 6.9 plots a series of calculated energy levels of N–phenylmaleimide. HOMO is 32nd state. LUMO is 33rd state. LUMO + 1 is 34th and 35th state. Figure 6.10 shows contour of molecular orbitals near HOMO and LUMO. π–bond orbitals are selected, taking account that, in the laser power range (1–10 mW) used for SHG measurement, a material does not undergo photolysis, indicating that chemically non-destructive optical transitions between

π–bond orbitals are main contributor to SHG. π–bond orbitals are selected by choosing orbitals that has a node at a plane of benzene ring and maleimide. For calculating β and γ, HOMO, LUMO, and LUMO+1 levels are selected as 4 states $|0\rangle$, $|1\rangle$, $|2\rangle$, $|3\rangle$ illustrated in Fig. 6.6. That is, optical transitions from HOMO to LUMO and HOMO to LUMO+1 are considered. In the calculation, 31st and 32nd states are used as HOMO, 33rd state is used as LUMO, and 34th and 35th states are used as LUMO+1. Among these 5 states, 6 transitions are dominant, *i.e.*, transitions from 31st to 33rd, 34th, 35th state, and transitions from 32nd to 33rd, 34th, 35th state. By using these 6 transitions, configuration interaction states are obtained with keyword CI = Read. Resulting CI states are CIS(n) n $= 1, 2, \ldots 6$. The amplitude of transition dipole moments from ground state to these CI states is calculated with energy for transition.

The result is plotted in Fig. 6.11(a). In vertical axis, the square of transition dipole moment for six transitions is plotted as a function of wavelength corresponding to transition energy. The calculated wavelength was shifted by 50 nm to longer wavelength for comparison with experiment. Figure 6.11(b) shows measured optical absorption spectrum of polyimide film. Here, polyimide film is prepared as follows: As a precursor of polyimide, polyamic acid (PAA) is synthesized from pyromelliticdianhydride (PMDA) and oxydianiline (ODA) in N, N'–dimethylacetoamide. The PAA film is formed from the solution, and imidized thermally or chemically to prepare polyimide film. The resulting polyimide is considered as a polymer of calculated molecule of N–phenylmaleimide. The calculated spectrum fits to measured spectrum. Note that a series of monomer, dimer, ... of N–phenylmaleimide, MOPAC calculation shows no significant change of energy levels, and only the number of orbitals close to each other increases by $1, 2, 3, \ldots$ times for monomer, dimer, and so on.

By using the same transition dipole moments used for reproducing optical absorption spectrum, multiplied transition moments in expression for second and third order nonlinear polarizability β and γ are calculated. The calculated result is shown in Fig. 6.11(c). The calculated result suggests that EFISHG is enhanced by non-zero γ in visible SHG wavelength region (SHG wavelength 400–600 nm, laser

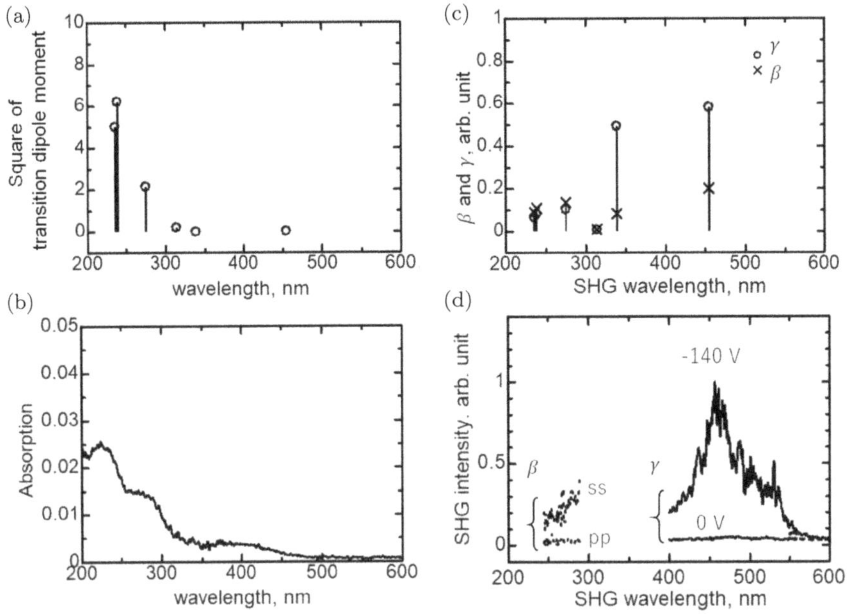

Figure 6.11. (a) Calculated linear optical spectrum. (b) Measured optical absorption spectrum. (c) Calculated SHG spectrum of β and γ. (d) Measured SHG spectrum of β and γ.

wavelength 800–1200 nm), while contribution from β is enhanced in ultra-violet SHG wavelength region (SHG wavelength 200–300 nm, laser wavelength 400–600 nm). In more detail, the calculated SHG spectrum shows large β at SH wavelength around 285 nm (laser wavelength 570 nm), whereas the EFISHG spectrum shows large γ at SH wavelength of 450 nm (laser wavelength 900 nm). The difference in resonant wavelength between SHG and EFISHG indicates that these peaks are originating from different optical transition contributions, which allows us to distinguish the SHG origins between charge displacement and dipolar orientation. In other words, difference in β and γ spectrum of the molecule gives an opportunity to probe charge displacement and dipolar motion in the material, individually.

Figure 6.11(d) is experimentally measured SHG spectrum of β and γ. SHG spectrum of β is obtained by using polyimide film that is mechanically rubbed by a cloth. The mechanical rubbing on polyimide causes polar ordering of molecules on polyimide surface. As a result, SHG is enhanced under laser irradiation with polarization parallel to the rubbing direction. In Fig. 6.11(d), ss denotes the experimental condition where s-polarized laser is a probe and s-polarized SHG is measured. pp indicates that p-polarized laser is used as a probe and p-polarized SHG is measured. The polyimide film is rubbed in the direction along s-polarization. The result shows that SHG intensity is enhanced with ss polarization condition where polarization direction of laser and SHG are pointing along the rubbing direction. The result is consistent with an idea that the rubbing of polyimide caused polar orientational ordering of molecules toward rubbing direction. Note that the SHG in ultra-violet region is active after triboelectric charge producing electrostatic field is decayed, supporting that SHG of β plays a main role while EFISHG of γ is minor in this wavelength region.

The EFISHG spectrum of γ is obtained experimentally by using sandwich type device. The polyimide film is sandwiched between parallel electrode of indium–zinc–oxide (IZO) and Ag. By applying external voltage to IZO electrode in reference to grounded Ag electrode, EFISHG is enhanced depending on the external voltage. That is, modulation of SHG spectrum in the presence of electrostatic field in polyimide indicates non-zero γ. In Fig. 6.11(d), SHG measurement is carried out for this polyimide device. The two EFISHG spectra is obtained for the device with applying 0 V and -140 V. The device with 0 V, *i.e.*, IZO and Ag electrode were connected through conducting wire without voltage source, produces no SHG. On the other hand, under external voltage application of -140 V, the polyimide radiates SHG at around 450 nm, in accordance with the calculation shown in Fig. 6.11(c). Note that SHG of β in visible wavelength region from rubbed polyimide is minor. A good fitting of calculated and measured SHG spectrum demonstrates that a way of quantum chemical calculation is effective to obtain nonlinear optical spectrum of β and γ. This calculation is useful to predict availability of SHG measurement to selectively probe charge displacement and dipole alignment. It is noteworthy that the spectrum peaks in SHG and

EFISHG spectrum of polyimide are broader than those in optical absorption spectrum. This is probably due to asymmetric vibration of electrons in molecule. In linear vibration, electrons go to and come back from an exited state at the identical frequency. This situation is similar to the linear vibration of a spring, resonantly enhanced at a specific frequency. As a result, electron vibration is resonantly enhanced infinitely as long as an external force continues. In nonlinear optical excitation, electrons go to an excited state more smoothly and come back from the state more slowly, due to deviation from analogy to a linear spring. This situation is like an air suspension. The resonant enhancement is distributed to a route of going to and backing from an excited state. Accordingly, resonant enhancement is suppressed, and a peak is broad.

6.3 SHG technique

In this section, SHG measurement system is described from technical view point. Nonlinear optical system such as multiphoton microscope, sum-frequency generation microscope and others, attracts much attention as conventional equipment in research and development (R&D) laboratories in the world [21–23]. The SHG and EFISHG measurement systems are the same as those employed in nonlinear microscope. In this chapter, the SHG techniques are described, on focusing key points for visualizing charge motion and dipole alignment in electronics devices and materials.

As described in Sec. 6.1 and Sec. 6.2, spectroscopic difference between SHG and EFISHG provides a way to selectively visualize electronic charge displacement and dipolar alignment. In other words, nonlinear polarization is utilized for probing charge displacement and dipolar alignment, by focusing on nonlinear polarizability of β and γ in energy space. We here introduce a typical SHG measurement system that utilizes the spectroscopic difference.

Figure 6.12(a) illustrates a conceptual idea of SHG measurement system. The pulsed laser is used as a probe light. The light travels to sample, and induces SHG. By measuring SHG intensity, charge displacement and dipole alignment are probed. As a pulsed laser

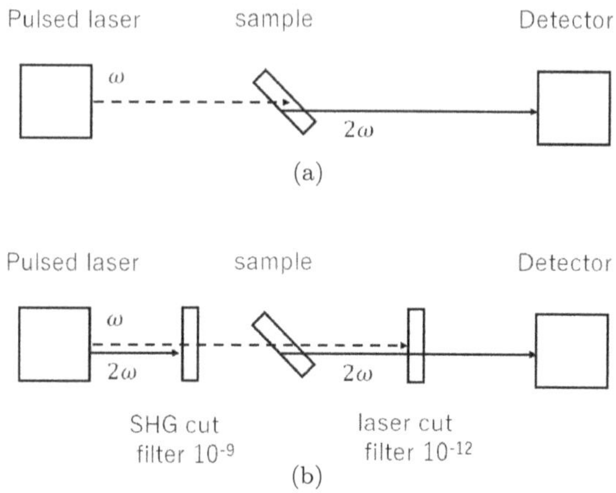

Figure 6.12. (a) Concept of SHG measurement system. (b) Implemented SHG measurement system.

source, Q–switched Nd:YAG laser equipped with an optical parametric oscillator (OPO) is used. OPO provides a way to scan laser wavelength. The laser wavelength is chosen for SHG measurement, depending on the microscopic origins of materials, *i.e.*, dipolar orientation and electronic charge displacement. Different material has different electronic energy structures, and gives rise to different SHG spectrum. SHG measurement system with an OPO is useful to find a proper laser wavelength that is specific to materials and devices to be probed. Shorter laser pulses can induce SHG with larger intensity. Femto-second laser beam comprised of Ti:Sappire and Nd:YLF coupled with optical parametric amplifier (OPA) is also available for the SHG measurement. OPA is used to carry out laser wavelength scan.

As a detector of SHG, photomultiplier tube (PMT) is useful, and imaging sensor such as charge-coupled device (CCD) camera is also useful to record an image of SHG. Figure 6.12(a) displays a concept of SHG measurement system where a sample is arranged in transmission style. The reflection style arrangement is useful when SHG in reflection direction is larger than SHG in transmission direction.

In Fig. 6.12(b), two optical filters are installed to implement SHG measurement system. The first filter removes SHGs from laser equipment and from optical components. Only laser beam at fundamental angular frequency ω is allowed to transmit the filter, and impinges on a sample. For example, the filter totally transmits laser with a wavelength of λ, but it blocks SHG passing through it with a wavelength of $\lambda/2$ by a factor of $a = 10^{-9}$. From the sample, transmitted laser ω and SHG 2ω are coming out. The transmitted laser is stopped by the second filter, meanwhile the SHG passes through this filter. Only SHG comes into the detector. The second filter reduces laser intensity by $b = 10^{-12}$ and allows SHG to pass through the filter. For example, Q-switched Nd:YAG laser, equipped with OPO, emits laser pulses with pulse energy 1 mJ/pulse, and sample radiates SHG with 100 fJ/pulse. Assuming that OPO emits SHG with a light intensity of 0.1% of laser beam, SHG with 1 μJ/pulse is emitted from OPO. After passing through the two filters, 1 fJ/pulse SHG is measured by a detector due to the SHG from OPO. In this situation, one can measure SHG from sample with a signal to noise (S/N) ratio of 100, with considering the SHG from laser light source is a noise. In a different idea of optical arrangement, a way of using a slit is available to improve detection of EFISHG [24].

A choice of polarization condition is key for visualizing charge motion and alignment of dipoles. Linear polarization, radially polarization, and other polarization conditions are used in the measurement. Figure 6.13 illustrates an example of polarization used in the SHG measurement. More details of the polarization of light are described in books [25]. *s*-polarization and *p*-polarization are produced by using linear polarizer such as wire-grid polarizer and others. Many organic electronic devices have an MIM structureMIM structure where active layer is sandwiched between parallel electrode. Among them are organic light-emitting diodes (OLED), organic solar cells (OSC), memory device, and so on. In these devices, electrostatic field is formed along the film thickness direction normal to electrode surface. Oblique laser incidence of *p*-polarization is selected to efficiently probe electrostatic field formed in film thickness direction [2]. Different direction of electrostatic field and dipolar alignment formed in material is, respectively, probed by using *p*- and *s*-polarized laser beam [26]. Azimuthal and radial polarizations are produced by using liquid crystal polarizers [27]. Focusing radially polarized laser beam

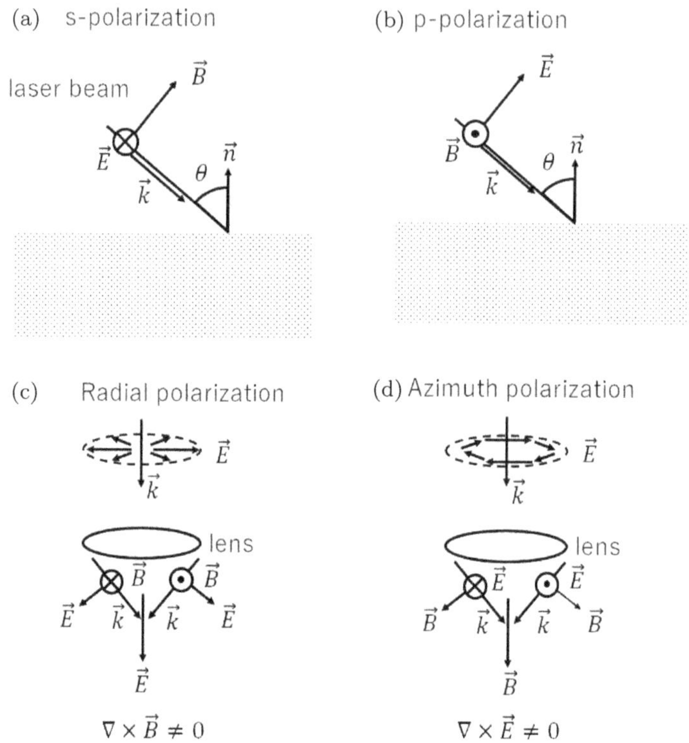

Figure 6.13. Polarization of light used in SHG experiment. (a) s-polarization. (b) p-polarization. $\vec{n}$ is an unit vector pointing in surface normal direction. $\vec{k}$ is an unit vector that is normal to electric field $\vec{E}$ and magnetic flux density vector $\vec{B}$. $\vec{n}$ and $\vec{k}$ form an incident plane. Incident angle is $\theta = \cos^{-1}(\vec{k} \cdot \vec{n})$ (c) Radial polarization. (d) Azimuth polarization.

by using lens, electric field of laser beam vibrates normal direction to focus plane. SHG microscope equipped with radially polarized laser beam allows us to probe electrostatic field formed in film thickness direction in organic devices [28–30]. It is noteworthy that the use of circularly polarization is also key for characterization of materials and device application, as described in Chap. 5.

To probe charge and dipole in electronic devices such as field-effect transistors, organic light-emitting diodes, and other devices, external voltage is applied to the device. Accordingly, optical part and electrical part are coupled in SHG measurement system to investigate materials and devices.

It is instructive for experiment to discuss SHG intensity measured. $\chi^{(3)}$ of pentacene film is $10^{-22}\,\text{m}^2/\text{V}^2$ at laser wavelength of $\lambda = 1120\,\text{nm}$, $\chi^{(3)}$ of C60 film is $10^{-18}\,\text{m}^2/\text{V}^2$ at laser wavelength of $\lambda = 1000\,\text{nm}$. More and more the electric field of laser E_ω is large, more and more nonlinear polarization P_{NL} increases. Along with introduction of pulsed lasers, more and more easy to access nonlinear optical phenomena through P_{NL} without damaging sample [31]. Q-switched Nd:YAG laser and OPO provide, for example, $t = 4\,\text{ns}$ pulse duration with $W = 1\,\text{mJ/pulse}$, focused on $S = 10^{-2}\,\text{mm}^2$ spot size on sample. The laser intensity on sample is $I_\omega = W/(tS) = 2.5 \times 10^{13}\,\text{W/m}^2$ ($2.5 \times 10^9\,\text{W/cm}^2$) that accompanies the vibrating electric field $E_\omega = 9.7 \times 10^7\,\text{V/m}$ ($9.7 \times 10^5\,\text{V/cm}$). Electrostatic field E_0 formed in electronics devices is in the order of $E_0 = 10^8\,\text{V/m}$ ($10^6\,\text{V/cm}$). In pentacene-OFET with a gate insulating layer whose thickness is $d = 100\,\text{nm}$, external voltage application of $V = 10\,\text{V}$ will create an electrostatic field in an FET channel as $E_0 = V/d = 10^8\,\text{V/m}$ in the pentacene film with $\chi^{(3)} = 10^{-22}\,\text{m}^2/\text{V}^2$, $P_{NL} = \varepsilon_0\chi^{(3)}E_0E_\omega E_\omega = 8.3 \times 10^{-10}\,\text{C/m}^2$ ($8.3 \times 10^{-14}\,\text{C/cm}^2$) vibrating with SHG wavelength of 560 nm. Using pointing vector $\vec{E}_{2\omega} \times \vec{H}$, the relationship between SHG light intensity $I_{2\omega}$ and SHG polarization P_{NL} is given by

$$I_{2\omega} = |\vec{E}_{2\omega} \times \vec{H}| = 2c_0\varepsilon_0\frac{|P_{NL}/\varepsilon_0|^2}{2}. \tag{6.31}$$

The SHG intensity is calculated using Eq. (6.31) as $I_{2\omega} = 23\,\text{W/m}^2$ ($2.3\,\text{mW/cm}^2$). This SHG in the area of $S = 1\,\text{mm}^2$ for $t = 4\,\text{ns}$ corresponds to an energy with 9.4×10^{-14} J/pulse ($= I_{2\omega}\,St$). That is, 92 fJ/pulse of SHG light is generated. This light intensity is easy to be detected by using a conventional photomultiplier tube. Note that 100 fJ corresponds to 2.8×10^5 photons at SHG wavelength of 560 nm. Such a large number of photons can be easily detected by using PMTs and high sensitivity CCD camera. Nd:YLF laser coupled with Ti:Sapphire laser, for example, 80 fs pulse duration, $100\,\mu\text{J/pulse}$, laser beam spot area $1\,\text{mm}^2$, conveys vibrating electric field whose amplitude is 7.0×10^8 V/m (7.0×10^6 V/cm) and generates $P_{NL} = 43\,\text{nC/m}^2$, resulting in 5 fJ/pulse at 1 kHz, giving average SHG power of 5 pW. Using CCD camera as a detector, SHG image is obtained in real time, for example, 10 fps with 50 pixels times 50 pixels has been achieved. In another system, the line scan, 50 mm by 1 mm width area, is continuously operated at 1 kHz for reconstruction of 2-dimensional (2D) image of charge displacement. Understanding these technical key points allows us to visualize carrier motion and dipolar motion.

The $\chi^{(2)}$ is related to dipolar alignment, and $\chi^{(3)}$ is enhanced due to electrostatic field originating from electronic charges. Accordingly, we can optically catch charge motion of electronic charge and dipoles. For optical imaging by means of SHG, microscope system is constructed and is available to visualize electrostatic field pointing in image plane direction [32] as well as the field directed in depth direction [28–30]. Of course, these measurement systems are advancing along with further development of optical and electrical techniques. Technical measures to construct a stably operating SHG measurement system is described in Appendix F.

It is instructive here to note that a variety of optical setups is available other than the system described here. Keeping into mind different aspects of dipolar alignment and charge displacement, we can find a way to construct a novel measurement system. The second and third nonlinear polarizability β and γ exhibit different time response. By focusing on this different time response, a novel SHG measurement system can be constructed. Different electromagnetic interaction between probe light and electrons in molecule is available in the construction of new systems. Direction of electric field and dipolar alignment is depending on a device and materials. By using a polarization of probe laser that is enhanced along with these directions, selective probing of charge displacement and dipolar alignment will be realized. A variety of SHG systems will be designed based on different concepts, and all of these systems will be helpful to investigate charge displacement and dipolar alignment in materials and devices.

6.4 Summary

The principle of SHG and experimental SHG technique were summarized. SHG microscope and other systems are available for visualizing carrier motion and dipolar motion in materials and devices. Idea of visualizing charge and dipole motions was summarized from the viewpoint of analysis and application of organic materials for organic electronics, on the basis of dielectric physics. The second order nonlinear molecular polarizability β contributes to identify dipoles being involved in electrostatic phenomena in device and materials. The third order molecular polarizability γ is responsible for visualizing charge displacement $qd\vec{r}$ through the measurement of electrostatic field $\vec{E}_0$. Examples of experimental results will be described in Chap. 7 and Chap. 8. In the development of quantum mechanics, idea of energy transitions between stationary states replaced the idea of classical view that the optical transition occurs through wave harmonics. This led to paradigm shift that accounts for microscopic structure of molecules [33]. On the other hand, SHG discovers microscopic dielectric nature, *i.e.*, electronic charge displacement and dipolar orientational change, on the basis of harmonics of quantum mechanical transitions between discrete levels. By measuring second-harmonic light, identifying electronic charge and dipole is allowed, and their motion in materials and devices is clearly visualized. Visualization technique is of course not limited to SHG, but SHG is one of the best ways for visualization, in terms of dielectric polarization phenomena. A variety of optical methods has been introduced along with technical development. Optical techniques that are available to visualize electrical phenomena have been referenced in [34–41].

References

1. Zhang, L., Taguchi, D., Manaka, T. and Iwamoto, M. (2011). *Appl. Phys. Lett.*, 99, 083301, https://doi.org/10.1063/1.3626851.
2. Taguchi, D., Weis, M., Manaka, T. and Iwamoto, M. (2009). *Appl. Phys. Lett.*, 95, 263310, https://doi.org/10.1063/1.3277155.
3. Chen, X., Taguchi, D., Manaka, T. and Iwamoto, M. (2014). *Appl. Phys. Lett.*, 104, 013306, https://doi.org/10.1063/1.4861620.

4. Boyd, R. W. Ed. (2008). *Nonlinear Optics*, 3rd ed. (Academic, London) Chapter 1, https://doi.org/10.1016/B978-0-12-369470-6.00001-0 pp. 1–67.
5. Popov, S. V., Svirko, Y. P. and Zheludev, N. I. (1995). *Susceptibility Tensors for Nonlinear Optics* (IOP Publishing) English translation by Kisin, V. I, https://doi.org/10.1201/9780203737699.
6. Fröhlich, H. (1958). *Theory of Dielectrics*, 2nd ed. (Oxford) pp. 26–30. (Nagamiya, T. and Nakai, Y. (1960). *Yudentai ron* (Yoshioka, Kyoto) [Japanese translation], https://doi.org/10.11501/1378529 Chapter II.6. p. 28–33.)
7. McConnel, J. (1980). *Rotational Brownian Motion and Dielectric Theory* (Academic, London New York Toronto Sydney San Fransisco) Chapter 1. pp. 1–11.
8. Kubo, R. (1971). *Toukei Rikigaku* (*Statistical Mechanics*), Kaitei Ed. (Kyoritsu, Tokyo) Sec. 4.7. pp. 116–119 (in Japanese).
9. Feynman, R. P., Leighton, R. B. and Sands, M. (1964). *The Fynman Lectures on Physics II, Mainly Electromagnetism and Matter* (Addison–Wesley, Massachusetts) pp. 27–6–7.
10. Nye, J. F. (1957). *Physical Properties of Crystals* (Oxford, Glasgow) Chap. VII.2, pp. 113–115.
11. Nye, J. F. (1957). *Physical Properties of Crystals* (Oxford, Glasgow) Chap. I.2, pp. 8–14.
12. Iwamoto, M. and C.-X., Wu (2001). *The Physical Properties of Organic Monolayers* (World Scientific, Singapore), https://doi.org/10.1142/4575 Chap. 7. pp. 140–141.
13. Verbiest, T., Clays, K. and Rodriguez, V. (2009). *Second-order Nonlinear Optical Characterization Techniques — An Introduction* (CRC, USA), https://doi.org/10.1201/9781420070736 Chap. 1.4. pp. 11–21.
14. Yamazaki, I. (2013). *Hikari Kagaku no tame no laser bunkou, hisenkei bunkouhou* (*Laser Spectroscopy and Nonlinear Spectroscopy for photo chemistry*) (Kodansha Scientific, Tokyo) Chap. 6.4, pp. 196–210 (in Japanese).
15. Kishida, H. *et al.* (1994). *Phys. Rev. B*, 50, 7786, https://doi.org/10.1103/PhysRevB.50.7786.
16. Stewart, J. J. P. (1999). *MOPAC 2000.00 Manual* (Fujitsu Limited, Tokyo) Chap. 8.21. Kurtz, H. and Korambath, P. Polarizability and hyperpolarizability calculation.
17. *Bunshi kido keisan program MOS–F V.4.2 User's manual*, 2nd Ed. (Fujitsu Limited, Japan) *Chap.* 5.7 and 5.8. pp. 82–87.
18. Tokita, S., Matsuoka, M., Kogo, Y. and Kihara, H. (1989). *Kinousei Shikiso no Bunshi Sekkei* (*Molecular Design of Functional Dyes*) (Maruzen, Tokyo) Chap. 7. pp. 155–162.
19. Clark, T. (1985). *A Handbook of Computational Chemistry — A Practical Guide to Chemical Structure and Energy Calculations* (John Wiley & Sons, New York); Japanese translation by Osawa, E., Tanabe, K., Mizuno, M. and Sugie, M. (1988). *Keisan Kagaku Guidebook* (Maruzen, Tokyo) Chap. 4 and Chap. 5. pp. 141–328 (in Japanese).

20. Kafafi, S. A. (1990). *Chem. Phys., Lett.*, 169, 561, https://doi.org/10.1016/0009-2614(90)85647-U.
21. Miyamae, T., Takada, N., Ohata, H. and Tsutsui, T. (2017). *Appl. Phys. Express*, 10, 102101, https://doi.org/10.7567/APEX.10.102101.
22. Hama, H., Hioki, H., Namiki, K., Hoshida, T., Kurokawa, H., Ishidate, F., Kaneko, T., Akagi, T., Saito, T., Saido, T. and Miyawaki, A. (2015). *Nature Neuroscience*, 18, 1518–1529, https://doi.org/10.1038/nn.4107.
23. Kawakami, R., Sawada, K., Kusama, Y., Fang, Y.-C., Kanazawa, S., Kozawa, Y., Sato, S., Yokoyama, H. and Nemoto, T. (2015). *Biomedical Optics Express*, 6, 891–901, https://doi.org/10.1364/BOE.6.000891.
24. Jašinskas, V., Gedvilas, M., Račiukaitis, G. and Gulbinas, V. (2016). *Opt. Lett.*, 41, 2759–2762, https://doi.org/10.1364/OL.41.002759.
25. Jenkins, F. A. and White, H. E. (1957). *Fundamentals of Optics*, 4th Ed. (McGraw–Hill, Tokyo) Chap. 24. pp. 497–522.
26. Taguchi, D., Manaka, T. and Iwamoto, M. (2019). *Appl. Phys. Lett.*, 114, 233301, https://doi.org/10.1063/1.5094171.
27. Yoshiki, K., Kanamaru, R., Hashimoto, N. and Araki, T. (2007). *Opt. Lett.*, 32, 1680, https://doi.org/10.1364/OL.32.001680; Hashimoto, N. *et al.* (2011). IEICE Technical Report, R2011–5, 19–22 (in Japanese).
28. Taguchi, D., Manaka, T., Iwamoto, M., Karasuda, T. and Kyomasu, M. (2013). *Jpn. J. Appl. Phys.*, 52, 04CK04, https://doi.org/10.7567/JJAP.52.04CK04.
29. Sadakata, A., Yano, R., Taguchi, D., Manaka, T. and Iwamoto, M. (2014). Probing of electric field distribution and carrier behavior in double-layer organic light-emitting diodes by a novel microscopic electric-field-induced optical second-harmonic generation measurement system, *Trans. Mat. Res. Soc.* Japan, 39, 443–447, https://doi.org/10.14723/tmrsj.39.443.
30. Sadakata, A., Taguchi, D., Manaka, T. and Iwamoto, M. (2017). A novel microscopic analyzing system for characterizing organic light-emitting diodes using EFISHG and LBIC measurements, *Electronics and Communications in Japan*, 100, 76–83, https://doi.org/10.1002/ecj.12017.
31. Ristau, D. ed. (2015). *Laser-induced Damage in Optical Materials* (CRC, Boca Raton London New York), https://doi.org/10.1201/b17722.
32. Manaka, T., Lim, E., Tamura, R. and Iwamoto, M. (2007). *Nat. Photon.*, 1, 581–584, https://doi.org/10.1038/nphoton.2007.172.
33. Tomonaga, S. (1969). *Ryoushi Rikigaku* (*Quantum mechanics*), vol. 1. 2nd Ed. (Misuzu, Tokyo) p. 195. Chap.5 (in Japanese).
34. Böer, K. W., Hänsch, H. J. and Kűmmel, U. (1959). *Z. Phys.*, 155, pp. 170–183, https://doi.org/10.1007/BF01337934.
35. Cola, A., Farella, I., Auricchio, N. and Caroli, E. (2006). *J. Opt. A*, 8, S467–S472, https://doi.org/10.1088/1464-4258/8/7/S24.
36. Shimizu, N., Katsukawa, H., Kosaki, M. and Horii, K. (1979). *IEEE Trans. Electr. Insul.*, EI–14, 256–263, https://doi.org/10.1109/TEI.1979.298229.
37. Otsuka, T., Taguchi, D., Manaka, T. and Iwamoto, M. (2017). *J. Appl. Phys.*, 122, 185501, https://doi.org/10.1063/1.5004002.

38. Lei Lei Yin Win, Taguchi, D. and Manaka, T. (2019). *Jpn. J. Appl. Phys.*, 58, SBBG18, https://doi.org/10.7567/1347-4065/ab027c.
39. Toriyama, Y. (1961). *Dust Figure of Surface Discharge and its Applications* (Kinokuniya, Tokyo).
40. Arita, Y., Shiratori, S. S. and Ikezaki, K. (2003). *J. Electrostaics*, 57, 263–271, https://doi.org/10.1016/S0304-3886(02)00166-3.
41. Taguchi, D., Kawate, T., Miyazawa, R., Weis, M., Manaka, T. and Iwamoto, M. (2009). *Mol. Cryst. Liq. Cryst.*, 512, 100–108, https://doi.org/10.1080/15421400903050608.
42. Verbiest, T., Clays, K. and Rodriguez, V. (2009). *Second-order Nonlinear Optical Characterization Techniques — An Introduction* (CRC, USA), https://doi.org/10.1201/9781420070736 pp. 76–77 Sec. 3.2.2 Maker fringe technique.
43. Chen, C. K., Heinz, T. F., Ricard, D. and Shen, Y. R. (1981). *Phys. Rev. Lett.*, 46, 1010–1012, https://doi.org/10.1103/PhysRevLett.46.1010.
44. Baba, A., Kanda, K., Ohno, T., Ohdaira, Y., Shinbo, K., Kato, K. and Kaneko, F. (2010). *Jpn. J. Appl. Phys.*, 49, 01AE02, https://doi.org/10.1143/JJAP.49.01AE02.
45. Nishi, S. *et al.* (2015). *J. Appl. Phys.*, 117, 245502, https://doi.org/10.1063/1.4923058.
46. Zhang, L., Taguchi, D., Li, J., Manaka, T. and Iwamoto, M. (2010). *J. Appl. Phys.*, 108, 093707, https://doi.org/10.1063/1.3483929.
47. Ohshima, Y., Lim, E., Manaka, T., Iwamoto, M. and Sirringhaus, H. (2011). *J. Appl. Phys.*, 110, 013715, https://doi.org/10.1063/1.3606423.

Chapter 7

Application of MDC and SHG for Modelling and Analysis of Organic Devices: Two-electrode System

The modelling and analysis of carrier motions in organic materials and in organic devices is an important topic of this book entitled "*Analysis and application of organic materials for organic electronics*". As described in earlier chapters, by probing the dielectric polarization phenomena caused by the electric field arising from electrons and dipoles, we can find a way to study carrier motions in materials. Maxwell displacement current (MDC) measurement allows us to probe orientational dipolar motion in films and in monolayers, whereas electric-field-induced optical second-harmonic generation (EFISHG) measurement allows long-range electron and hole transport in solids to be directly probed. This chapter is devoted to the demonstration of MDC and EFISHG measurements for analysis and modelling of carrier motions in solids and in organic devices, where the carrier motions in solids and in devices are driven by external stimuli such as applied d.c. voltage, photo-radiation, and others. Note that these external stimuli are most often used in the study of electrical properties of materials and in the field of materials' application. In this chapter, analysis and modelling of carrier process in two-electrode systems are described, where organic single and multilayer are sandwiched between two electrodes. Examples are metal–insulator–semiconductor (MIS) diodes, organic light-emitting diodes (OLEDs), organic memory devices, and organic solar cells (OSCs).

In the next Chap. 8, analysis and modelling of carrier process in three-electrode systems will be discussed.

7.1 Introduction

Analysis and modelling of carrier processes in materials is a fundamental research subject in electronics and materials science, and the knowledge obtained from this research is useful for developing device technologies. As we discussed in earlier chapters, total current flowing across solids is the sum of the conduction current and the Maxwell displacement current (MDC), according to Maxwell's electromagnetic field theory. The conduction current flows when charged particles, *e.g.*, electrons, holes, ions, *etc.*, are conveyed across solids. The external electric field is used to convey charged particles in solids, and we usually apply d.c. voltages on electrodes which sandwich the solids to form this external field.

The current–voltage (I–V) characteristics describes the steady flow of charges in solids, and gives us information on carrier injection, carrier transport, carrier trapping, and other carrier processes. We conventionally pay attention to the I–V characteristics to study carrier transport mechanisms in solids and in devices. There are many books that describe the I–V characteristics of films sandwiched between two electrodes [1, 2]. Carrier behaviours in electronic devices are discussed using the theories and analyses described in these books. For example, carrier transport across metal–insulator–metal (MIM) elements are analysed using the theories on electron tunnelling, Schottky current, Space charge limited current, and others [2]. Similarly, carrier transport across p–n diodes, MIS diodes, and other electronic semiconductor devices are discussed using the energy-band theory of semiconductors [3].

However, we should note that this approach is not so successful when we talk about the carrier transport in systems comprised of organic layers whose electrical properties are unknown, in particular, in the case of organic multilayer systems. Owing to the ambiguity of their electrical properties, charge accumulation at the interface, the presence of defects and impurities, and others, discussions on the carrier transport become controversial ones. One of the main reasons is that the *electric field* in each layer cannot be determined definitely, in particular, in the case of organic multilayer systems. It will be helpful to get information on the electric field distribution in the multilayer systems under the external applied voltage, because carriers will be conveyed under the established electric field distribution in multilayers. In other words, "*the static electric field distribution*" is playing an important role to transport charged particles.

Nevertheless, it is quite a task to determine the electric field distribution in multilayer systems only by looking at I–V characteristics, without knowledge of the fundamental electrical properties of materials used in multilayer systems. If we can directly measure the electric field distribution in multilayer systems, it will be of great help for analysing and modelling carrier behaviours. EFISHG measurement allows us to probe the electric field distribution in multilayer systems. This can help for understanding the steady-state and transient carrier transport in multilayer systems comprised of materials with unknown electrical properties, as will be demonstrated in this chapter.

By contrast, MDC is generated when the distribution of electric flux density in solids changes with time, where electrons, holes, and dipoles are the main source of the electric flux density. This MDC is a transient current, and we can study the dynamical carrier motion, in particular, the motion of carriers injected from electrode and orientational change of dipoles in solids, by analysing the trace of MDCs using an appropriate model [4, 5]. The dynamical carrier motion can be triggered by a variety of external stimuli, such as pulse voltages, light illumination, heating, pressure, and so forth. For example, mechanical compression and expansion can trigger orientational motion of dipoles in monolayers on water surface (see Chap. 5), alternating UV and Visible-light irradiation can directly induce reversible molecular conformational change in films (Chap. 2), *e.g.*, *cis–trans* conformational change in azobenzene monolayer, thermal heating can cause dielectric relaxational motion of dipoles in solids and release of trapped charges in solids (Chap. 3 and Chap. 4), light irradiation can excite electrons in solids (Chap. 2 and Chap. 3), and so forth.

It is instructive here to note that MDC is generated as the result of dynamical short-range and long-range carrier motions in solids and it can be recorded by using an ammeter. What we measure in the MDC measurement is the change of amount of induced charge on the electrode, where "*the electric field diverging from carriers*" makes a main contribution to the amount of induced charge. For example, MDC measurement is used for studying the dynamical behaviour of monolayers on water surface during the course of mechanical monolayer compression, where the change of amount of induced charge on an electrode suspended above the water surface is recorded using an ammeter. That is, dynamical rotational motion of dipoles in monolayer is indirectly probed in the MDC measurement, as a result of change of the amount of induced charge on the suspended electrode.

Time-of-flight (TOF) technique is often used to probe carrier transfer in solids sandwiched between two electrodes, to determine the carrier mobility (Chap. 2). What we measure by using this TOF technique is also the change of amount of induced charge on the electrode connected to an ammeter. The principle of the TOF measurement is basically the same with the MDC measurement used for the study of dipolar rotation in monolayers on water surface, in that change of amount of induced charge on the electrode connected to an ammeter is recorded. Consequently, the TOF measurement is also a technique that probes the carrier motion indirectly. In the actual TOF measurement, a long-range carrier motion in solids is recorded as a transient current with time, from which we determine the transit time of carriers to estimate the carrier mobility, by assuming that carriers move at a constant speed through solids under a constant external electric field formed in solids by application of d.c. voltage.

Noteworthy that the actual location of carriers traveling across solids cannot be determined directly from the trace of transient current. An appropriate model is necessary to estimate the location of carriers. Consequently, the TOF measurement exerts an elaborate mathematical approach on us for analysing the actual location of traveling carriers, but this analysis leads to a puzzling situation in which there are many possible solutions that fit observed transient currents, in particular, in the case of carriers being transported in the presence of multiple carrier traps. This means that it is quite a task to trace actual carrier behaviour in solids by the TOF method.

The idea of using EFISHG is available for probing the static electric field E formed in solids. By the way, there are two important directions to understand the role of "*electric field*" in materials and devices. One is as the "*static electric field*" that can give mechanical force to move charged particles, leading to conduction current. In this case, information on the distribution of static electric field formed in single and multilayer systems is key for analyzing I–V characteristics. The other is as "*the electric field arising from charged particles*" according to the Gauss law. Owing to the presence of this field, materials surrounding charged particles can be dielectrically polarized. In the observation of carrier motions using EFISHG measurement, the propagation of this polarization is effectively detected with time. Using EFISHG is a way for directly pursuing carriers that are traveling in devices, and this approach contrasts to TOF that indirectly traces carrier behavior (see also Chap. 8).

In the following Sec. 7.2, we go back to the principle of EFISHG measurement, and summarize the basics of EFISHG measurement as a tool for probing the above two kinds of "*electric field*". One is to understand a role of "*electric field*" leading to conduction current. The other one is "*electric field*" arising from charged particles distributed in organic devices. In the succeeding sections, typical examples of the EFISHG measurement are demonstrated on focusing the role of two kinds of "*the electric field*". In Sec. 7.3, on paying attentional to the role of electric field in organic devices, modelling of device is described in two approaches, *i.e.*, dielectric physics approach and semiconductor physics approach. Introduction of these two approaches will make clear a concept of dielectric time constant that is a key to device-modelling in terms of electronic material property. In Sec. 7.4, electric fields in single and multilayer device is summarized, and they are shown to be probed by using EFISHG. In Sec. 7.5, conduction mechanism of single layer device is described. In Sec. 7.6 to Sec. 7.9, two-layer devices are described as examples of analyzing and modelling of devices. The topics demonstrated here are typical ones in electrical and electronics engineering, and the

EFISHG measurement will be widely used for investigating biological, physical, electrochemical and other topics.

7.2 EFISHG as a tool for viewing "electric field"

To understand the EFISHG, we go back to Faraday's idea [4–6] formulated as the Gauss law in the electro-magnetic theory. According to this law, the static electric field arises from charges. We postulate that solids surrounding electrons and holes can be polarized due to the presence of this static electric field. Actually, if electrons and holes are injected into organic solids, non-zero static electric field $\vec{E}(0)$ is additionally and freshly formed according to the Gauss law, and nonlinear polarization $\vec{P}_{NL}$ can be freshly produced by laser irradiation, due to the presence of this static electric field $\vec{E}(0)$. The induced nonlinear polarization $\vec{P}_{NL}$ is due to the interaction between electromagnetic fields of the laser light and electrons of molecules in organic materials. Figure 7.1 illustrates the basic concept of EFISHG measurement [7–9]. The non-zero static electric field $\vec{E}(0)$ diverging from charges serves to distort the electron cloud of molecules surrounding them. This kind of nonlinear polarization $\vec{P}_{NL}$ is induced even in centrosymmetric molecular systems [10], such as phthalocyanine,

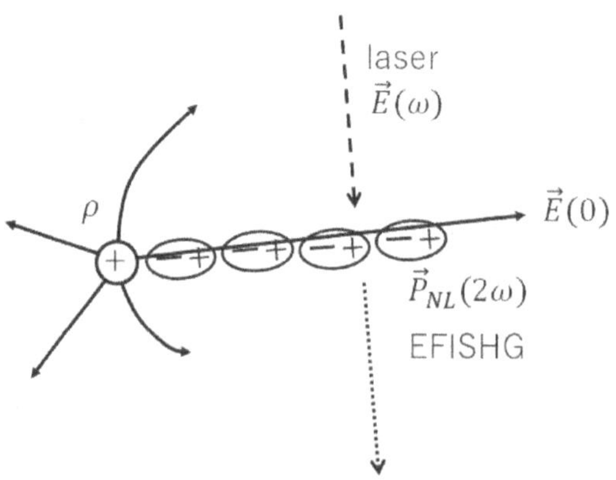

Figure 7.1. Basic concept of EFISHG measurement. Electric field $\vec{E}(0)$ is formed in a space surrounding charge ρ, according to Gauss law. $\vec{E}(0)$ induces polarization in materials, and nonlinear polarization $\vec{P}_{NL}(2\omega)$ is induced at double frequency by irradiating laser beam with vibrating electric field $\vec{E}(\omega)$ at angular frequency ω.

pentacene, C60 and others [11–13]. Induced nonlinear polarization $\vec{P}_{NL}$ in the presence of $\vec{E}(0)$ is given as

$$\vec{P}_{NL}(2\omega) = \varepsilon_0 \chi^{(3)}(-2\omega; 0, \omega, \omega) \vdots \vec{E}(0)\vec{E}_i(\omega)\vec{E}_j(\omega) \tag{7.1}$$

where $\chi^{(3)}(-2\omega; 0, \omega, \omega)$ represents nonlinear optical susceptibility, and ω stands for the angular wave frequency of the incident electromagnetic wave. $\vec{E}_i(\omega)$ and $\vec{E}_j(\omega)$ represent the electric fields of the incident laser, and $\vec{E}(0)$ is the static electric field arising from injected carriers, which satisfies the relation given by the Gauss law as [6]

$$\nabla \cdot \vec{E}(0) = \frac{\rho}{\varepsilon_s \varepsilon_0}, \tag{7.2}$$

where ρ is the charge density. ε_s and ε_0, respectively, are the relative permittivity of the organic materials and the absolute permittivity of free space.

It is instructive here to note that there are two important aspects of charges in terms of the analysis and modelling of organic devices. One aspect is that the charge distribution plays an important role in modeling carrier behavior in materials installed in devices. For example, for analyzing Si-semiconductor devices, understanding of depletion layer development at the Si–SiO_2 interface in terms of the charge distribution in the layer is key, on the basis of the established semiconductor device physics [3]. By contrast, we face a difficulty in analyzing the function of organic semiconductor device, in particular, for analyzing the charge distribution around the interface, owing to a lack of established organic semiconductor physics. But if we can experimentally probe the electric field distribution in organic device, we can make it clear the charge distribution used for modeling and analyzing organic devices.

The other aspect is that the behavior of charges directly impacts on the conduction current flowing through devices. For example, for p–n junction of Si-diode, an electrostatic potential distribution is built at the p–n junction interface and this potential distribution governs the number of charges crossing the junction. In other words, the electrical current flowing across the interface is regulated by the electrostatic potential distribution around the interface. In this situation, the *I–V* characteristics of p–n diode is theoretically analyzed, using semiconductor device physics.

On the other hand, for analyzing organic diodes, we do not have such established theory based on organic device physics that accounts for the function of interface. It is therefore difficult to say how the interface governs carrier conduction of organic diode. As a result, the analytical solution of organic p–n diodes become speculative, owing to the use of a model with many assumptions. Nevertheless, if we can probe the electric field $\vec{E}(0)$ of Eq. (7.2) experimentally, this situation will be improved, and we can explore the mechanism of the *I–V* characteristics, in terms of the probed potential distribution around interface. It is therefore quite important to find a way for measuring the electric field $\vec{E}(0)$ in devices. This is the reason why we here pay attention to the nonlinear dielectric polarization of organic materials in the presence of the electric field $\vec{E}(0)$, which is induced due to the coupling with incident electromagnetic waves.

The polarization $\vec{P}_{NL}(2\omega)$ induced by the coupling of electromagnetic wave, *e.g.* laser, is a source of the second-harmonic (SH) signal, and results in the enhancement of SH light intensity $I(2\omega)$ in proportion to the square of $\vec{P}_{NL}(2\omega)$ as

$$I(2\omega) \propto |\vec{P}_{NL}(2\omega)|^2 = |\varepsilon_0\chi^{(3)}(-2\omega; 0, \omega, \omega) \vdots \vec{E}(0)\vec{E}_i(\omega)\vec{E}_j(\omega)|^2 \tag{7.3}$$

From Eq. (7.2) and Eq. (7.3), it is clear that the SH intensity in solids changes in accordance with the nonlinear polarization distribution, which is generated in proportion to the static electric field distribution $\vec{E}(0)$ caused by injected charges. Consequently, by using EFISHG technique, we can directly observe the injected charge distribution, and if the injected carriers move in solids, we can catch the carrier transport in solids by monitoring the propagation of polarization $\vec{P}_{NL}(2\omega)$ induced in the solids along with the carrier motion [14] (see also Chap. 8).

It is noteworthy that, in the EFISHG measurement, laser irradiation is used to generate nonlinear polarization, but it can be postulated that no photo carriers are generated. This is because the basic second-harmonic generation process is a two-photon process, and the EFISHG process follows this two-photon process. For example, in SHG measurement of OLED with α–NPD transport layer, probe laser beam at wavelength of 820 nm is incident to molecules and EFISHG emitted at 410 nm is probed. These two wavelengths, 820 nm and 410 nm, are corresponding to one-photon and two-photon energy quanta of laser beam, respectively, in EFISHG process. As described in Chap. 6, two-photon emission at 410 nm is a result of resonant enhancement at energy gap of a molecule, and the molecule has no coupling with one-photon absorption of laser at 820 nm. Consequently, we can monitor carrier transport in solids and others in a nondestructive way. This is the advantageous point of the EFISHG measurement.

By the way, in electronic devices as well as in solids sandwiched between electrodes, there are many sources of static electric fields. Among them are applied external voltage, moving carriers, trapped charges, work function difference, and so forth. Consequently, the $\vec{E}(0)$ in Eq. (7.3) is rewritten by the sum of the electric fields from many sources as [7, 8]

$$\vec{E}(0) = \vec{E}_{ext}(0) + \vec{E}_m(0) + \vec{E}_t(0) + \cdots . \tag{7.4}$$

Here, $\vec{E}_{ext}(0)$ represents the external electric field formed by applied external voltage, $\vec{E}_m(0)$ is the electric field diverging from moving carriers, and $\vec{E}_t(0)$ is the electric field formed by trapped charges. Equation (7.3) with Eq. (7.4) suggests that we can probe electric fields $\vec{E}_{ext}(0)$, $\vec{E}_m(0)$, $\vec{E}_t(0)$, and others by using the EFISHG measurement. The time-resolved EFISHG method is particularly effective for probing carrier motion in solids, as described in Chap. 8, where the propagation of the polarization is monitored along with the carrier transport. It is noteworthy that $\chi^{(3)}(-2\omega; 0, \omega, \omega)$ is a material dependent parameter, and specifies the feature of materials. As a result, the SHG is activated at a specific wave length $\lambda_{2\omega} = 2\pi c/2\omega(c$, velocity of light) of the materials, and this fact allows us to selectively probe SHG generated from a certain layer of the multi-layers by choosing the laser wavelength appropriately [15].

The magnitude of $\chi^{(3)}(-2\omega; 0, \omega, \omega)$ is very important in the EFISHG measurement. For example, the values of pentacene and C60 are in the order of $10^{-22}\,\mathrm{m}^2\,\mathrm{V}^{-2}$ (10^{-14} esu) at a laser wavelength of $\lambda = 1120$ nm and $10^{-18}\,\mathrm{m}^2\,\mathrm{V}^{-2}$ (10^{-10} esu) at $\lambda = 1000$ nm, respectively, and EFISHG signals generated by laser light irradiation from Q–switched YAG laser with an energy of 1 mJ/pulse are probed as shown in Chap. 6. Note that these $\chi^{(3)}(-2\omega; 0, \omega, \omega)$ values are very large at an angular frequency ω due to the resonant coupling between the laser beam and the electrons in the molecules [16]. It should be noted that there are many reports on SHGs since the pioneering works by Bloembergen and Shen [17]. Those works mainly focus on crystal and molecular structures and arrangement of molecules from viewpoints of spectroscopy for characterizing material structure. That is, analysis of the SHG signal from non-symmetric crystal is a main topic. The idea of SHG measurement employed in these works is different from that of the EFISHG measurement which we discuss in this book. EFISHG measurement is argued from viewpoints of carrier motion, where *electrostatic field* shall be essential for analysis and modelling of electronic devices.

In more detail, spectroscopic SHG experiments are carried out on focusing the SHG signal generated from materials in a form below,

$$I(2\omega) \propto |\vec{P}_{NL}(2\omega)|^2 = |\varepsilon_0 \chi^{(2)}(-2\omega; \omega, \omega) : \vec{E}_i(\omega)\vec{E}_j(\omega)|^2, \quad (7.5)$$

where a key parameter is the material dependent $\chi^{(2)}(-2\omega; \omega, \omega)$ that depends on the arrangement of molecules and symmetry of molecular structure, but this is independent of the presence of the static d.c. field, $\vec{E}(0)$. This situation is very helpful to characterize material structures, *e.g.*, for characterization of organized molecular films (see Chap. 5). Noteworthy that $\chi^{(2)}(-2\omega; \omega, \omega) = 0$ for symmetric molecular systems such as phthalocyanine, pentacene, C60, in dipole approximation, and the SHG ruled by Eq. (7.5) is not generated even when the external d.c. electric field is non-zero.

Generally, the angular frequencies $\omega = \omega_1$ in $\vec{E}_i(\omega)$ and $\omega = \omega_2$ in $\vec{E}_j(\omega)$ are chosen to be the same value by using the same light source in the ordinary SHG measurement, but it is possible to choose two different angular frequencies by using two light sources with angular frequencies ω_1 and ω_2 for $\vec{E}_i(\omega)$ and $\vec{E}_j(\omega)$, respectively. This is the technique called as the sum frequency generation (SFG) method, and it is used to precisely characterize non-symmetric material structures, the dynamical change of the orientational order of polar molecules, the interaction between molecules and surrounding solutions, and others (see text books [17, 18] and review papers [19–21]).

On the other hand, the situation is quite different for the case of the electric-field-induced optical second-harmonic generation (EFISHG) described by Eq. (7.3), where the non-zero d.c. field $\vec{E}(0)$ makes a significant contribution and this field couples with $\chi^{(3)}(-2\omega; 0, \omega, \omega)$. In other words, EFISHG is not activated if $\vec{E}(0) = 0$. Probing $\vec{E}(0)$ in Eq. (7.3) by means of EFISHG is indeed helpful for analyzing carrier behaviors and related electronic phenomena in organic devices. That is, we need to focus on the EFISHG process given by Eq. (7.3) to probe carrier motions in electronic organic devices, where $\vec{E}(0) = \vec{E}_m(0)$. All electric phenomena have something to do with carrier behaviors. EFISHG provides a way for analyzing carrier behaviors in devices, *e.g.*, organic field effect transistors, organic solar cells, organic light-emitting diodes, organic memory devices, *etc.* It should be noted that EFISHG is also available to probe $\vec{E}(0) = \vec{E}_{ext}(0)$, $\vec{E}(0) = \vec{E}_t(0)$ and others. That is, EFISHG is capable of probing the distribution of electrostatic field E formed in multilayer systems and in devices. Consequently, it will be effective to study the I–V characteristics of multilayer systems comprised of layers whose electrical properties are unclear, in coupling with the current–electrostatic field (I–E) characteristics of individual layer of multilayer.

7.3 Dielectric physics approach and semiconductor physics approach

EFISHG provides a powerful way to probe carrier motion in organic device, meanwhile the I–V characteristics is the trace of carrier transport in devices. Consequently, for analysing the I–V characteristics of organic devices, we focus on the electric field probed by the EFISHG. In the light of the dielectric nature of organic materials, this approach is quite reasonable in comparing with the approach which is adopted when we talk about the I–V characteristics of conventional semiconductor devices such as Si-devices. We generally focus on the nature of semiconductors and focus on the potential distribution in conventional semiconductor devices. In the followings, the difference of the two approaches are briefly described. This section will be of help as the guide to read the following sections.

A. Dielectric Physics Approach

As you know, a variety of new materials and novel processing technologies have been developed since the discovery of organic semiconductors, where the fundamental carrier process is found to be different from the conventional inorganic semiconductors, owing to the dielectric nature of organic semiconductor materials. For organic material, dielectric relaxation time $\tau = \varepsilon/\sigma$ is long in the region $10^0 \sim 10^{15}$ s (see Fig. 7.2 [22]; for dielectric relaxation time, see

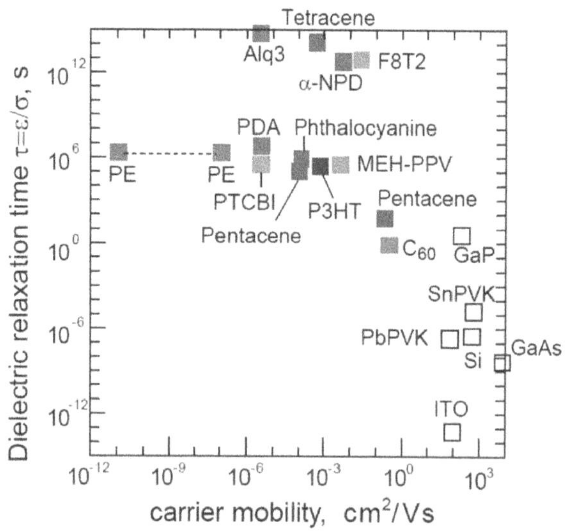

Figure 7.2. Dielectric relaxation time and carrier mobility of materials. PE: polyethylene, Alq3: Tris(8–hydroxyquinolinato)aluminium(III), PDA: polydiacetylene, PTCBI: 3,4,9,10–Perylenetetracarboxylic Bisbenzimidazole, α–NPD: N, N'–di–[(1–naphthyl)–N, N'–diphenyl]–(1,10–biphenyl)–4,4′–diamine, MEH–PPV: poly[2–methoxy–5–(2–ethylhexyloxy)–1,4–phenylenevinylene], F8BT: poly(9,9–dioctylfluorene–*alt*–*co*–benzothiadiszole, P3HT: regio–regular poly(3–hexylthiophene–2,5–diyl), C60: fullerene, ITO: Indium–tin–oxide, PbPVK: lead perovskite, SnPVK: tin perovskite.

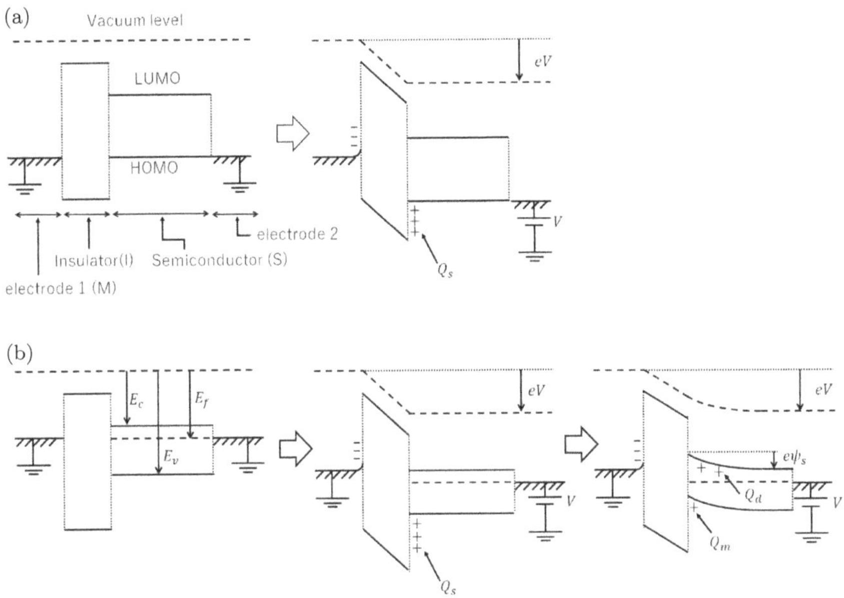

Figure 7.3. A concept of organic device. Example of metal (M)–insulator (I)–semiconductor (S) device. (a) Organic semiconductor devices. (b) Conventional semiconductor, *e.g.*, Si single crystal, devices.

Appendix B). By contrast, the relaxation time of the conventional semiconductor, *e.g.*, Si, is much shorter, in the region $10^{-15} \sim 10^{0}$ s. Figure 7.3 illustrates energy diagram of metal–insulator–semiconductor (MIS) device comprised of organic semiconductor and the conventional n-type semiconductor (Si-single crystal). In organic MIS device (Fig. 7.3(a)), insulator layer (I) and semiconductor layer (S) are sandwiched between two identical metal electrodes 1 and 2. Electrodes 1 and 2 are electrically connected and shorted in an external circuit. The external voltage ($V > 0$) is applied to electrode 2 in reference to electrode 1 (M) grounded. Upon positive voltage application to electrode 2, positive charges are conveyed from electrode 1 to electrode 2 through the external circuit, on leaving negative charges on electrode 1 (M). At the same time, holes are smoothly injected into the semiconductor (S) from the electrode 2, and they are conveyed to insulator (I)–semiconductor (S) interface. Finally, holes are accumulated at the interface due to the Maxwell–Wagner effect (see Chap. 4). That is, relaxation time τ_1 of organic semiconductor (S) is generally long, but the relaxation time τ_2 of insulator (I) is much longer, *i.e.*, $\tau_1 \ll \tau_2$ and infinite $\tau_2 \to \infty$. As the result, $Q_s = I(\tau_2 - \tau_1) = C_i V$ is accumulated at the interface (C_i: capacitance of insulator). This accumulated charge Q_s serves as a source of carriers in electronic devices. For example, by making use of alternatingly switching between on-state ($Q_s \neq 0$) and off-state ($Q_s = 0$) in MIS structure, we can drive organic field-effect transistor (organic-MIS FET) as an electrical switching element.

B. Semiconductor Physics Approach

General conventional semiconductor transistors, *e.g.*, Si-transistor, are available as switching elements, but the switching mechanism is different from organic transistors [3]. Figure 7.3(b) illustrates an energy diagram of Si-semiconductor of n-type MIS diode. By applying positive voltage to electrode 2 in reference to electrode 1 (M) grounded, holes are injected into semiconductor layer (S) from electrode 2, and they are then quickly traveling through the semiconductor (S). After that, they are accumulated at the insulator–semiconductor interface. This is possibly due to the Maxwell–Wagner effect in the same way as that seen for organic MIS device (Fig. 7.3(a)), but the accumulation process further proceeds until thermo-dynamic equilibrium is established in semiconductor (S). That is, at the final state, Fermi energy level alignment is established over the entire region of semiconductor (S), as illustrated in the right-hand-side figure of Fig. 7.3(b). This final state is established very quickly after the application of the positive voltage, and the accumulated holes Q_s, are re-distributed into depletion layer Q_d and into inversion layer Q_m. Q_m is mobile charge and serves as a source of carriers of transistor function in metal–oxide–semiconductor field effect transistor (MOSFET), whereas Q_d are fixed charges. By switching on-state ($Q_m \neq 0$) and off-state ($Q_m = 0$) alternatingly, electrical switching behaviour is realized by using MOSFET.

As mentioned above, the function of switching is the same between organic semiconductor MIS FET and the conventional semiconductor MOSFET in that Maxwell–Wagner-type charging Q_s is original source charges of switching of these devices, but the actual device operation is different. For organic MIS FETs, Q_s directly contributes to the on-off switching, if we cannot find inversion layer in the device. By contrast, for the conventional MOSFETs, the contribution of Q_s to the switching is indirect, but Q_m governs the on-off switching.

In the following subsection C, based on the common concept of Maxwell–Wagner effect in the charging of MIS diodes, charges Q_s in organic MIS diodes and Q_m in the Si-MIS diodes are derived, respectively.

C. Maxwell–Wagner effect in the charging of MIS diodes

C.1 Organic MIS diodes case

In organic MIS diodes, semiconductor layer with dielectric relaxation time τ_1 and insulator with τ_2, are connected in series, and an electrical conduction route is constructed. The relaxation times are given by

$$\begin{aligned} \tau_1 &= \frac{C_1}{G_1} = \frac{\varepsilon_1}{\sigma_1} \\ \tau_2 &= \frac{C_2}{G_2} = \frac{\varepsilon_2}{\sigma_2} \end{aligned} \tag{7.6}$$

C_1, C_2, and G_1, G_2 are capacitance and conductance of semiconductor layer and insulator layer, respectively. ε_1, ε_2, and σ_1, σ_2 are dielectric constant and conductivity of semiconductor and insulator, respectively. Under voltage application V to electrode 2 of the device (see Fig. 7.3(a)), conduction current I flows and steady state is eventually established. The charge Q_s is accumulated at the two-layer interface according to the Maxwell–Wagner effect as,

$$Q_s = I(\tau_2 - \tau_1) = I\left(\frac{C_2}{G_2} - \frac{C_1}{G_1}\right) \tag{7.7}$$

Assuming insulating layer is ideal with $\tau_2 \to \infty$ and $\tau_1 \ll \tau_2$,

$$Q_s = C_2 V = \varepsilon_2 E_2. \tag{7.8}$$

Q_s in Eq. (7.8) is a source of carriers in device function of organic MIS devicc. ε_2 and E_2 are dielectric constant and electric field in the insulating layer. Equation (7.8) indicates that the electric field E_2 formed in the insulating layer serves as a linker between Q_s at the interface and $Q_1 = -Q_s$ at the electrode 1, and changes in proportion to the amount of interface charge Q_s. In other words, probing electric field E_2 is identical to directly probing charge Q_s. This is the reason why probing electric filed is key to analyse and model carrier behaviours in organic MIS diode. Electromagnetic theory suggests us two ways to probe Q_s, as described in Chap. 6. One way is directly probing charge Q_s itself, the other way is probing "*electric field*" instead of charge Q_s. EFISHG measurement is corresponding to the latter case.

C.2 Si-MIS diode case

For the conventional Si-MIS diodes, Q_s is supplied in the way as given by Eq. (7.8), but Q_s is redistributed into depletion layer charge Q_d and inversion layer charge Q_m. Q_m and Q_d are mobile charges and fixed charges, respectively, and they are given as a function of surface potential ψ_s, established at the insulator (I)–semiconductor (S) interface as illustrated in the energy band diagram of Fig. 7.3(b). That is, the surface potential ψ_s, is a key parameter for describing Q_d and Q_m, and they are given as below

$$Q_d = C_d\psi_s \tag{7.9}$$

$$Q_m = C_2(V - \psi_s) - C_d\psi_s \tag{7.10}$$

Here, C_d is capacitance of depletion layer. ψ_s changes in accordance with the applied voltage V. As illustrated in Fig. 7.3(b), the depletion layer is developing into the semiconductor layer (S) from the I–S interface. Accordingly, in the semiconductor layer (S), there are at least two regions with different dielectric relaxation time τ_d (depletion layer region) and τ_1 (bulk region of semiconductor (S)). Consequently, Maxwell–Wagner-type charging is anticipated at the depletion layer/bulk semiconductor interface Q'_s and also at insulator/depletion layer interface Q''_s as follows:

$$Q'_s = I(\tau_d - \tau_1) \tag{7.11}$$

$$\sim I\tau_d = C_d\psi_s(\tau_d \gg \tau_1)$$

$$Q''_s = I(\tau_2 - \tau_d) = C_2\frac{I}{G_2} - C_d\frac{I}{G_d} = C_2(V - \psi_s) - C_d\psi_s$$

$$= C_2V - (C_d + C_2)\psi_s \tag{7.12}$$

G_2 and G_d are conductance of insulator and depletion layer, respectively. Tentatively, $Q_m = Q''_s$ are available charges for the device function. In the voltage region $0 < V < V_{th}$, external voltage V drops across the depletion layer ψ_s as

$$\psi_s = \frac{C_2}{C_d + C_2} V \tag{7.13}$$

and, from Eq. (7.12) and Eq. (7.13) the available charge is zero as below,

$$Q''_s = 0. \tag{7.14}$$

In the voltage region $V_{th} < V$, inversion layer appears at the insulator/semiconductor interface. Depletion layer is stable in this voltage region, and $\psi_s = \psi_{th}$ and $C_d = C_{th}$, are kept constant. The available charge is given as

$$Q''_s = C_2(V - V_{th}) \quad \text{with} \quad \frac{C_2}{C_{th} + C_2} V_{th} = \psi_{th}. \tag{7.15}$$

In summary, in the conventional semiconductor device such as Si-device, surface potential ψ_s plays a main role for driving the device function.

7.4 "Electric field" in organic devices

The electric field significantly effects on carrier injection, transport, and other carrier processes in organic devices. Many experimental methods have been developed to study these electronic processes in organic materials, with consideration of the role of "*electric field*" in the measurement. TOF measurement is utilized as a conventional way to study carrier transit, where d.c. field is utilized to control the direction of the carrier transit [2, 5, 23–25]. I–V measurement is the most conventional way to study single- and double-carrier injection process, and carrier transport in steady state, and others, where the relationship between the "average external electric field" and the current density is utilized to explore these carrier processes [2]. It is therefore very helpful to understand the electric field formed in organic device. In this section, we briefly discuss the electric field distribution formed in devices. This knowledge will be of help for deep understanding of the EFISHG measurement employed for double layer diodes, organic field effect transistors (OFET), organic light-emitting diode (OLED) devices and others.

a. "Electric field" in single layer MIM devices

As the first step for the study of carrier process, a single layer device with a metal–insulator–metal (MIM) structure is very helpful. The external electric field is uniformly formed in the single layer, which helps to carry the analysis easy. Figure 7.4(a) illustrates a single layer device, where a single-layer is sandwiched between two identical parallel electrodes with a spacing of L. The external voltage V is applied to the device, where the constant external electric field is formed uniformly in the single layer. The average electric field $\bar{E}$ in the film is given by

$$\bar{E} = \frac{V}{L} \tag{7.16}$$

Upon square-wave voltage application, the transient electric field is formed in the film in the manner as

$$E(t) = \bar{E}(1 - e^{-\frac{t}{RC}}) \tag{7.17}$$

Here, R is electrical resistance of circuit and C is capacitance of the single layer. Similarly, a variety of transient electric fields can be formed depending on the applied voltages, *e.g.*, Large amplitude AC voltage, double-pulse, ramp voltage, and others. Accordingly,

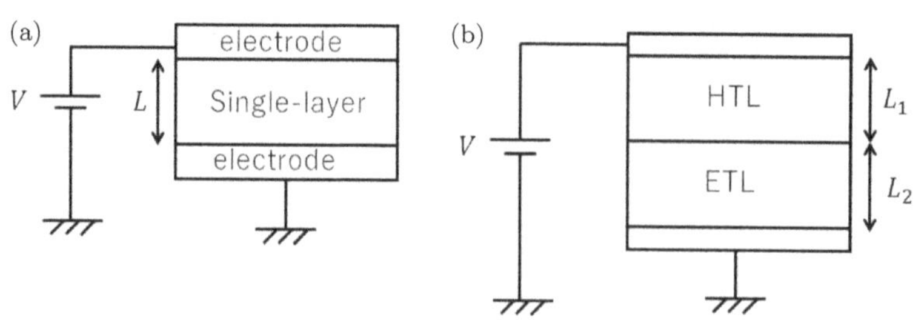

Figure 7.4. (a) A single layer device. (b) Two-layer OLED as an example of multilayer device. HTL and ETL denote hole and electron transport layer, respectively.

many kinds of electrical responses of carriers are anticipated, and they are used for analysing carrier processes. In this chapter, response of carriers against square wave voltage application is mainly described, under assumption that the electric field changes according to Eq. (7.17). Note that in actual experiment, the wave form will slightly deviate from the form of Eq. (7.17), owing to the performance of the electrical instruments employed for the generation of square wave voltage. However, by using a set of calibrated voltage source V, we can carry reliable experiments for analysing the carrier processes in the single layer. Note that in actual device, single layer is sandwiched between two dissimilar electrodes, with consideration of work function of electrodes for hole and electron injection, *etc.* Owing to the work function difference between two dissimilar electrodes, internal field E_i adds to Eq. (7.16) as described in the Appendix G.

b. "*Electric field*" in multi-layer organic devices

The electric field $E(t)$ directly gives impact on electronic phenomena in single-organic layer, *e.g.*, carrier injection, carrier transport, carrier detrapping from traps, dipolar alignments, *etc.*, and we can explore the electric phenomena in single organic layers, in terms of the electric field $E(t)$. Meanwhile, this situation is slightly different when we explore the electronic phenomena observed in multi-layer organic devices, owing to the presence of interfaces in multilayers, such as insulator/organic semiconductor, organic semiconductor/ organic semiconductor, *etc.* Basically, Maxwell–Wagner type charging occurs at these interfaces, and this charging influences the electric field distribution in the layers of organic devices. We therefore must consider the contribution of accumulated charge Q_s at these interfaces, in order to analyse the electric phenomena generated in multi-layers and organic devices.

Figure 7.4(b) shows a structure of two-layer organic light-emitting diode (OLED). The two layers are hole transport layer (HTL) and electron transport layer (ETL), and they are sandwiched between top and bottom electrodes. The hole–electron recombination leading to the EL emission occurs around the HTL/ETL interface. As that, understanding of carrier behaviours at the interface is quite important. In two-layer system, charges Q_s are accumulated at the interface due to the Maxwell–Wagner effect, and it is given as

$$Q_s = I(\tau_2 - \tau_1). \tag{7.18}$$

Here I is steady-state current flowing across the interface. Under the operation of OLED, large current I flows to emit bright light through hole–electron recombination at the interface. Assuming G_1 and G_2 are conductance of the two layers, steady-state current I is given by Eq. (7.19) with Eq. (7.18) as follows:

$$I = \frac{G_1 G_2}{G_1 + G_2} V \tag{7.19}$$

$$= \frac{Q_s}{\tau_2 - \tau_1}. \tag{7.20}$$

Equation (7.20) suggests that the interface charge Q_s governs the I–V characteristics of OLED device function. Therefore, it is very important to pay attention to the "*electric field*" arising from Q_s, for understanding the operation of organic devices, *e.g.*, OLED.

By contrast, in the case of conventional p–n diodes, the accumulation of excess charges at the interface due to the Maxwell–Wagner effect is minor, owing to the very short relaxation time of p- and n-type layers, *e.g.* $\sim 10^{-13}$ s for Si-single crystal. In this situation, the I–V characteristics of p–n diode is described as a function of the interface potential barrier height that changes according to the applied voltage V, where the electric field in the bulk of the adjacent two semiconductor layers is nearly zero due to the establishment of thermodynamic equilibrium (see Fig. 7.3(b)). The current I flowing across the p–n junction is described as

$$I = \int_{E_n}^{\infty} env dE = I_0(e^{\frac{eV}{kT}} - 1) \quad \text{where} \quad \int_{E_n}^{\infty} nv dE = \frac{I_0}{e}(e^{\frac{eV}{kT}} - 1) \tag{7.21}$$

I_0 is diffusion current at external voltage $V = 0$, and it balances with drift current in p–n junction. Therefore, in this case, it is not necessarily important to pay attention to the electric field, while external voltage V is important.

As described in Eq. (7.20), we can sketch the operation of organic devices by using the Maxwell–Wagner (MW) charges Q_s. This accumulated charge Q_s is excess charge and is a source of space charge field in organic semiconductor layers. Therefore, it is quite reasonable to probe the *electric field* in organic electronics devices under operation to investigate the carrier processes. The electric field in the first layer $\bar{E}$ and in the second layer $\bar{E}'$ in steady state is (see Chap. 4).

$$
\begin{aligned}
&\bar{E} = \bar{E}_e(V) - \bar{E}_s(Q_s) \\
&\bar{E}' = \bar{E}'_e(V) + \bar{E}'_s(Q_s) \\
\text{with}\quad &\bar{E}_e(V) = \frac{1}{L_1}\frac{C_2}{C_1+C_2}V, \quad \bar{E}_s(Q_s) = \frac{1}{L_1}\frac{Q_s}{C_1+C_2} \\
\text{and}\quad &\bar{E}'_e(V) = \frac{1}{L_2}\frac{C_1}{C_1+C_2}V, \quad \bar{E}'_s(Q_s) = \frac{1}{L_2}\frac{Q_s}{C_1+C_2}
\end{aligned}
\tag{7.22}
$$

$\bar{E}_e$ and $\bar{E}'_e$ are external electric fields, given by external voltage, $\bar{E}_e = \bar{E}_e(V)$ and $\bar{E}'_e = \bar{E}'_e(V)$. $\bar{E}_s$ and $\bar{E}'_s$ are space charge field originating from MW charge Q_s at the interface, $\bar{E}_s = \bar{E}_s(Q_s)$ and $\bar{E}'_s = \bar{E}'_s(Q_s)$. Using SHG probes $\bar{E}$ and $\bar{E}'$, and allows us to estimate Q_s.

The transient electric fields are

$$
\begin{aligned}
E(t) &= \bar{E}_e(1 - e^{-\frac{t}{RC}}) - \bar{E}_s\left(1 - e^{-\dfrac{t}{\tau}}\right) \\
E'(t) &= \bar{E}'_e(1 - e^{-\frac{t}{RC}}) + \bar{E}'_s\left(1 - e^{-\frac{t}{\tau}}\right)
\end{aligned}
\tag{7.23}
$$

Here C is total capacitance of the device, *i.e.*, $C = C_1C_2/(C_1 + C_2)$, RC is time constant of external circuit, and τ is relaxation time τ_{MW} of Maxwell–Wagner effect,

$$\tau_{MW} = \frac{C_1 + C_2}{G_1 + G_2}. \tag{7.24}$$

Different time constants RC and τ_{MW} allows us to evaluate $\bar{E}_e$ and $\bar{E}_s$, by means of the time-resolved EFISHG measurement.

In conclusion, Sec. 7.4 is summarised as follows:

(1) For single layer device, the electric field is constant and given by a ratio of applied voltage to layer thickness ($\bar{E} = V/L$). The relaxation time of the transient state is the same with the response time of the external circuit RC. Accordingly, electronic phenomena are well analysed using the average electric field $\bar{E}$ that is only a function of external voltage V.

(2) For two-layer device, the first term $\bar{E}_e$ and $\bar{E}'_e$ of Eq. (7.22) are defined by using external voltage V, and their time-constant RC corresponds to the response time of external circuit. The second term, $\bar{E}_s$ and $\bar{E}'_s$, of Eq. (7.22) is a function of Q_s. The time-constant of accumulating Maxwell–Wagner charge Q_s is $\tau = \tau_{MW}$. By measuring electric field E_s (E'_s) in the layer, we can explore the carrier process Q_s that governs the device function. This is a key reason why we probe the *"electric field"* in electronic devices. EFISHG is one of the powerful tools to probe the *"electric field"*. Sum-frequency generation (SFG) [26, 27], electro-absorption measurement [28, 29], and others are also available for probing these electric fields.

7.5 Metal–insulator–metal (MIM) device

As described in previous section, the electric field $\vec{E}(0)$ formed in solids and in organic devices can be probed by measuring nonlinear polarization $\vec{P}_{NL}$. The source of EFISHG signals is nonlinear polarization, and it is generated in proportion to $\left|\vec{E}(0)\right|^2$ ((see Eq. (7.3)). Further by paying attention to the source of the electric field $\vec{E}(0)$, carrier motions, distribution of carrier traps, *etc.*, can be explored (see Eq. (7.2) and Eq. (7.4)). In the following sections, some examples are demonstrated, and carrier distribution and carrier transport *etc.*, are modeled, on the basis of EFISHG measurement. Noteworthy, there are two important points in the probing of electric field by EFISHG, for analysis and modelling of organic devices. One is that we can probe the location of charges in device on-time, *e.g.*, mobile charges, trapped charges, and others. Accordingly, the charge behaviors in device can be modeled. The other is that we can argue carrier conduction mechanism in devices, with taking account of the space charge field caused from charges. We show that the probing electric field is key for modeling I–V characteristics and others, in particular, for the case of multilayer systems.

Before going into the discussion on carrier behaviors in organic semiconductor layers in organic devices, we focus on the carrier transport mechanism of ideal insulator sandwiched between identical metal electrodes, to make clear the importance of the *"electric field"*.

The key function of electrical insulator is to block carriers crossing the insulator. The electrical resistivity of the insulator is very high and the free carrier density of the insulator is very low. Consequently, the carrier transport is dependent on the film thickness d, applied voltage V, and the work-function ϕ of metal electrodes.

For electrical insulating films whose thickness is less than several nano-meters, tunneling current is allowed to flow according to the quantum effect [30–32]. We can see such examples in MIM diodes and Josephson junctions with ultrathin organic layers, *e.g.*, polyimide, polyethylene, and others. For films whose thickness is more than several nanometers, carrier conduction is ruled by intrinsic carrier transport process, carrier injection process, and others.

The mechanism is dependent on the applied voltage to the film. In the low electric field region (region 1 of Fig. 7.5), the carrier conduction is ruled by the transport of intrinsic carriers in the film, where ohmic conduction current I flows in proportion to the applied voltage V in a manner as $I = en_0v = \kappa E$. Here, e is elemental charge, n_0 is intrinsic carrier density. v is carrier velocity given by $v = \mu E$ where μ is carrier mobility with $E = V/d$. κ is the electrical conductivity given by $\kappa = en_0\mu$. E is external electric field and acting to transport intrinsic carriers n_0. The intrinsic carrier density n_0 is determined in thermodynamic equilibrium condition, and it depends on the energy gap E_g in the way as $n_0 \propto \exp(-E_g/2kT)$. This value is generally very small for insulators owing to the large energy gap, *e.g.*, the energy gap of polyethylene is larger than 5.0 eV.

In the higher electric field region (region 2 of Fig. 7.5), the carrier conduction is ruled by the injected carrier density n. This

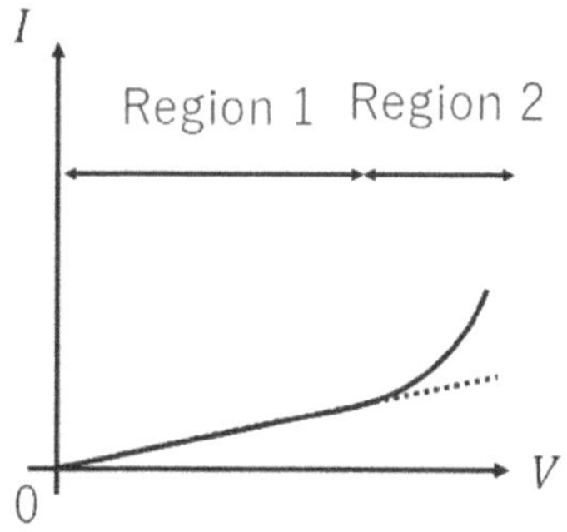

Figure 7.5. I–V of single layer sandwiched between identical electrode, *i.e.*, metal–insulator–metal (MIM) device. In low voltage region (Region 1), Ohmic current flows, and I–V is linear. In high voltage region (Region 2), current injection initiates, and I–V is nonlinear.

situation occurs when the physical condition $end > en_0d$ is established by applying the external voltage V. The carrier transfer process governed by injected carriers consists of two electronic processes, *i.e.*, carrier injection from electrode and the succeeding carrier transit across the film. If the carrier transit time across the film $t_r = d/v = d^2/\mu V$ is much longer than the time required for carrier injection, injected charges accumulate in the film. Consequently, the carrier conduction process is limited by the carrier transport process. The most typical one is the space charge limited current (SCLC) process, where the SCLC is given as follows (Child's law);

$$I = \frac{9}{8}\varepsilon_s\varepsilon_0\mu\frac{V^2}{d^3} \tag{7.25}$$

The SCLC current flows in proportion to V^2, but it is easily rewritten as

$$I \approx env = \varepsilon_s\varepsilon_0\mu\frac{V^2}{d^3} = \frac{\varepsilon_s\varepsilon_0\mu}{d}E^2,$$

$$\text{with } en = \frac{\varepsilon_s\varepsilon_0\frac{V}{d}}{d} > en_0 \quad \text{and} \quad v = \mu E. \tag{7.26}$$

Here $E = V/d$. That is, SCLC flows under the condition when injected carrier density exceeds the intrinsic carrier density. The amount of injected charges *end* corresponds to the amount of induced charge on electrode of the capacitor, $\varepsilon_s\varepsilon_0\frac{V}{d}$, and this amount of injected charge *end* are conveyed at a constant speed v under the external electric field E in steady-state. Noteworthy that the local electric field at the injection electrode, where carrier injection happens, becomes zero, owing to the space charge field created by the injected charge *end* in the film. In other words, carrier injection is regulated by the space charge field due to carriers accumulated in the film and this regulation by the space charge field proceeds until the zero-field position (see Chap. 3) is to be created at the injection electrode (see Fig. 7.6).

On the other hand, if the injected-carriers transit across the film in much shorter time after the carrier injection, injected carriers do not accumulate in the film. Consequently, the process is limited by the carrier injection. The most typical one is the Schottky-type injection, and the current injection is governed by the effective barrier height, $\phi-\Delta\phi$, built at the electrode–insulator interface, due to mirror image charge effect. This effective barrier height varies depending on the applied external electric field (see Fig. 7.7), and the Schottky current

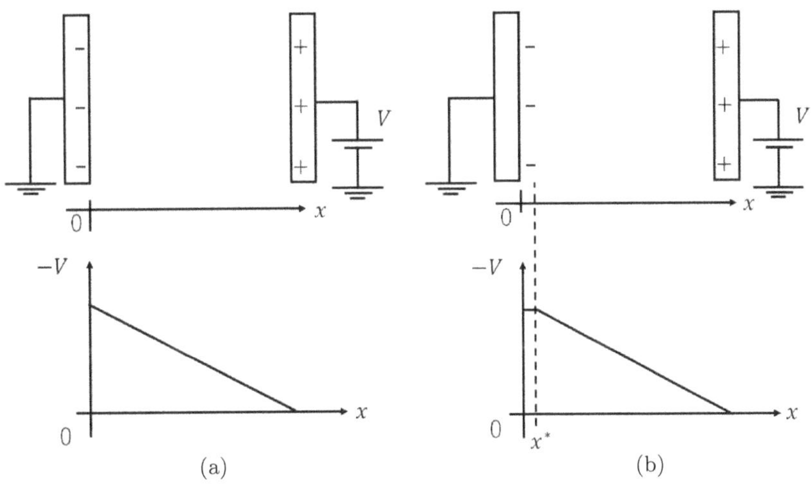

Figure 7.6. SCLC current in single insulating layer sandwiched between two identical electrodes. (a) In low electric field region where injected charge is minor. (b) In high electric field region where zero-field position x^* is created with SCLC condition.

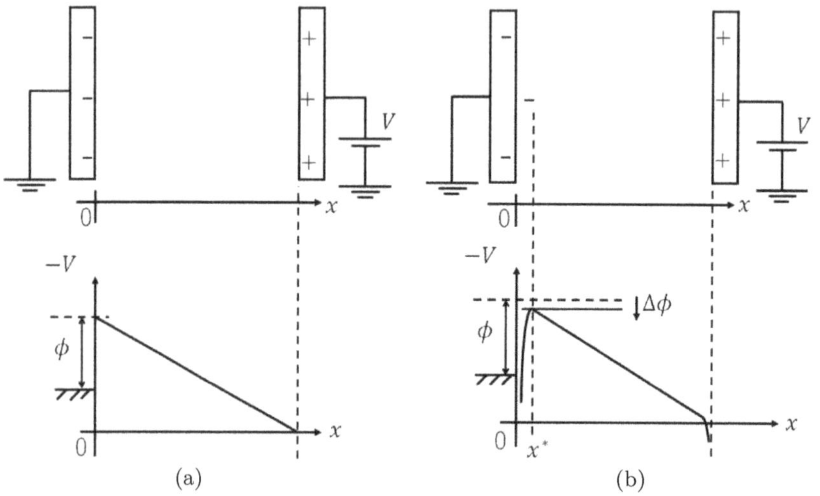

Figure 7.7. Schottky current in single insulating layer sandwiched between two identical electrodes. (a) In low electric field region where injected charge is minor. (b) In high electric field region where zero-field position x^* is created, where carriers cross over the effective injection barrier height $\phi - \Delta\phi$. $\Delta\phi$ is caused by the mirror image charge effect.

I flows in the way as

$$\ln I \propto const + \frac{\Delta\phi}{kT} = const + \frac{\beta\sqrt{E}}{kT} \quad \text{with} \quad \beta = \sqrt{\frac{e^3}{4\pi\varepsilon_s\varepsilon_0}}. \tag{7.27}$$

That is, the external electric field E is acting to vary the effective barrier height, and it works to collect injected carriers to the counter electrode. Note that the x-axis of Fig. 7.7 is the location of charge carrier in single layer insulators injected from the left-hand-side electrode [3]. As mentioned above, for understanding the electrodes, the measurement of "*electric field*" is very important.

The electric field distribution in multilayer systems will be more complex, owing to the accumulation of excess charges at the interface, difference in electrical resistivity and dielectric constant, and others. Therefore, it will be of help to directly determine the electric field distribution by experiments for discussing the carrier transport mechanism of the multilayer systems.

7.6 MIS structure device

a. Steady state

Along with the discovery of new synthetic organic semiconductor materials, preparation methods, characterization methods, organic-based circuit techniques have been intensively proposed for the use of organic semiconductor materials in electronics. Among electronic devices, Metal–Insulator–Semiconductor (MIS) diode is the most typical ones.

Conventionally, the current–voltage (I–V) and capacitance–voltage (C–V) measurements are carried out to analyse the carrier transport. The I–V and C–V characteristics are analysed using a RC equivalent circuit model, *e.g.*, in a study on metal–insulator–semiconductor (MIS) capacitors incorporating a polyimide (PI) layer as the gate insulator and *regio*-regular P3HT as the active semiconductor layer. A model of energy diagram for organic devices that illustrates the molecular energy level of organic semiconductors and the Fermi energy level of electrodes has been developed. I–V and C–V measurements are effectively employed for studying organic devices, however, the actual experimental results often lead us to a puzzling situation; the carrier injection and transport cannot be fully discussed on the basis of semiconductor device physics, even in the case of simple double layer devices. Main reason is that interfacial charge Q_s accumulates at the two-layer interface in organic device, and the space charge field from the charge Q_s deforms the electric field in the layer. The deformed electric field directly gives effect on carrier injection and transport process as described in Sec. 7.5. Therefore, a technique that can probe the electric field is necessary. EFISHG measurement is capable of probing the electric field distribution in organic devices, and the electric field of the two layers in double layer device is determined one by one.

For example, using EFISHG measurement coupled with the conventional current–voltage (I–V) measurement, the carrier transport

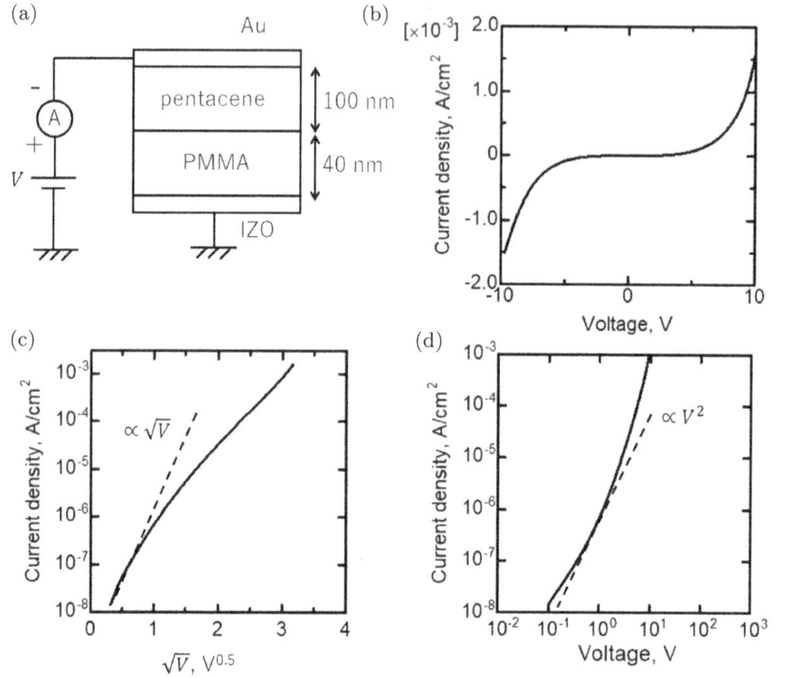

Figure 7.8. I–V measurement of MIS diode. (a) IZO/PMMA/pentacene/Au diode structure. (b) I–V characteristics of the diode. (c) and (d) are conventional $\ln I - \sqrt{V}$ plot and $\ln I - \ln V$ plot ($V > 0$) for Schottky and SCLC analysis. Dotted lines represent a fitting by Schottky and SCLC conduction mechanisms, but these are unsatisfactory to fit the whole region of I–V [34].

of an organic double-layer diode with a MIS structure of indium zinc oxide (IZO)/polymethylmethacrylate (PMMA)/pentacene/Au diode has been successfully clarified. Figure 7.8 shows a structure of IZO/PMMA/pentacene/Au diode and I–V characteristics of the diode. Here the pentacene layer is used as semiconductor layer, and PMMA layer is used as insulator. As shown in Fig. 7.8(c)(d), the conventional Schottky and SCLC analyses of I–V characteristics is unsatisfactory to fit the whole region of I–V. The effect of space charge field originating from insulator–semiconductor interface charge accumulation is one of the candidates that make the analysis more complex. EFISHG is available to probe the electric fields of insulator and semiconductor one-by-one to overcome this situation.

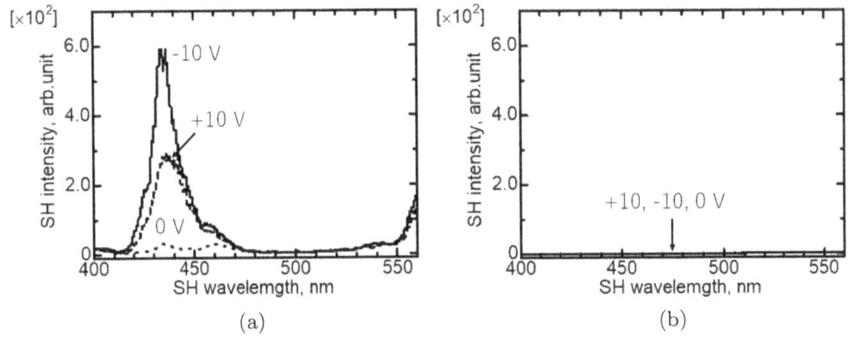

Figure 7.9. EFISHG spectrum of (a) pentacene and (b) PMMA [34].

Figure 7.9 shows EFISHG spectrum of pentacene and PMMA. By scanning probe laser wavelength from 1300 nm to 800 nm, SHG intensity is recorded in the wavelength region from 650 nm to 400 nm under external voltage application. Pentacene actively emits EFISHG but PMMA is very silent in this wavelength region. Accordingly, by probing EFISHG of pentacene/PMMA diode, electric field in the pentacene layer $\bar{E}$ is selectively probed. Electric field in PMMA layer $\bar{E}'$ is calculated using $\bar{E}$ with the relation $V = \bar{E}d_1 + \bar{E}'d_2$ (V: external voltage, d_1: pentacene thickness, d_2: PMMA thickness). The laser wavelength is set at 860 nm and SHG wavelength is selected at 430 nm, to probe electric field $\bar{E}$ in the pentacene layer of pentacene/PMMA diode.

Figure 7.10 is EFISHG of the diode, measured under sweeping voltage in the region of $-10\,\mathrm{V} < \mathrm{V} < +10\,\mathrm{V}$ [33]. The voltage is applied to Au electrode in reference to IZO electrode grounded. One of the two EFISHG curves is measured under pulse voltage application with a delay time $t_d = 300\,\mathrm{ns}$ in the time-resolved EFISHG measurement. This delay time t_d was chosen so that the electrode charging completes at t_d but interfacial charging does not initiate. This EFISHG curve reflects the electric field in the absence of interfacial charge Q_s. That is, $\bar{E}_e(V)$ of Eq. (7.22) is probed. The EFISHG intensity is calibrated to electric field strength based on calculation of $\bar{E}_e(V)$ in the way as described in Appendix F (ii).

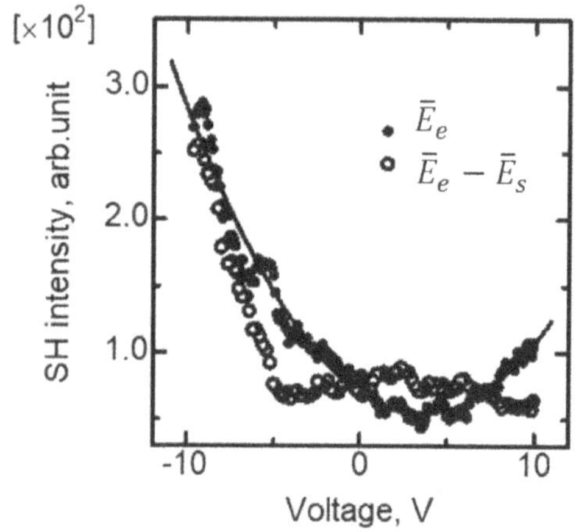

Figure 7.10. EFISHG measurement of Au/pentacene/PMMA/IZO diode. SHG in the pentacene layer is selectively probed at laser wavelength 860 nm and SHG wavelength at 430 nm. The two SHG curves are reflecting electric field with and without charge Q_s, *i.e.*, $\bar{E}_e$ and $\bar{E}_e - \bar{E}_s$ [34].

The other EFISHG curve in Fig. 7.10 is SHG measured at $t_a = 1\,\mathrm{s}$ after d.c. voltage application V. The time t_a is chosen so that SHG probes electric field in steady state of diode where conduction current I is flowing through the diode, after completion of charge accumulation Q_s. This EFISHG curve reflects electric field in pentacene layer in steady state. That is, the measured electric field is $\bar{E} = \bar{E}_e(V) - \bar{E}_s(Q_s)$ in Eq. (7.22). The accumulation of interfacial charge Q_s is suggested by non-zero $\bar{E}_s(Q_s)$. The two SHG curves in Fig. 7.10 do not coincide, indicating non-zero $\bar{E}_s(Q_s)$. SHG measurement in steady state evidently shows that the electric field in pentacene layer and PMMA layer of MIS diode is deformed by the presence of charge accumulation in the device. The electrical conduction in the layers is governed by $\bar{E} = \bar{E}_e(V) - \bar{E}_s(Q_s)$, not governed by an average external field $\bar{E}_e(V)$. The current–voltage (I–V) characteristics in Fig. 7.8(b) is measured simultaneously with the EFISHG measurement under d.c. voltage application, and is analysed as a function of electric field in pentacene and PMMA. In Fig. 7.11(a), the electric field in pentacene $\bar{E}$ and electric field in PMMA $\bar{E}'$ from the EFISHG is plotted as a function of external

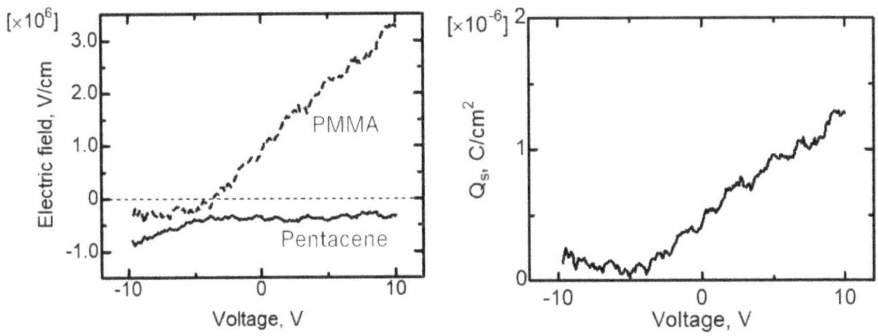

(a) (b)

Figure 7.11. EFISHG measurement of Au/pentacene/PMMA/IZO diode. (a) Electric field in pentacene $\bar{E}$ (solid curve) and in PMMA $\bar{E}'$ (dotted curve) in steady state. (b) Charge Q_s at the pentacene/PMMA interface [34].

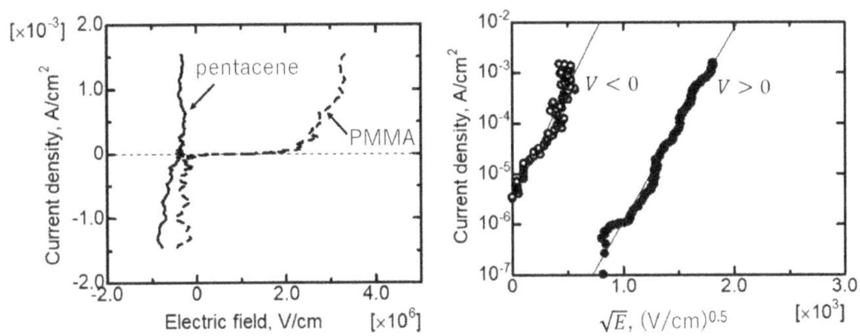

(a) (b)

Figure 7.12. (a) Current–electric field (I–E) characteristics of Au/pentacene/PMMA/IZO diode. (b) Schottky plot ($\ln I - \sqrt{E}$) of PMMA layer by using electric field E' of PMMA layer in the pentacene/PMMA double-layer [34].

voltage V. Figure 7.11(b) shows a plot of interfacial charge Q_s at the pentacene/PMMA interface calculated by using Eq. (7.22).

To analyse conduction mechanism of pentacene and PMMA layer, installed in the diode, current I is plotted as a function of electric field $\bar{E}$ and $\bar{E}'$, *i.e.*, I–E characteristics$I$–E characteristics. Figure 7.12(a) is I–E characteristics of pentacene and PMMA layer installed in the MIS diode. It is clear that the conduction mechanisms of pentacene and PMMA are different. Pentacene shows ohmic behavior and current increases with increasing electric field with conductivity $\sigma = 3.0 \times 10^{-7} \sim 3.0 \times 10^{-6}\,\text{S/m}$. PMMA shows injection-type behaviour, where I–E characteristics is no longer linear function of electric field but current increases nonlinearly. Figure 7.12(b) is Schottky plot of I–E characteristics of PMMA layer. A fairly good fitting to Schottky plot allows one to analyse conduction mechanism of PMMA layer in terms of Schottky injection current. By fitting Eq. (7.27) to I–E characteristics, for hole injection, relative dielectric constant is $\varepsilon_r = 3.4$, and barrier height is $\phi = 1.0\,\text{eV}$, for electron injection, relative dielectric constant is $\varepsilon_r = 4.5$, and barrier height is $\phi = 1.2\,\text{eV}$. Note that in the Schottky current mechanism, ε_r and ϕ reflect dielectric constant and barrier height that are responsible to a single particle injected at the interface.

As mentioned above, we can find a way to analyse the details of the I–V behavior of MIS diode by using the EFISHG measurement in combination with the I–V measurement. MIS with different organic materials shows different I–V curves. These rectifying behaviour of MIS diodes are well analyzed by using I–V measurement in combination with EFISHG measurement [34–36]. Hysteresis behaviour in the I–V curve of Au/α-NPD/PMMA/IZO diode is successfully analyzed by using EFISHG measurement coupled with I–V measurement [37]. Stress-biasing effect and other behaviors of 6,13–Bis (triisopropylsilylethynyl)pentacene (TIPS–pentacene)/polyimide MIS diodes are studied and modelled by using EFISHG results [38–40].

b. Transient state

The time-resolved EFISHG measurement allows us to measure the transient electric field $\bar{E}(t)$ and $\bar{E}'(t)$ with time. That means we can study the changing of electric field distribution in the diode in the course of carrier transport in MIS device [41–45]. Figure 7.13(a) illustrates a MIS diode with indium tin oxide (ITO)/polyimide/pentacene/Au structure [46]. Pentacene accepts hole and electron injection under voltage application. In the MIS diode, pentacene is expected to serve as a conductor after hole (or electron) injection from Au electrode. Polyimide is assumed to serve as an insulator that blocks carrier injection from ITO electrode.

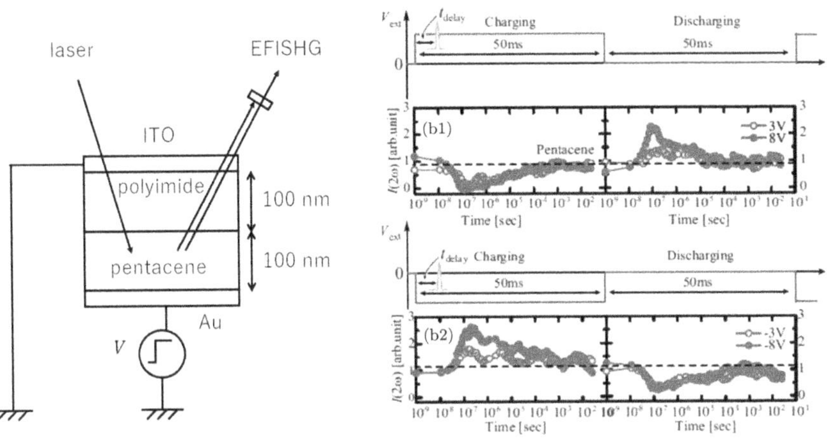

(a) (b)

Figure 7.13. Time-resolved EFISHG measurement of MIS device. (a) Device structure of ITO/polyimide/pentacene/Au MIS diode. (b) Time-resolved EFISHG measurement of the MIS diode. By setting laser wavelength at 860 nm (SHG wavelength 430 nm), changing of electric field $\bar{E}(t)$ in pentacene layer is selectively probed. Reprinted from L. Zhang, D. Taguchi, J. Li, T. Manaka, M. Iwamoto, *J. Appl. Phys.*, Vol. 108, 093707, 2010, with the permission of AIP Publishing [46].

But these are merely a speculation. We need to know actual carrier behavior in MIS diodes. A transient behavior in MIS diode with single carrier injection, *i.e.*, hole or electron injection, has been modeled using Eq. (7.23). A transient electric field in pentacene $\bar{E}(t)$ and in polyimide $\bar{E}'(t)$ is coming from electrode charging $\bar{E}_e$ with time constant RC of external circuit and Maxwell–Wagner charging $\bar{E}_s$ with time constant $\tau = (C_1 + C_2)/(G_1 + G_2)$ as follows.

$$\begin{aligned} E(t) &= \bar{E}_e(1 - e^{-\frac{t}{RC}}) - \bar{E}_s(1 - e^{-\frac{t}{\tau}}) \\ E'(t) &= \bar{E}'_e(1 - e^{-\frac{t}{RC}}) + \bar{E}'_s(1 - e^{-\frac{t}{\tau}}) \end{aligned} \tag{7.28}$$

Assuming that the polyimide layer totally blocks carrier injection from ITO electrode, Eq. (7.28) is calculated with $Q_s = C_2 V$ as

$$\begin{aligned} &E(t) = \bar{E}_e(1 - e^{-\frac{t}{RC}}) - \bar{E}_s(1 - e^{-\frac{t}{\tau}}) \\ &\text{with} \quad \bar{E}_e = \bar{E}_s = \frac{1}{L_1}\frac{C_2}{C_1 + C_2}V \end{aligned} \tag{7.29}$$

$$E'(t) = \bar{E}'_e(1 - e^{-\frac{t}{RC}}) + \bar{E}'_s(1 - e^{-\frac{t}{\tau}}) \quad \text{with}$$
$$\bar{E}'_e = \frac{1}{L_2}\frac{C_1}{C_1 + C_2}V \quad \text{and} \quad \bar{E}'_s = -\frac{1}{L_2}\frac{C_2}{C_1 + C_2}V$$

Upon application of positive voltage to Au bottom electrode, positive charges are induced on Au electrode, meanwhile negative charges are induced on ITO electrode grounded. The presence of induced electrode charges is an origin of electric field transient in the first term $\bar{E}_e$ of Eq. (7.29), and this electric field is formed in pentacene layer with time constant RC. The induced charge on electrodes is followed by carrier injection, transport and accumulation processes. Positive charges on Au electrode are injected into pentacene layer, and they travel across the layer and accumulate at the pentacene/polyimide interface, according to the Maxwell–Wagner effect. The accumulated charge is a source of space charge field, and generates the second term of the transient electric field $\bar{E}_s$ expressed in Eq. (7.29). If no carriers cross polyimide layer, the electric field $\bar{E}_e$ and $\bar{E}_s$ in the first and second transient, respectively, should be the same (see Eq. (7.29)). Consequently, in steady state at $t \gg RC$ and $t \gg \tau$, electric field is zero in the pentacene layer. That is, electric field $E(t)$ in pentacene layer increases and decreases in a two-step manner, corresponding to the two different transients due to electrode charging and Maxwell–Wagner charging, respectively. Their time-constants are given by RC and τ based on the Maxwell–Wagner effect model. For electron injection from Au to pentacene layer, similar two-step transient is expected. These behaviours were experimentally confirmed by the time-resolved EFISHG measurement.

Figure 7.13(b) shows the result of time-resolved EFISHG measurement. The wavelength of probe laser beam was set at 860 nm (SHG wavelength 430 nm) to selectively measure the electric field of pentacene layer. No SHG is generated from the polyimide layer at this wavelength. The laser beam is incident on MIS device from ITO electrode side, where transparency of ITO electrodes allows probe-laser light to transmit into pentacene layer. SHG light is coming out through the transparent ITO electrode. The incident laser angle was set at 45° from substrate surface normal direction. The p-polarized laser beam was used as a probe and p-polarized SHG was measured. Combination of oblique laser incidence and p-polarization allows one to efficiently probe electrostatic field of sandwiched-type device such as MIS diodes, where the electrostatic field is mainly formed in the direction along the layer thickness direction.

Result showed the two-step transient behaviour in the way as expected from Maxwell–Wagner effect model in Eq. (7.29). Upon negative voltage application to Au electrode, electrons are injected from Au electrode to pentacene, and accumulated at the pentacene/polyimide interface. After positive voltage application to Au electrode, with reference to ITO electrode grounded, holes are injected from Au to pentacene, and accumulated at the pentacene/polyimide interface. Accordingly, after the time constant τ of Maxwell–Wagner effect type charging, electric field in pentacene layer is zero. The analysis of time constant from transient electric field confirmed that the time constant RC of the first transient is independent of voltage as expected while that of second transient is reflecting carrier process in pentacene layer.

It is noteworthy that the discharging process is a reversal process of this MIS diode, as result shows in the right-hand-side of Fig. 7.13(b). That is, after hole (electron) accumulation completed at Time = 50 ms, the applied voltage is removed and the device is short-circuited. Positive and negative charges induced on the two electrodes recombine and disappear with the time constant RC. The accumulated holes (electrons) discharge through the pentacene layer, going back to Au electrode with the time constant $\tau = \tau_d$(=τ_c: time constant in charging process). Afterwards, the MIS diode returned to the initial state, *i.e.*, the state before voltage application. This simple reversal carrier injection and extraction process is a case of single carrier injection of MIS diodes. Example of non-reversal process is described in Sec. 7.7 for OLED and Sec. 7.9 for organic solar cell where interaction between holes and electrons is an origin of device function.

Figure 7.14 portrays another EFISHG measurement results that demonstrate electric field transients in a double-layer MIS diode.

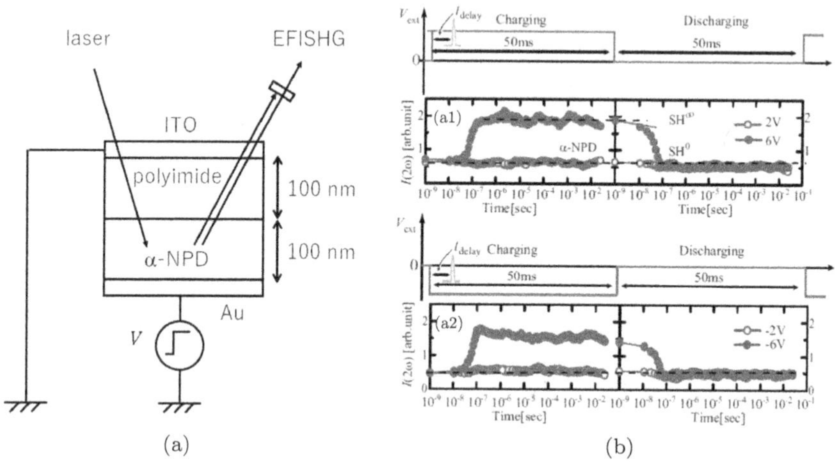

Figure 7.14. Time-resolved EFISHG measurement of MIS device. (a) Device structure of ITO/polyimide/α-NPD/Au MIS diode. (b) Time-resolved EFISHG measurement of the MIS diode. By setting laser wavelength at 820 nm (SHG wavelength 410 nm), electric field in α-NPD layer is selectively probed. Reprinted from L. Zhang, D. Taguchi, J. Li, T. Manaka, M. Iwamoto, *J. Appl. Phys.*, Vol. 108, 093707, 2010, with the permission of AIP Publishing [46].

The device structure is illustrated in Fig. 7.14(a), *i.e.*, ITO/ polyimide/*N,N′–di*–[(1–naphthyl)–*N,N′*–diphenyl]–(1,10–biphenyl)–4,4′–diamine (α–NPD)/Au structure. In SHG measurement, laser wavelength was set at 820 nm (SHG wavelength 410 nm) to selectively probe the electric field in α–NPD layer. The time-resolved EFISHG showed that hole and electron injection is minor in comparison with that of polyimide/pentacene MIS diodes. The SHG transients is a single transient reflecting electrode charging. That is, in Eq. (7.29), only the first term $\bar{E}_e$ contributed to the electric field transient, whereas the contribution of the second term $\bar{E}_s$ due to interfacial charge accumulation is minor $\bar{E}_s \sim 0$. The different carrier behaviours in polyimide/pentacene and polyimide/α–NPD MIS diodes suggest that the different carrier process proceeds in pentacene and α–NPD. The EFISHG measurements are useful to analyse these carrier process in MIS diode, and this is of help for modelling carrier behaviours in organic materials in it, such as *regio*-regular poly(3–hexylthiophene–2,5–diyl) [47], organic photoconductors [48], electron-blocking hole-transport layers [49–53], and others.

7.7 Double-layer EL diode

OLEDs have attracted much attention in electronics. Basically, OLEDs are injection-type devices with two different electrodes that are separated by an organic double-layer. The double layer consists of hole transport layer and electron transport layer. For example, in IZO/α–NPD/Alq3/Al diode, under positive voltage application to IZO electrode in reference to Al electrode, bright green light is coming out from the diode. α–NPD and Alq3 are used as hole and electron transport layers, respectively. Under the voltage application, holes are injected from IZO electrode, and they are transported through α–NPD layer. Meanwhile, electrons are coming into Alq3 from Al electrode, and they are conveyed across the Alq3 layer. The transported holes and electrons meet and recombine at the α–NPD/Alq3 interface, and bright EL light is emitted from Alq3 molecules.

As described in Sec. 7.4a, space charge Q_s accumulates at the interface in double-layer device, and forms space charge field $\bar{E}_s$ and $\bar{E}'_s$ in the layers. In the case of single-injection MIS diode, holes are only injected to and extracted from interface, in a reversal way. That is, a model of conventional Maxwell–Wagner effect is enough to describe carrier behavior in MIS diodes. By contrast, in OLED, double-injection, *i.e.*, simultaneous hole and electron injection is key in device operation.

SHG measurement provides a way to probe carrier process in double-layer OLED to model carrier processes in this kind of double-layer diode. Figure 7.15 shows an example of double layer OLED device. The top electrode is indium–zinc–oxide (IZO) which serves as hole injecting electrode, bottom electrode is Al serving as an electron injecting electrode. *N,N′–di–*[(1–naphthyl)–*N,N′*–diphenyl]–(1,10–biphenyl)–4,4′–diamine (α–NPD) is a hole transport layer and tris (8–hydroxy–quinolinato) aluminum (III) (Alq3) is an electron transport material. Alq3 adapts hole–electron recombination to emit light with electroluminescence peak at the wavelength 520 nm. To probe

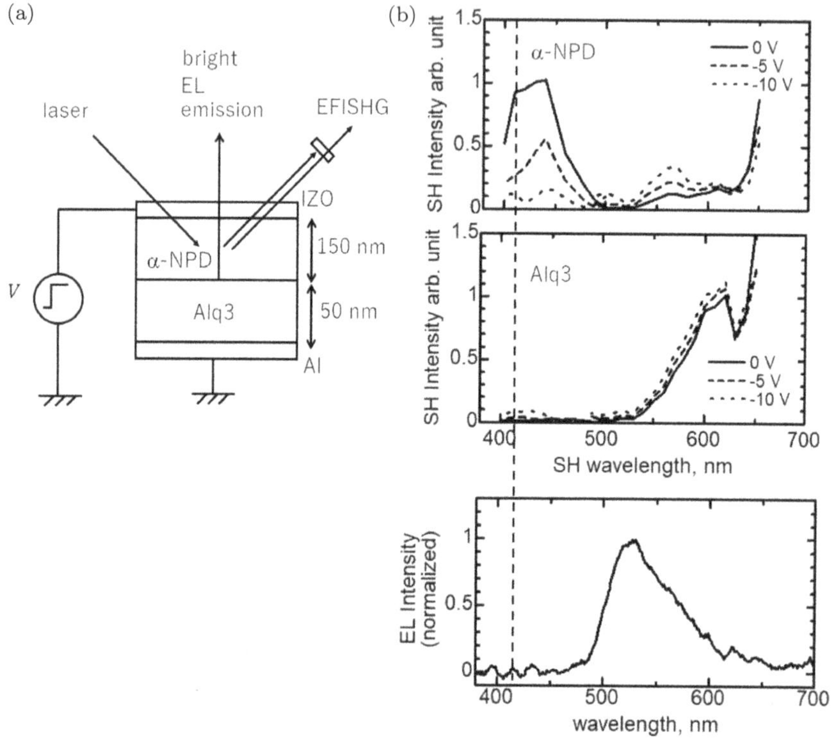

Figure 7.15. (a) Double-layer light-emitting diode device structure. (b) From top to bottom, EFISHG spectra of α–NPD and Alq3, and EL spectrum of α–NPD/Alq3 OLED [54].

the electric field in α–NPD layer selectively, laser wavelength is chosen at 820 nm, and EFISHG activated at the wavelength of 410 nm is measured. The spectrum of EFISHG of α–NPD and Alq3 layer, and EL spectrum of α–NPD/Alq3 OLED is displayed in Fig. 7.15(b) [54]. As shown in Fig. 7.15(b), bright EL emission is coming out from OLED with a peak at wavelength 520 nm. This light emission is due to hole–electron recombination in Alq3. But hole and electron motion in OLED leading to this recombination is unclear, and modelling carrier behavior is carried out based on time-resolved EFISHG measurement. EFISHG at laser wavelength of 820 nm (SHG wavelength 410 nm) is available (see Fig. 7.15(b)) to selectively probe electric field in α–NPD layer in α–NPD/Alq3 double layer OLED.

By setting the laser wavelength at 820 nm (SH wavelength 410 nm), the probe laser beam impinged onto device obliquely to substrate surface with p-polarization, to probe electrostatic field in sandwich-type device. Time-resolved EFISHG measurements are carried out to probe transient electric field and interfacial charge behaviors $E(t)$ and $Q_s(t)$ to model carrier behavior. Transient of electroluminescence intensity is recorded to know the relationship between carrier behaviour and recombination [55].

Figure 7.16 is the result of time-resolved SHG and transient EL measurements [56–58]. Upon application of the external voltage V, the electric field in α–NPD layer $E(t)$ is obtained from Eq. (7.23),

$$E(t) = \bar{E}_e(1 - e^{-\frac{t}{RC}}) - \bar{E}_s(1 - e^{-\frac{t}{\tau}}) + E_b$$
$$\text{with} \quad \bar{E}_e(V) = \frac{1}{L_1}\frac{C_2}{C_1 + C_2}V \quad \text{and} \quad \bar{E}_s(Q_s) = \frac{1}{L_1}\frac{Q_s}{C_1 + C_2}. \tag{7.30}$$

C_1 and C_2 are the capacitance of the α–NPD layer and Alq3 layer, respectively, L_1 is the thickness of the α–NPD layer. In Eq. (7.30), E_b is internal field due to the work function difference between the two dissimilar electrodes, *etc.*, which corresponds to other terms in Eq. (7.4). Q_s is the interface charge accumulated in steady state owing to the Maxwell–Wagner effect. The EFISHG is given by Eq. (7.3) with $E(0) = E(t)$ of Eq. (7.30). Under application of a square wave voltage with an amplitude of $V = +5\,\text{V}$ on IZO electrode, with reference to Al electrode grounded, time-resolved EFISHG measurement and transient electroluminescence measurement have been carried out simultaneously. Figure 7.16(a)–7.16(d), respectively, show the SHG intensity $I(2\omega)$, electric field $E(t)$, interfacial charge $Q_s(t)$, and EL intensity. The electric field in α–NPD, $E(t)$, is calibrated with reference SHG measurement in a way as described in Appendix F (ii).

In Fig. 7.16, regions 1–3 and 4–6, respectively, correspond to charging and discharging processes, as displayed on top of the figure. Regions 1–3, origin of "Time" is at rising edge of square wave voltage. Regions 4–6, origin of "Time" is at the beginning of falling edge of square wave voltage. In region 1, $E(t)$ increases with a

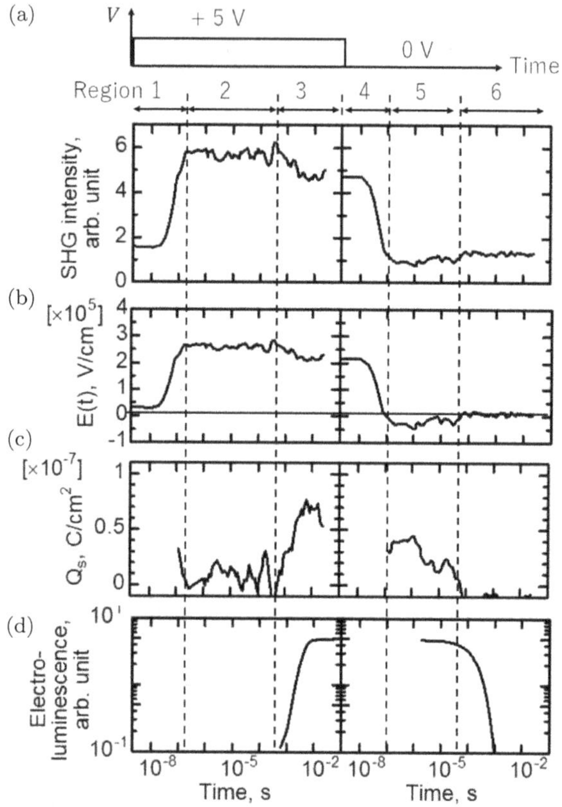

Figure 7.16. Time-resolved EFISHG measurement of double-layer (α–NPD/Alq3) EL device. (a) SHG intensity, (b) electric field in α–NPD layer, (c) interfacial charge Q_s at α–NPD/Alq3 interface, and (d) transient electroluminescence (EL) under square voltage application. In EFISHG measurement, laser wavelength is 820 nm and SHG wavelength is 410 nm. This setup allows one to selectively probe electric field in α–NPD layer only [56, 57].

response time of $RC = 7 \times 10^{-8}$ s. This value agrees well with a calculated value $RC = 7.7 \times 10^{-8}$ s with a series capacitance of $C = C_1C_2/(C_1 + C_2) = 4.5 \times 10^{-10}$ F comprised of α–NPD, C_1, and Alq3, C_2, layers and $R = 170\,\Omega$ of lead resistance connecting IZO and Al electrode. In this OLED, $RC \ll \tau$, and interfacial charging is minor at the end of region 1. The $E(t)$ at the end of region 1, *i.e.*, $E(t) = \bar{E}_e$, is 2.7×10^5 V/cm, and corresponds to the external electric field in α–NPD layer ($\bar{E}_e$ in Eq. (7.30)). Over the entire region 1, EL enhancement was not observed. This result shows that region 1 reflects a process of electrode charging by external voltage V. Under voltage application to OLED, positive charges are induced on IZO electrode, whereas negative charges are induced on Al electrode. This is the carrier process in region 1. In region 2, the SHG intensity $I(2\omega)$ and $E(t)$ are nearly constant, and EL enhancement is not observed. In this region, the amount of positive and negative charges induced on electrodes are kept constant, meanwhile positive and negative charges are injected into α–NPD and Alq3 from electrodes. In region 3, the electric field $E(t)$ decays with a relaxation time of 2×10^{-3} s, and finally saturates. This process evidently indicates that positive charge Q_s is accumulated at the α–NPD/Alq3 interface. The accumulated positive charge $Q_s > 0$ produces electric field $\bar{E}_s(>0)$ pointing from interfacial charge to IZO electrode in α–NPD layer. As Eq. (7.30) shows, this space charge field $\bar{E}_s(>0)$ cancels external field $\bar{E}_e$ formed in process 1. Consequently, the electric field in α–NPD layer decreases owing to positive charge accumulation $Q_s > 0$. Note that this charge accumulation is due to the Maxwell–Wagner effect, and it is accompanied by the EL emission. At the end of region 3, the electric field is $E(t) = 2.2 \times 10^5$ V/cm and the accumulated charge density is $Q_s = 6.8 \times 10^{-8}$ C/cm^2. The relaxation time corresponds to the Maxwell–Wagner relaxation time of a double layer system $\tau_c = 2 \times 10^{-3}$ s. After the hole accumulation at the interface, the electric field in Alq3 layer is increased, and electron injection is enhanced. Accordingly, EL intensity is enhanced after the hole accumulation at the interface. This is the carrier behaviour in region 3, which is concluded based on the results of EFISHG and EL transient measurement.

Regions 4–6 are discharging processes. At the beginning of region 4, external voltage $V = +5$ V is removed and returned to zero. Accordingly, in region 4, electric field $E(t)$ decreases with the time constant $RC = 7 \times 10^{-8}$ s, in the way same as in region 1. This result indicates that the charge induced on electrodes at the end of region 1 discharges through the external circuit. This electrode charging and discharging behavior is the same as that is seen in MIS diode (Fig. 7.13 and Fig. 7.14). On the other hand, $Q_s(>0)$ remains at the α–NPD/Alq3 interface, which gives rise to an electric field $E(t) = -\bar{E}_s$ in α–NPD layer and $E'(t) = \bar{E}'_s$ in Alq3 layer. At the end of region 4, the electric field saturated at $E(t) = -3.1 \times 10^4$ V/cm, suggesting that most of charge Q_s remains at the interface. The enhanced EL intensity is nearly constant over the entire region 4. In region 5, the electric field $E(t)$ in the α–NPD layer decreases with time constant $\tau_d = 5 \times 10^{-6}$ s and finally reaches $E(t) = 0$, while the enhanced EL intensity does not change, keeping bright emission. In region 6, $E(t) = 0$ and EL gradually decays with relaxation time $t_d = 3 \times 10^{-4}$ s.

Results showed that the charging and discharging processes are non-reversible in OLED. The time-constant RC is the same in charging and discharging processes, but the Maxwell–Wagner charging and discharging at the interface is non-reversible $\tau_c \neq \tau_d$. This carrier behavior is different from charging and discharging behavior in MIS diodes. That is, the relaxation time τ in charging and discharging is the same in MIS diodes where single carrier injection governed carrier process in the diode, but in OLED with double-injection, relaxation time in charging is $\tau_c = 2 \times 10^{-3}$ s and in discharging $\tau_d = 5 \times 10^{-6}$ s. The non-reversal behavior is mainly due to the difference in the contribution of electric field arising from accumulated charges at the interface. In charging process, accumulated positive charges disturb hole injection from IZO electrodes to α–NPD layer, and balance hole injection current crossing α–NPD layer with electron injection current crossing Alq3 layer. On the other hand, in discharging process, accumulated positive charge $Q_s > 0$ keeps assisting electron injection from Al electrode via Alq3 layer.

This model well accounts for the appearance of two-EL modes generated in this sandwich-type EL devices [59–61]. As mentioned above, we can monitor dynamical carrier behaviours in charging and discharging processes in EL devices, by means of EFISHG. To further clarify the mechanism of EL enhancement, experiments have been carried out. Figures 7.17 and 7.18 show, respectively, EL and EFISHG measurements under application of a large AC square voltage to the ITO/α–NPD/Alq3/Al OLEDs. Two kinds of AC square

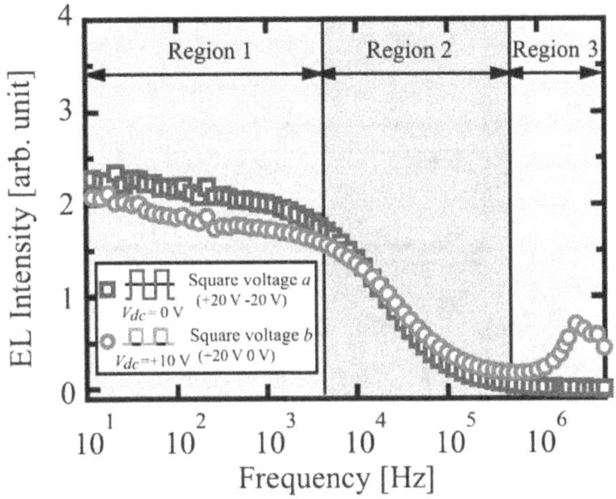

Figure 7.17. Frequency response of EL intensity from double-layer (α–NPD/Alq3) EL device where large AC square voltage application with no d.c. component (square voltage *a*) and with d.c. component (square voltage *b*) is utilized. Reprinted from A. Sadakata, K. Osada, D. Taguchi, T. Yamamoto, M. Fukuzawa, T. Manaka, M. Iwamoto, *J. Appl. Phys.*, Vol.112, 083723, 2012, with the permission of AIP Publishing [59].

voltage waveforms are chosen, one with no d.c. components (square voltage *a*), and the other with d.c. components (square voltage *b*). As Fig. 7.17 shows, EL is generated in accordance with the applied AC square voltage, where the EL intensity is dependent on the frequency of square wave voltage and also the waveform. In more detail, the EL intensity gradually decays with the increase of the frequency of the applied AC square voltage, and reaches a minimum at the frequency corresponding to the carrier transit time of the Alq3 layer. Interestingly, the EL intensity again increases by the application of further higher frequency AC square voltages, relying on the d.c. component of the AC square voltages [59]. Results suggest that the presence of two EL modes, where the space charge field caused by the interface charge makes a significant contribution. That is, in the low frequency region, accumulated holes at the interface suppress hole injection, and lead to the decrease of the EL intensity activated by the recombination of holes and electrons injected from opposite electrodes. On the other hand, in the high frequency region, the accumulated holes

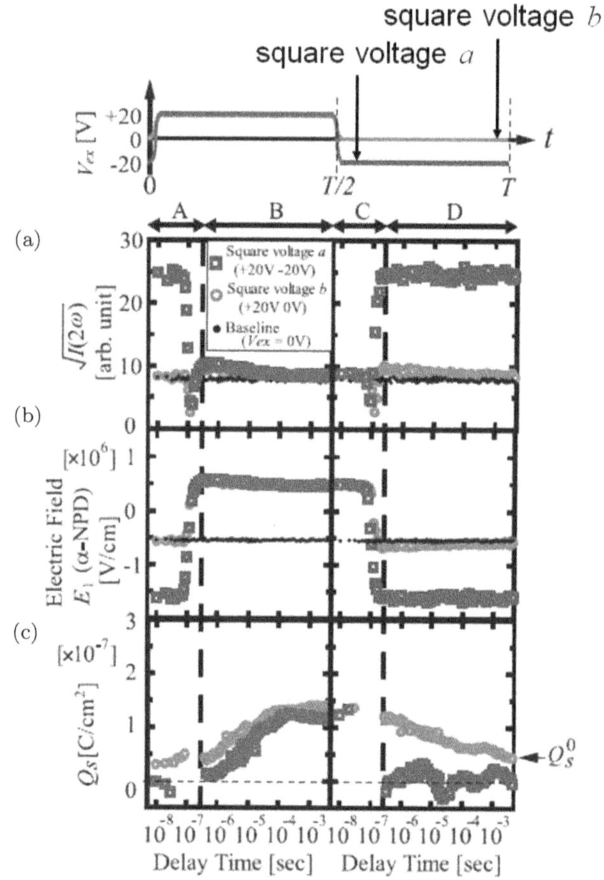

Figure 7.18. Time-resolved EFISHG measurement of double-layer (α–NPD/ Alq3) EL diode under large AC square voltage a and b. (a) The square-root of SHG intensity $\sqrt{I(2\omega)}$, (b) electric field E_1 $(= E(t))$ in α–NPD layer, and (c) interfacial charge Q_s at the α–NPD/Alq3 interface. Reprinted from A. Sadakata, K. Osada, D. Taguchi, T. Yamamoto, M. Fukuzawa, T. Manaka, M. Iwamoto, *J. Appl. Phys.*, Vol. 112, 083723, 2012, with the permission of AIP Publishing [59].

assist electron injection and result in the increase of EL intensity activated by the recombination of the interfacial accumulated holes and injected electrons from Al electrode.

However, this is merely speculation, and we need to make clear the presence of this space charge field caused from accumulated interface charges and its contribution. For this purpose, the EFISHG measurement is very useful. Figure 7.18 shows the EFISHG measurements in the low frequency region, where two kinds of AC square voltage waveforms a and b are chosen, *i.e.*, one with no d.c. component (square voltage a), the other with a d.c. component (square voltage b). From the top to bottom, the applied external voltage, V, generated EFISHG intensity $I(2\omega)$, electric field $E(t)$ and accumulated interface charge Q_s are plotted. Here the generated EFISHG intensity $I(2\omega)$ is plotted by $\sqrt{I(2\omega)}$ with consideration of the relation between the electric field and the generated EFISHG (see Eq. (7.3)). Q_s is calculated in the same way as Fig. 7.16.

Under the AC voltage application with no d.c. component (square voltage a), in region A, Q_s is zero, in prior to voltage application $V = +20\,\text{V}$. In region B, external voltage $V = +20\,\text{V}$ induces charges on the ITO and Al electrodes with a circuit response time $RC \sim 10^{-7}\,\text{s}$, and forms an electric field in the α–NPD layer as high as $E(t)(= \bar{E}_e) = 5.6 \times 10^5\,\text{V/cm}$. After that, holes are injected from the ITO to α–NPD layer and electrons are injected from the Al to Alq3 layer for radiative EL recombination. The holes smoothly transport in the α–NPD layer (hole mobility: $\mu_h \sim 10^{-4}\,\text{cm}^2/\text{Vs}$) but electrons go through the Alq3 layer slowly (electron mobility: $\mu_e \sim 10^{-5}\,\text{cm}^2/\text{Vs}$), due to the imbalance of hole and electron transport in the layers. As the result, holes accumulate excessively at the α–NPD/Alq3 interface, and the electric field in the α–NPD layer decreases. The EFISHG shows that the electric field actually decreases from $E(t) = 5.6 \times 10^5\,\text{V/cm}$ to $E(t) = 4.8 \times 10^5\,\text{V/cm}$ at $t_d = 10^{-5}\,\text{s}$ due to excess hole accumulation $Q_s = 6.5 \times 10^{-8}\,\text{C/cm}^2$.

On the other hand, under AC voltage application with a d.c. component (square voltage b), electrode charges are induced on the ITO and Al electrodes and holes accumulated excessively at the α–NPD/Alq3 interface, in a way similar to the carrier processes under the square voltage a. However, excess holes stay at the α–NPD/Alq3 interface in region A and region D by charge density $Q_s = Q_s^0 = 3.0 \times 10^{-8}\,\text{C/cm}$, as evidenced in the EFISHG result (see Fig. 7.18(c)). These results support the prediction that there are accumulated holes at the interface, and these charges suppress hole injection in the low frequency region and lead to the decrease of the EL intensity. Using the EFISHG results, we can carry out a similar discussion on the presence of interface charge and its contribution, in terms of the EL in the high frequency region. Under large AC voltage application with no d.c. component (square voltage a), in discharging process in region D, interfacial charges totally discharge, and initial OLED state without charge accumulation $Q_s = 0$ is recovered. Under AC square wave form with d.c. component (square wave voltage b), non-zero Q_s remained at the interface after discharging process. As the non-zero Q_s forms space charge electric field in Alq3 layer, this field keeps assisting electron injection from Al to Alq3 layer, resulting in EL enhancement in high frequency region (Fig. 7.17, region 3). In this way, we can monitor dynamical carrier behaviors in charging and discharging processes in EL devices, by means of EFISHG, and model OLED behaviors successfully [62–65].

7.8 Double-layer memory devices

Ferroelectric polymers are widely used as memory materials, where the copolymer of vinylidene fluoride (VDF) and trifluoroethylene (TrFE), P(VDF–TrFE) are the most popular materials. Remanent polarization can be repeatedly turned-over by an external electric field [66–68] to make stable two-states alternately. In organic devices, the remanent polarization is utilized as a source of internal electric field and interfacial charge accumulation in OFET [69–72], organic photovoltaic cells (OPVs) [73, 74], and other devices [75]. As a result, carrier behaviors in the active semiconductor layer, *e.g.*, pentacene, have been deeply studied in terms of the turn-over of the spontaneous polarization of ferroelectric layers. Electrical measurements such as displacement current measurement (DCM) using a ramp voltage are available, but these are no longer sufficient to account for carrier behaviors in organic multilayer devices with ferroelectric layers. Electrical measurement is the most conventional way to study polarization reversal of ferroelectric polymer, but it is indirect as described in Sec. 6.1. EFISHG directly probes the electric field change induced in organic devices with a ferro-electric layer, and allows one to analyze and model carrier behaviors in organic memory devices.

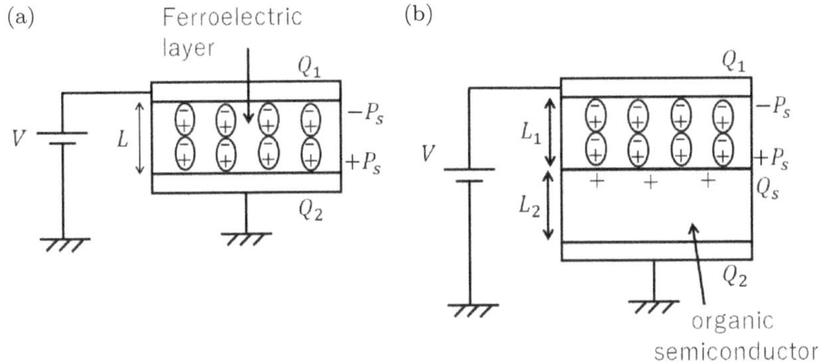

Figure 7.19. Ferroelectric memory device structure. (a) Metal–ferroelectric–metal (MFM), single layer device. (b) Metal–ferroelectric–organic semiconductor–metal (MFSM), double-layer device.

Figure 7.19 illustrates a single ferroelectric layer and double layer memory device with ferroelectric layer, where the layer is sandwiched between two identical electrodes [76, 77]. In ferroelectric layer, spontaneous polarization is modeled using polarization charge $\pm P_s$. With consideration of non-zero spontaneous polarization, $P_s \neq 0$, the electric field and charges in device are modeled. The internal electric field is freshly introduced due to the presence of spontaneous polarization P_s, and the electric field in single- and double-layer device in steady state is described as

$$\bar{E} = \frac{V}{L}, \tag{7.31}$$

in single layer device, and in double-layer device

$$\begin{aligned}
\bar{E} &= \bar{E}_e(V) - \bar{E}_s(Q_s, P_s) \\
\bar{E}' &= \bar{E}'_e(V) + \bar{E}'_s(Q_s, P_s) \\
\text{with} \quad \bar{E}_e(V) &= \frac{1}{L_1}\frac{C_2}{C_1 + C_2}V \\
\bar{E}_s(Q_s, P_s) &= \frac{1}{L_1}\frac{Q_s + P_s}{C_1 + C_2} \\
\bar{E}'_e(V) &= \frac{1}{L_2}\frac{C_1}{C_1 + C_2}V \\
\bar{E}'_s(Q_s, P_s) &= \frac{1}{L_2}\frac{Q_s + P_s}{C_1 + C_2}
\end{aligned} \tag{7.32}$$

For single layer MFM device (Eq. (7.31)), the electric field in the layer is the same as Eq. (7.16). By contrast, the electric field in double-layer MFSM device (Eq. (7.32)) is different from Eq. (7.22) due to the presence of the space charge fields $\frac{1}{L_1}\frac{P_s}{C_1+C_2}$ and $\frac{1}{L_2}\frac{P_s}{C_1+C_2}$ originating from the spontaneous polarization P_s at the ferroelectric–organic semiconductor interface.

The charges Q_1 and Q_2 induced on top and bottom electrodes are described as below:

(1) single layer device case

$$\begin{aligned} Q_1 &= CV + P_s \\ Q_2 &= -CV - P_s \end{aligned} \tag{7.33}$$

(2) double layer device case

$$\begin{aligned} Q_1 &= CV - C_1\frac{Q_s + P_s}{C_1 + C_2} + P_s \\ Q_2 &= -CV - C_2\frac{Q_s + P_s}{C_1 + C_2} \end{aligned} \tag{7.34}$$

with $C = \frac{C_1C_2}{C_1+C_2}$, where C_1 is capacitance of ferroelectric layer, and C_2 is capacitance of organic semiconductor layer. Note that charge conservation law is satisfied in Eq. (7.33) and Eq. (7.34), in the manner as $Q_1 + Q_2 + P_s - P_s = Q_1 + Q_2 = 0$ and $Q_1 + Q_2 + Q_s + P_s - P_s = Q_1 + Q_2 + Q_s = 0$ in MFM and MFSM device, respectively.

The turn-over of polarization undergoes when the electrostatic field in ferroelectric layer is over coercive field E_c. In MFM device, this condition is

$$\begin{aligned} \bar{E} &= \frac{V_c}{L} = E_c \\ V_c &= E_cL. \end{aligned} \tag{7.35}$$

External voltage V is a governor of turn-over of spontaneous polarization. Turn-over voltage $V = V_c$ is given as the product of coercive field E_c and layer thickness L. In MFSM device without interfacial charge, $Q_s = 0$, this condition is given by

$$\bar{E} = \frac{Q_1 - P_s}{\varepsilon_1} = \frac{1}{L_1}\frac{C_2}{C_1 + C_2}V_c - \frac{1}{L_1}\frac{P_s}{C_1 + C_2} = E_c \tag{7.36}$$

and is a function of external voltage V $(= V_c)$ and spontaneous polarization P_s.

In single layer device, internal field formed by spontaneous polarization is cancelled by the field due to induced charges on top and bottom electrode, which are facing to the spontaneous polarization. By contrast, internal field originating from P_s at the ferroelectric–organic semiconductor is active in the layer, in MSFM device. As a result, the turn-over voltage V_c is a function of E_c and P_s.

In MFSM device illustrated in Fig. 7.19(b), charge injection is allowed from the bottom electrode to organic layer. For example, in MFSM device with pentacene as an organic semiconductor layer, holes are injected from the bottom electrode and accumulated at the interface. The accumulated interfacial charge Q_s is Maxwell–Wagner charge, and makes contribution to electric field. In this case, the condition for turn-over of spontaneous polarization changes from Eq. (7.36) to

$$\begin{aligned} \bar{E} &= \frac{Q_1 - P_s}{\varepsilon_1} = \frac{1}{L_1}\frac{C_2}{C_1 + C_2}V_c - \frac{1}{L_1}\frac{Q_s + P_s}{C_1 + C_2} = E_c \\ V_c &= E_c L_1 \frac{C_1 + C_2}{C_2} + \frac{Q_s + P_s}{C_2}. \end{aligned} \tag{7.37}$$

In a special case where semiconductor layer is acting as a conductor, *i.e.*, $Q_s = -C_1 V_c - P_s$, turn-over voltage V_c in Eq. (7.37) is given in a simple form like Eq. (7.35) as

$$V_c = E_c L_1. \tag{7.38}$$

Clearly, the case of MFM layer is simple as in Eq. (7.35). DCM is very useful to investigate coercive field E_c and other ferroelectric characteristics. By contrast, in OFET, solar cells, and other organic devices, multilayer structure is utilized for the device performance.The turn-over happens depending on Q_s and P_s, as expressed in Eq. (7.37).

To investigate polarization reversal process in MFSM device, both charge Q_s and polarization P_s makes a significant contribution to electric field in the device, and to polarization reversal process. DCM indirectly probes these carrier and polarization behaviors as a superposition of displacement current induced in a course of charge motion and polarization motion. Accordingly, analysing DCM of MFM is simple but that of MFSM is rather complex. EFISHG allows one to probe charge behavior and polarization reversal process by the direct probing of electrostatic field induced by Q_s and P_s, providing a very powerful way to probe and model carrier behaviours in these devices.

Figure 7.20 shows a typical example of DCM measurements for metal–ferroelectric layer–metal (MFM) and metal–ferroelectric layer–semiconductor–metal (MFSM) (the inset of Fig. 7.20(b)) devices [76]. The MFM device has a structure of indium zinc oxide (IZO)/P(VDF–TrFE) (VDF:TrFE = 72: 28 mol. %)/Au and MFSM device has a structure of IZO/P(VDF–TrFE) (VDF:TrFE = 72: 28 mol. %)/pentacene/Au. P(VDF–TrFE) is a ferroelectric layer with a thickness 200 nm. Pentacene is utilized as an organic semiconductor layer with a thickness of 200 nm. In DCM and EFISHG measurement, external voltage is applied to IZO electrode in reference to Au grounded electrode. For the DCM measurement, a ramp voltage was applied. Figure 7.20(a) shows the results of the DCM for the single layer device, *i.e.*, MFM device. Two peaks (peaks A and B) are clearly observed at symmetric positions at voltages corresponding to the coercive electric field E_c of 0.6 MV/cm, and show no dependence on the sweeping frequencies of the ramp voltage. These results suggest that the appearance of peaks A and B are due to the turn-over of spontaneous polarization.

On the other hand, for MFSM devices, three peaks A, B, and C are clearly observed. The appearance of the third peak, peak C, at the non-symmetric position of peak A is a typical characteristic of the MFSM device. Results suggest a two-step polarization reversal of the ferro-electric layer in the pentacene/P(VDF–TrFE) double-layer device. However, this is merely speculation. For studying carrier behaviors with the turn-over of polarization, it is very helpful to

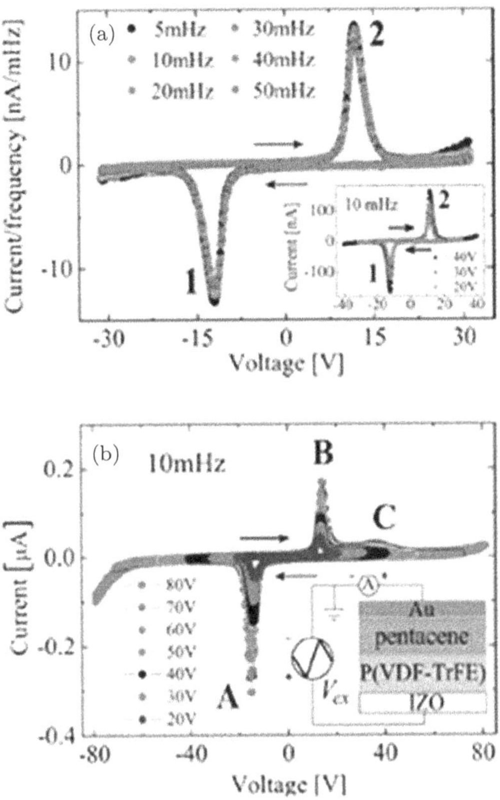

Figure 7.20. Displacement current measurement (DCM) of MFM and MFSM memory devices, where displacement current flowing through the device is measured under triangular voltage application. (a) MFM structure device (IZO/P(VDF–TrFE)/Au). (b) MFSM structure device (IZO/P(VDF–TrFE)/pentacene/Au. Reprinted from J. Li, D. Taguchi, W. OuYang, T. Manaka, M. Iwamoto, *Appl. Phys. Lett.*, Vol. 99, 063302, 2011, with the permission of AIP Publishing [76].

measure the electric field formed in pentacene and P(VDF–TrFE) layers, that are changed by Q_s and P_s as described in Eq. (7.32).

Figure 7.21 shows the result of EFISHG measurement, where the electric field across the pentacene layer is recorded [77]. Results clearly verify that the proposed two-step polarization reversal process by the two different hysteresis loops: one loop in Fig. 7.21 is corresponding to peak A and B, and the other loop corresponding to

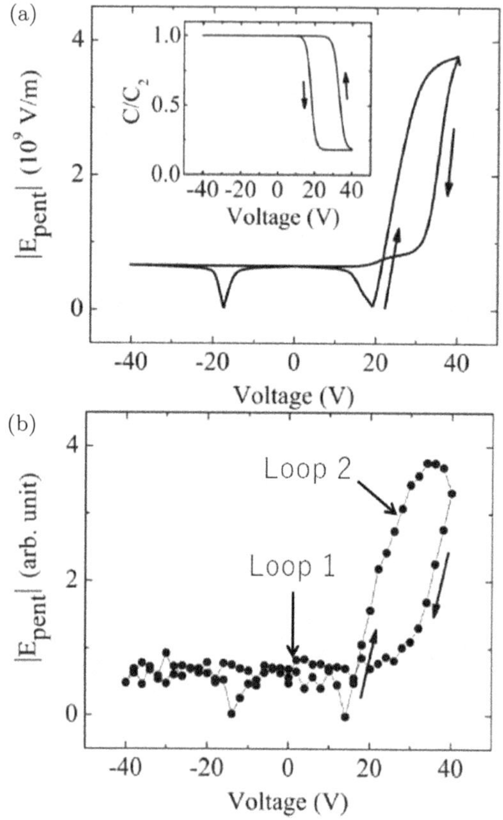

Figure 7.21. EFISHG measurement of MFSM memory device with IZO/P(VDF–TrFE)/pentacene/Au structure. (a) Electric field in pentacene layer of the MFSM memory device expected from conventional capacitance–voltage (C–V) measurement. (b) Electric field change directly probed by the EFISHG measurement. Loop 1 is the first-step polarization reversal corresponding to peak A and peak B in the DCM (see Fig. 7.20), while loop 2 is the second-step polarization reversal of peak C. Reprinted from J. Li, M. Weis, D. Taguchi, T. Manaka, M. Iwamoto, *J. Appl. Phys.*, Vol.111, 023706, 2012, with the permission of AIP Publishing [77].

peak C. Results of EFISHG suggest that the interaction of interfacial charge Q_s and ferroelectric polarization P_s is responsible for this two-step process.

It is noteworthy that the situation of the two-step turn-over strongly depends on the organic semiconductor materials. The charge accumulation Q_s is different for different set of double-layer materials as MIS diodes and OLEDs. Different Q_s gives an additional polarization reversal like peak C. Interaction between the ferroelectric layer and semiconductor layer is quite different depending on the contacting material. Accordingly, a new DCM peak like peak C can appear. For example, for P(VDF–TrFE)/C60/Au (or Al) double layer devices, only two peaks appear, but the peak position appears at quite different voltages from those of the IZO/P(VDF–TrFE)/pentacene/Au double layer device, owing to the potential drop across the C60 layers [78]. Using the EFISHG measurement is an effective way for clarifying the contribution of organic semiconductor layers in MFSM-type memory devices [79–81].

It is noteworthy that there are two origins for memory function from dielectric physics point of view. One is spontaneous polarization due to aligned dipoles, as described here for memory devices with ferroelectric layer. The other one is electronic charge displacement due to trapped electrons and holes. EFISHG also provides a powerful way for probing and modelling carrier behaviors in electronic charge-based memory devices, using nanoparticles, natural synthesized graphene as a trapping center [82, 83], *etc.*

As Fig. 7.20 and Fig. 7.21 show, external turn-over voltage V_C depends on Maxwell–Wagner charge Q_s accumulated at the interface. The accumulated charge Q_s is governed by electrode material if the injection process dominates device performance. For example, Al is low work function material. If Al is used as a hole injection electrode to pentacene OFET, carrier injection is limited due to high injection energy barrier from Al to pentacene. In this case, accumulation of holes Q_s at pentacene/gate insulator interface proceeds very slowly, and turn-over voltage V_c depends on a voltage sweeping speed in DCM.

Figure 7.22 illustrates DCM of Au/pentacene/poly(vinylidene fluoride–tetrafluoroethylene) (P(VDF–TeFE))/Al and Al/pentacene/P(VDF–TeFE)/Al memory OFET [69]. The P(VDF–TeFE) is a gate insulator and ferroelectric layer. Turn-over of P(VDF–TeFE) governs switching on-state and off-state of the memory OFETs. In a DCM that measures displacement current flowing through the

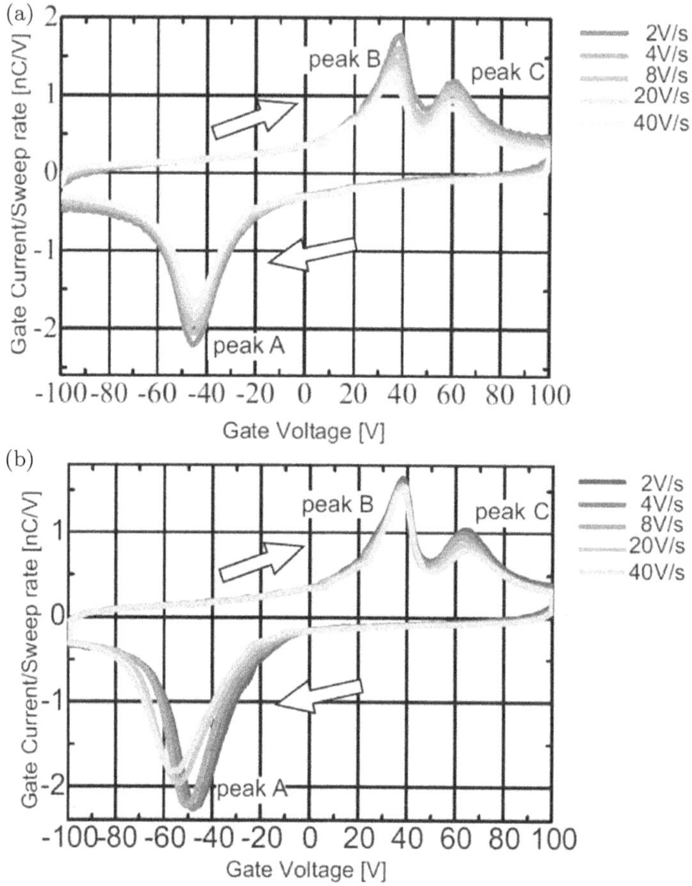

Figure 7.22. Memory OFET where carrier injection process governs polarization reversal. (a) In Au/pentacene/P(VDF–TeFE)/Al device, hole injection from Au electrode is smooth and turn-over voltage is independent of voltage sweeping rate in DCM. (b) In Al/pentacene/P(VDF–TeFE)/Al device, hole injection from Al electrode is limited and turn-over voltage (peak A) shifts to higher voltage due to insufficient hole injection at higher voltage sweeping rate in DCM. Reprinted from S. Yoshita, R. Tamura, D. Taguchi, M. Weis, E. Lim, T. Manaka, M. Iwamoto, *J. Appl. Phys.*, Vol. 106, 024505, 2009, with the permission of AIP Publishing [69].

gate electrode, a part of OFET around Source electrode is considered as a MFSM device. In DCM, triangular AC voltage is applied to the gate electrode in reference to Source electrode grounded. The hole injection barrier height from Au electrode to pentacene is low and that from Al electrode to pentacene is high. Accordingly, injection time for hole injection is very short in Au/pentacene/P(VDF–TeFE)/Al memory OFET while the injection time is very long in Al/pentacene/P(VDF–TeFE)/Al memory OFET. In Au/pentacene/P(VDF–TeFE)/Al memory OFET, due to smooth hole injection from electrode at peak A of DCM (see Fig. 7.22), Maxwell–Wagner charge Q_s increases and decreases on keeping the steady state. As the result, turn-over voltage V_c is independent of frequency of triangular AC voltage in the frequency region from 5×10^{-3} to 10^{-1} Hz. That is, hole injection time from Au electrode to pentacene is approximately zero, *i.e.*, $\tau = \tau_{MW} = 0\,\text{s}$ in Eq. (7.23). The turn-over voltage V_c is well described by using Eq. (7.38).

By contrast, in Al/pentacene/P(VDF–TeFE)/Al memory OFET, hole injection from Al electrode to pentacene is slow, owing to high injection barrier at Al–pentacene interface. Accordingly, turn-over voltage V_c shifts to higher voltage at higher frequency. In more detail, the hole injection time from Al to pentacene is $\tau = \tau_{MW} = 1.59\,\text{s}$. The time τ_{MW} is negligibly small at low frequency 5×10^{-3} Hz, and turn-over voltage for peak A of DCM is very similar to that in Au/pentacene/P(VDF–TeFE)/Al memory OFET. But the time τ_{MW} cannot be omitted at high frequency 10^{-1} Hz. Consequently, turn-over voltage V_c is given by Eq. (7.37) and not by Eq. (7.38). The results in Fig. 7.22 indicates the importance of analyzing and modelling MSFM devices, by taking account of carrier behavior Q_s. EFISHG measurement is available to directly probe carrier behavior Q_s for this purpose.

7.9 Organic solar cells

There are many kinds of organic solar cells (OSCs), *e.g.* double-layer type, bulk-heterojunction (BHJ) type, *etc.* [84, 85]. By using donor and acceptor material interfaces, hole–electron pairs are broken and separated into free holes and electrons. This is the origin of electromotive force for the power generation. To increase power conversion efficiency, the key is to find new materials and to develop new processing techniques that can enhance free carrier production in

OSCs [86]. Solar cell parameters such as open-circuit voltage, short-circuit current, and others are available as the guide for this purpose. However, probing and modelling of carrier behaviors in OSCs is also very important to know key material parameters and device parameters that are responsible for the enhancement, as shown in earlier sections. In this section, double-layer type solar cell case is analyzed, but this analysis can be extended to other type of organic solar cells.

For describing carrier behaviors in OSC, the basic model of two-layer organic device presented in Sec. 7.4b is modified. Free holes $+Q_p$ and free electrons $-Q_p$ are freshly considered in the model, to find a way to describe solar cell device function. The details are as below [87]. Figure 7.23 illustrates a two-layer organic solar cell consisting of two organic layers sandwiched between parallel electrodes. Electron donating material, *i.e.*, donor, and electron accepting material, *i.e.*, acceptor, are stacked and installed as the two layers of OSCs.

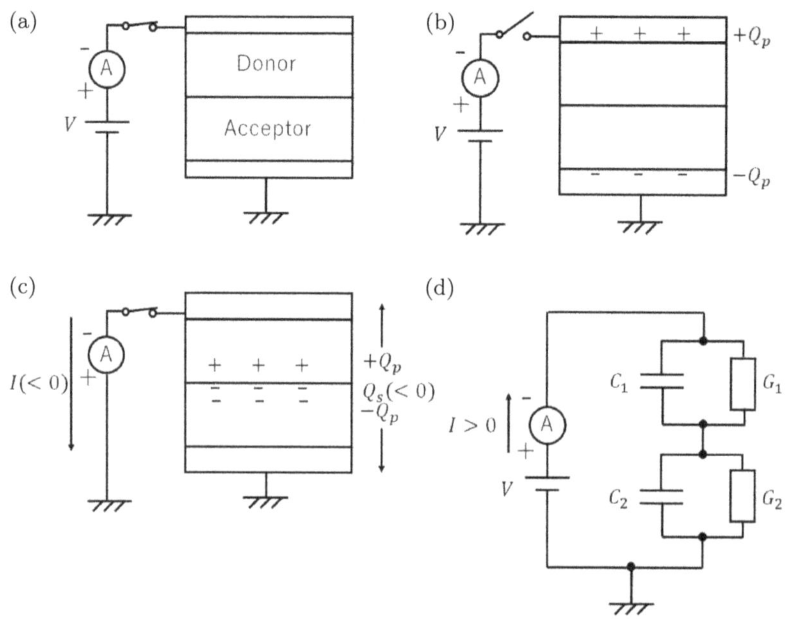

Figure 7.23. Double-layer organic solar cell (OSC). (a) I–V measurement arrangement of OSC. (b) Open-circuit condition. (c) Short-circuit condition. (d) Equivalent circuit model of double-layer OSC.

Under photoillumination, photons are optically absorbed in donor and acceptor, and excitons are created in the layers. Successively, the excitons diffuse to donor–acceptor interface. At the interface, HOMO and LUMO energy difference between donor and acceptor materials allows excitons to be separated to free holes and electrons. The created holes $+Q_p$ and electrons $-Q_p$, respectively, move through donor toward top electrode and through acceptor toward bottom electrode. Consequently, holes and electrons are separated to top and bottom electrode. These charges on electrode are extracted and make electrical work in the external circuit, connected to top and bottom electrode. It is instructive here to note that the concept of donor and acceptor described in this section is different from idea in conventional silicon solar cells. In organic solar cells, donor transfers electrons to acceptor meanwhile acceptor receives electrons from donor. After the electron transfer between donor and acceptor, free holes remain in donor and free electrons are in acceptor. By contrast, in silicon solar cells, donor provides electrons to Si while acceptor supplies holes to Si, forming n-type and p-type Si. After the electron transfer from p-type Si to n-type Si at p–n junction, free holes spread to p-type Si where acceptor is active, and free electrons goes to n-type Si where donor dominates.

In open-circuit condition, steady state current I flowing through the device is zero, $I = 0$. That is, the separated holes $+Q_p$ and electrons $-Q_p$, respectively, move to top and bottom electrode as shown in Fig. 7.23(b). In this situation, interfacial charge Q_s generated due to the Maxwell–Wagner effect is zero. That is, $Q_s = I(\tau_2 - \tau_1) = 0$ with $I = 0$. On the other hand, in short-circuit condition, where top and bottom electrodes are connected by a conducting wire as illustrated in Fig. 7.23(c), short-circuit current I flows though the two layer interface, thereby Maxwell–Wagner charge Q_s is accumulated in proportion to the difference of dielectric relaxation time τ_1 and τ_2 of donor and acceptor material, respectively. Under photoillumination, with the presence of interfacial charge Q_s, free holes and electrons $\pm Q_p$ are created at the donor–acceptor interface, and short-circuit current $I = I_s < 0$ flows through the external circuit. As described in

Eq. (7.23), the electric field transient in two-layer device is given as

$$E(t) = \bar{E}_e(1 - e^{-\frac{t}{RC}}) - \bar{E}_s(1 - e^{-\frac{t}{\tau}})$$

$$\text{with} \quad \bar{E}_e(V) = \frac{1}{L_1}\frac{C_2}{C_1 + C_2}V$$

$$\text{and} \quad \bar{E}_s(Q_s) = \frac{1}{L_1}\frac{Q_s}{C_1 + C_2},$$

in the top layer of the two-layer system, whereas in the bottom layer of the two-layer, the electric field is given by

$$E'(t) = \bar{E}'_e(1 - e^{-\frac{t}{RC}}) + \bar{E}'_s(1 - e^{-\frac{t}{\tau}})$$

$$\text{with} \quad \bar{E}'_e(V) = \frac{1}{L_2}\frac{C_1}{C_1 + C_2}V \quad \text{and} \quad \bar{E}'_s(Q_s) = \frac{1}{L_2}\frac{Q_s}{C_1 + C_2}.$$

In short-circuit condition of OSC, external voltage V is zero in above equations, *i.e.*, $V = 0$, and electric field changes upon photoillumination as

(1) In donor layer in short-circuit condition

$$E(t) = -\bar{E}_s(1 - e^{-\frac{t}{\tau}}) \quad \text{with} \quad \bar{E}_s(Q_s) = \frac{1}{L_1}\frac{Q_s}{C_1 + C_2}. \tag{7.39}$$

(2) In acceptor layer in short-circuit condition

$$E'(t) = +\bar{E}'_s(1 - e^{-\frac{t}{\tau}}) \quad \text{with} \quad \bar{E}'_s(Q_s) = \frac{1}{L_2}\frac{Q_s}{C_1 + C_2}. \tag{7.40}$$

This electric field change $E(t)$ and $E'(t)$ is to be probed by using MDC and EFISHG measurement.

For example, copper-phthalocyanine (CuPc) is used as a donor layer and C60 is used as an acceptor layer [88–90]. The two-layer OSC is prepared in the structure of IZO/CuPc/C60/Al, where the thickness of CuPc is 60 nm and thickness of C60 is 40 nm. To selectively probe electric field in the two-layer OSC, EFISHG spectra of two materials are explored.

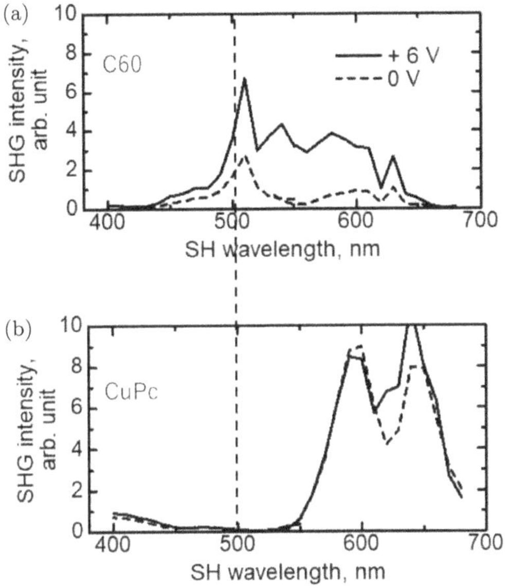

Figure 7.24. EFISHG spectrum of (a) C60 and (b) CuPc used as donor and acceptor materials in double-layer (CuPc/C60) OSC [87].

Figure 7.24 shows EFISHG spectra of CuPc and C60. It is seen that, by setting the laser wavelength at 1000 nm and SHG at 500 nm, we can selectively probe electric field $\bar{E}'(t)$ in C60 layer. That is, at probe laser wavelength 1000 nm, SHG from C60 is activated in the presence of electrostatic field while SHG from CuPc is minor. Figure 7.25 shows a time-resolved EFISHG measurement under photoillumination and transient current measurement of IZO/CuPc/C60/Al OSC. In this time-resolved EFISHG measurement, the OSC is in short-circuit condition. At "Time" = 0 in the left-hand-side panel, red light at wavelength of 650 nm begins to illuminate OSC to induce photovoltaic effect. At "Time" = 0 in the right-hand-side panel, red light is switched off and transient behavior is monitored in dark condition. The changing of EFISHG signal $I(2\omega)$ in Fig. 7.25 evidently indicates that interfacial charge Q_s in Eq. (7.40) is induced at the donor–acceptor interface, and freshly establish electrostatic field in the C60 layer, causing changing of SHG intensity. This is a direct evidence of the source of electromotive force in two-layer

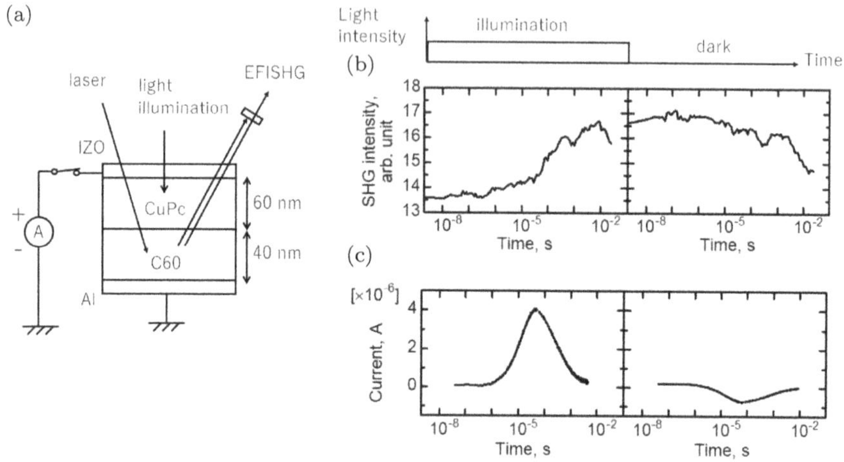

Figure 7.25. EFISHG measurement of two-layer (CuPc/C60) organic solar cell (OSC). (a) EFISHG and transient current measurement arrangement and double-layer OSC with IZO/CuPc/C60/Al structure. (b) Time-resolved EFISHG and (c) transient current measurement of double layer OSC (short-circuit condition). By setting laser wavelength at 1000 nm (SHG wavelength 500 nm), electric field in C60 layer is selectively probed [89].

OSC induced by photoillumination [88]. That is, photoillumination induced Maxwell–Wagner charges accompanied by steady state current, *i.e.*, short-circuit current, flowing through the closed circuit.

The relationship between short-circuit current $I < 0$ and interfacial charge is

$$Q_s = I(\tau_2 - \tau_1). \tag{7.41}$$

In the donor layer, hole current is sum of the motion of photo generated charge $+Q_p$ and a part of interfacial charge Q_{s1}. In acceptor layer, electron current is sum of the motion of charge $-Q_p$ and the other part of interfacialcharge Q_{s2} $(= Q_s - Q_{s1})$. Assuming that zero-field position is on the donor–acceptor interface, these interfacial charges are described as

$$Q_{s1} = C_1 \frac{Q_s}{C_1 + C_2} \tag{7.42}$$

$$Q_{s1} = C_2 \frac{Q_s}{C_1 + C_2}. \tag{7.43}$$

Accordingly, hole current I_h and electron current I_e is described as

$$I_h = \frac{Q_p}{\tau_1} + \frac{Q_{s1}}{\tau_1} \tag{7.44}$$

$$I_e = \frac{Q_p}{\tau_2} - \frac{Q_{s2}}{\tau_2} \tag{7.45}$$

with a condition in steady state $I_h = I_e = I_s$ (I_s: short-circuit current). After calculation using Eq. (7.42)–Eq. (7.45), free holes and electrons $\pm Q_p$ is expressed in a term of interfacial charge Q_s as

$$Q_p = \frac{Q_s}{\tau_1 - \tau_2} \frac{\tau_1 \tau_2}{\tau_{MW}} \tag{7.46}$$

and open-circuit voltage is described as

$$V_{oc} = \frac{Q_p}{C} \quad \text{with total capacitance} \quad C = \frac{C_1 C_2}{C_1 + C_2} \tag{7.47}$$

It is instructive here to mention that right-hand-side of Eq. (7.46) is the product of two dielectric relaxation times τ_1 and τ_2. The appearance of the term $\tau_1 \tau_2$ is obvious, with taking into account that the short-circuit current flows through the donor and acceptor layer in series. Technically Eq. (7.46) gives an idea that larger production of photogenerated charge is possible with a condition $\tau_1 \sim \tau_2$. That is, mixture of semiconducting materials with a matched dielectric relaxation time is one of the best ways for OSC. If imbalance of dielectric relaxation time is too big, $Q_p = 0$.

The relationship Eq. (7.46) is also derived from basic equivalent circuit analysis [87]. Figure 7.23(d) is the circuit model where donor and acceptor layer are modeled as a Maxwell–Wagner effect element (see Chap. 4). C_1 and C_2 are capacitance of donor and acceptor layer, respectively. G_1 and G_2 are conductance of donor layer and acceptor layer, respectively. Total capacitance C and conductance G are a series capacitance and series conductance of donor and acceptor layer. Based on electric circuit analysis, they are

$$G = \frac{G_1 G_2}{G_1 + G_2} \quad \text{and} \quad C = \frac{C_1 C_2}{C_1 + C_2}. \tag{7.48}$$

In short-circuit condition, short-circuit current I_s flows though the circuit where open-circuit voltage is mainly drops across the OSC. Consequently, short-circuit current is expressed in terms of open-circuit voltage as

$$I_s = -GV_{oc}. \tag{7.49}$$

In open-circuit condition, photo-generated charge Q_p is a source of photo-voltage, and described as

$$Q_p = CV_{oc}. \tag{7.50}$$

By substituting V_{oc} in Eq. (7.49) to Eq. (7.50) with the relation $Q_s = I_s(\tau_2 - \tau_1)$ in Maxwell–Wagner effect, the relationship between Q_p and Q_s is derived as

$$Q_p = \frac{Q_s}{\tau_1 - \tau_2} \frac{\tau_1 \tau_2}{\tau_{MW}} \tag{7.51}$$

which is the same as Eq. (7.46).

For IZO/pentacene/C60/Al OSC, quantitative analysis of EFISHG measurement results has been carried out [87]. By increasing illumination intensity, interfacial charge$|Q_s|$ increases in accordance with increase of short-circuit current. At $1\,\mathrm{mW/cm^2}$ illumination intensity, interfacial charge saturates at $Q_s = -3.6 \times 10^{-9}\,\mathrm{C/cm^2}$, establishing potential hill at the pentacene/C60 interface. Time-constant of excess charge generation τ_{MW} was analyzed as a function of illumination intensity. The time-constant τ_{MW}, similarly to charge density Q_s, changes to shorter time as the illumination intensity increases, and saturates at $\tau_{MW} = 1.4 \times 10^{-5}\,\mathrm{s}$. By using Q_s and τ_{MW} values from EFISHG measurement, short-circuit current $I = I_s = 0.72\,\mathrm{mA/cm^2}$ in I–V measurement, capacitance of pentacene and C60 layer $C_1 = 4.0 \times 10^{-8}\,\mathrm{F/cm^2}$ and $C_2 = 5.0 \times 10^{-8}\,\mathrm{F/cm^2}$, we can estimate conductance value G_1 of pentacene layer and G_2 of C60 layer installed in OSC. That is, by solving two equations Eq. (7.18) and Eq. (7.24) with these experimental values, conductance of donor and acceptor layer is obtained as $G_1 = 1.2 \times 10^{-4}$ S and $G_2 = 1.0 \times 10^{-4}$ S, respectively. Accordingly, photogenerated charges $Q_p = 9.4 \times 10^{-9}$ C/cm^2 is obtained by substituting these values to Eq. (7.46), and open-circuit voltage based on this model is estimated as $V_{oc} = 0.41$ V. The I–V measurement shows $V'_{oc} = 0.26$ V, showing a fairly good agreement between the model and V_{oc} predicted by the model. The discrepancy of V_{oc} and V'_{oc} suggests that non-zero effect of recombination loss, conductance change due to photoconduction, *etc.*, should be possible as an extension of the model. In the same way, pentacene/C60 two-layer [91–94], three-layer [95], multilayer cell [96], and inverted structure solar cell were analyzed [97]. Irreversible charging and discharging behavior is identified by using EFISHG. Conductance change under illumination and in dark caused this irreversible behavior [96]. Also, polymer bulk heterojunction solar cell was analyzed by using EFISHG [98–101].

The transient current measurement in Fig. 7.25(b) is conventional to probe carrier displacement in OSC. In accordance with the generation and decay of interfacial charge Q_s in EFISHG measurement, Maxwell displacement current (MDC) flows due to a change of induction charge on the top and bottom electrode. But as described in Sec. 6.1, MDC is indirectly probing carrier motion between two electrodes. The charge amount Q' calculated from time-integral of MDC is not identical to change of charge Q_s itself (see Chap. 2 and Chap. 3). That is, Q' is always smaller than Q_s, *i.e.*, $|Q'| < |Q_s|$. Analysis of Fig. 7.25(b) shows that the charge generation under illumination is $Q_s = 2.5 \times 10^{-7}$ C/cm^2 from EFISHG measurement. The amount of charge Q_s is of course the same in charging and discharging process. The same amount of Q_s decays in dark in EFISHG measurement. That is, SHG intensity increases upon illumination and recovers its original intensity in dark. On the other hand, MDC flows depending on the direction of charge motion in device, reflecting the zero-field position created in devices. Accordingly, from transient current, charge Q' calculated from an integral of MDC is $Q'_i = 9.0 \times 10^{-8}$ C/cm^2 under photoillumination and $Q'_d = 4.0 \times 10^{-8}$ C/cm^2 in dark. The two apparent charge Q'_i and

Q'_d is different to each other, implying that motion of charge in OSC is non-reversal due to charge recombination, traps, and other sources. As mentioned in Chap. 3, inequality relationship between charge is that anticipated based on Maxwell–Wagner effect, *i.e.*, $|Q'_i| < |Q_s|$ and $|Q'_d| < |Q_s|$. The result confirmed this inequality relationship by experiment. Using EFISHG coupling with MDC is very effective to analyze and model a device. The level of analytical technique in MDC such as TOF, TSC, *etc.*, is very high, owing to cumulative advances in measurement instruments, idea of device structure, knowledge of materials, *etc.* By using EFISHG, we can know the charge motion itself behind the MDC waveform implying charge motion and dipolar rotation in materials and devices.

Idea of bulk heterojunction (BHJ) is developed in a research field of OSC to enhance power conversion efficiency. In BHJ OSCs, electron donor and acceptor molecules are mixed to introduce larger contacting area for the creation of excitons to be separated into electrons and holes efficiently. The generation of the EFISHG depends on the material, and EFISHG spectrum of donor and acceptor materials are different. Consequently, EFISHG is capable of probing electric field in different material components one-by-one in the BHJ layer, by choosing appropriate laser wavelengths. This is an advantage of EFISHG for the study of carrier path in mixed material system such as BHJ OSCs. For example, for OSCs co-deposited with pentacene and C60, EFISHG measurement has been carried out at two laser wavelengths of 1000 nm (SHG 500 nm) and 860 nm (SHG 430 nm), by which carrier behaviors inside the C60 and pentacene have been selectively probed (Fig. 7.26). As a result, the presence of two carrier paths for electrons and holes has been identified from the EFISHG response by application of a laser pulse to the OSCs [102]. Furthermore, recently attention has been focused on perovskite solar cells, where the dissociation of excitons is postulated to happen in the bulk of the solar cells. EFISHG can be employed to make clear the contribution of the internal electric field, and thus the technique is available for studying the carrier process [103–106].

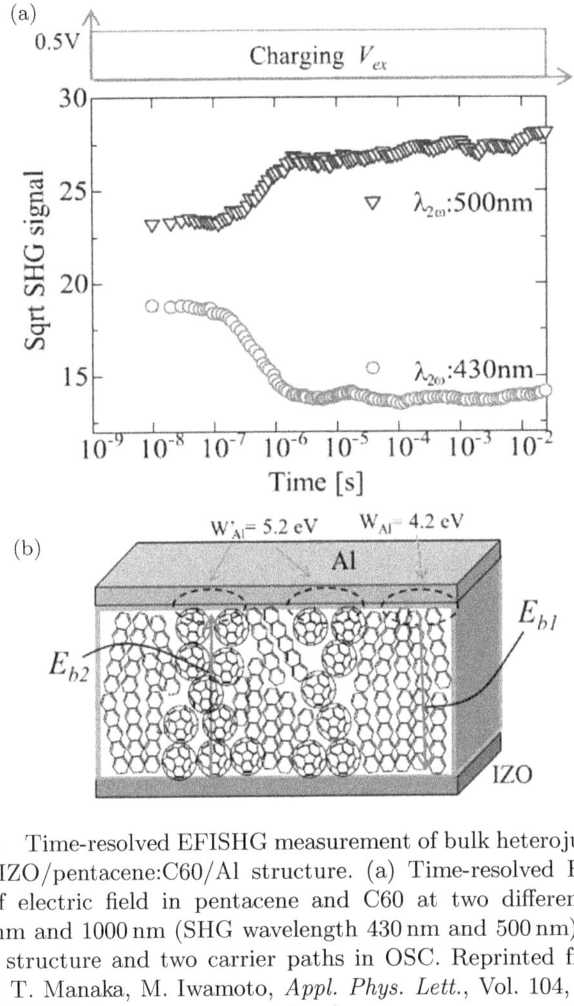

Figure 7.26. Time-resolved EFISHG measurement of bulk heterojunction (BHJ) OSC with IZO/pentacene:C60/Al structure. (a) Time-resolved EFISHG measurement of electric field in pentacene and C60 at two different laser wavelength 860 nm and 1000 nm (SHG wavelength 430 nm and 500 nm), respectively. (b) Sample structure and two carrier paths in OSC. Reprinted from X. Chen, D. Taguchi, T. Manaka, M. Iwamoto, *Appl. Phys. Lett.*, Vol. 104, 013306, 2014, with the permission of AIP Publishing [102].

7.10 Summary

In this chapter, analysis and modelling of sandwich-type organic electronic devices was described. On paying attention to "*electric field*" formed in single layer and multilayers, optical nonlinear polarization and Maxwell displacement current were probed, and the carrier behaviors in organic devices were analysed and modelled. Examples were MIS diode, OLED, MFSM memory diode, and OSC. The probing methods were Maxwell displacement current (MDC) measurement and electric-field-induced optical second-harmonic generation (EFISHG) measurement. In both measurements, *polarization* is a key parameter for the detection of dynamical carrier motions. In the MDC measurement, *polarization* induced by electronic charge displacement, *e.g.*, holes, electron, *etc.*, was electrically probed, meanwhile in the EFISHG measurement, *nonlinear polarization* induced in solids was optically probed with consideration of the presence of d.c. electric field caused from electronic carriers. The I–V characteristics of sandwich-type device were analysed and modelled on the basis of electric field probed by using EFISHG measurement. *"Electric field"* is the key to understand device operation of organic devices. The validity of these analyses is not limited to the case of organic devices. As described in this chapter, it is valid when materials installed in the device exhibit long dielectric relaxation time.

References

1. Mott, N. F. and Gurney, R. W. (1948) *Electronic processes in ionic crystals*, 2nd ed. (Oxford, Oxford University).
2. Lampert, M. A. and Mark, P. (1970) *Current injection in solids* (New York and London, Academic).
3. Sze, S. M (1981) *Physics of semiconductor devices*, 2nd ed. (John Wiley & Sons, New York Chichester Brisbane Toronto Singapore).
4. Iwamoto M. and Wu, C.-X. (2001) *The Physical Properties of Organic Monolayers* (World Scientific, Singapore), https://doi.org/10.1142/4575.
5. Iwamoto, M. and Taguchi, D. (2014) Thermally Stimulated Current in Electrical and Electronic Materials, Analysis and Application (Tokyo, Corona–sha). (in Japanese)
6. Purcell, E. M. (1965) *Electricity and magnetism* (McGraw–hill, New York St. Louis San Francisco Toronto London Sydney).
7. Iwamoto, M., Manaka, T. and Taguchi, D. (2015) Modeling and visualization of carrier motion in organic films by optical second harmonic generation and Maxwell-displacement current, *J. Phys. D*, 48, 373001, https://doi.org/10.1088/0022-3727/48/37/373001.
8. Iwamoto, M., Manaka, T., Taguchi, D. (2014). Probing and modeling of carrier motionin organic devices by electric-field-induced optical second-harmonic generation, Jpn. *J. Appl. Phys.*, 53, 100101, https://doi.org/10.7567/JJAP.53.100101.

9. Iwamoto, M., Manaka, T. and Taguchi, D. (2016). Denki Gakkai Ronbun Shi A, 136, pp. 678–684, https://doi.org/10.1541/ieejfms.136.678.
10. Boyd, R. W. (2008) *Nonlinear optics*, 3rd ed. (Academic, Amsterdam Boston Heidelberg London New York Oxford Paris San Diego San Francisco Singapore Sydney Tokyo) Sec.1.1, https://doi.org/10.1016/B978-0-12-369470-6.00001-0.
11. Manaka, T., Li, C.-Q., Cheng, X.-M. and Iwamoto, M. (2004). Spectroscopic consideration of the surface potential built across phthalocyanine thin films on a metal electrode, *J. Chem. Phys.* 120, 7725, https://doi.org/10.1063/1.1665676.
12. Manaka, T., Suzue, Y., and Iwamoto, M. (2005). Investigation of the Electrostatic Phenomena at Pentacene/Metal Interface by Second-Harmonic Generation, *Japanese Journal of Applied Physics*, Volume 44, Part 1, Number 4S, 2818, https://doi.org/10.1143/JJAP.44.2818.
13. Kajzar, F., Taliani, C., Danieli, R., Rossini, S. and Zamboni, R. (1994). Wave-dispersed third-order nonlinear optical properties of C60 thin films, *Chem. Phys. Lett.*, 217, 418–422, https://doi.org/10.1016/0009-2614(93)E1396-X.
14. Manaka, T., Lim, E., Tamura, R. and Iwamoto, M. (2007). *Nat. Photon.*, 1, 581, https://doi.org/10.1038/nphoton.2007.172.
15. Shibata, Y., Nakao, M., Manaka, T., Lim, E. and Iwamoto, M. (2009). Probing Electric Field Distribution in Underlayer of an Organic Double-Layer System by Optical Second-Harmonic Generation Measurement, *Jpn. J. Appl. Phys.*, 48, 021504, https://doi.org/10.1143/JJAP.48.021504.
16. Verbiest, T., Clays, K. and Rodriguez, V. (2009) *Second-order nonlinear optical characterization techniques: an introduction* (CRC, New York), https://doi.org/10.1201/9781420070736 Chapters 5 and 6.
17. Shen, Y. R. (1984) *The principles of nonlinear optics*(New York, Wiley).
18. Motti, S. G., Maia F. C. B. and Miranda, P. B. (2014) *Nonlinear Spectroscopy of Interfaces and Its Application to Organic Electronics Ultrafast Dynamics in Molecules, Nanostructures and Interfaces*, Series in Optics and Photonics, Vol. 8, G. G. Gurzadyan, G. Lanzani, C. Soci and T. C. Sum eds. (World Scientific , Singapore), https://doi.org/10.1142/9789814556927_0011.
19. Sung, W., Kim, D., Shen, Y. R. (2013). Sum-frequency vibrational spectroscopic studies of Langmuir monolayers, *Curr. Appl. Phys.*, 13, 619, https://doi.org/10.1016/j.cap.2012.12.002.
20. Miranda, P. B. and Shen, Y. R. (1999). Liquid Interfaces: A Study by Sum-Frequency Vibrational Spectroscopy, *J. Phys. Chem. B*, 103, 3292, https://doi.org/10.1021/jp9843757.
21. Tian, C. S. and Shen, Y. R. (2014). Recent progress on sum-frequency spectroscopy, *Surf. Sci. Rep.*, 69, 105, https://doi.org/10.1016/j.surfrep.2014.05.001.
22. A part of Fig. 7.2 is based on a survey by Dr. Yuki Ohshima and by Dr. Taishi Noma.

23. Köhler, A. and Bässler, H. (2015) *Electronic Processes in Organic Semiconductors: An Introduction* (Wiley , Weinheim), https://doi.org/10.1002/9783527685172 Chap. 3, pp. 193–306.
24. Basu, D. and Dodabalapur, A. (2010) *Drift Velocity and Drift Mobility Measurement in Organic Semiconductors Using Pulse Voltage Organic Electronics*, Advances in Polymer Science, Vol. 223, G. Meller, T. Grasser eds.(Springer, Heidelberg, Dordrecht, London, New York), https://doi.org/10.1007/12_2009_4.
25. Kepler, R. G. (1960). Charge Carrier Production and Mobility in Anthracene Crystals, *Phys. Rev.*, 119, 1226, https://doi.org/10.1103/PhysRev.119.1226.
26. Miyamae, T., Takada, N., Ohata, H. and Tsutsui, T. (2017). Direct probing of charge carrier behavior in multilayered organic light-emitting diode devices by time-resolved electric-field-induced sum-frequency generation spectroscopy, *Appl. Phys. Express*, 10, 102101, https://doi.org/10.7567/APEX.10.102101.
27. Sato, T., Miyamae, T., Ohata, H. and Tsutsui, T. (2019). Direct observations of the charge behavior of a high-efficiency blue organic light-emitting diode under operating conditions using electric-field-induced doubly resonant sum-frequency-generation vibrational spectroscopy, *Org. Electron.*, 74, 118–125, https://doi.org/10.1016/j.orgel.2019.07.008.
28. Rohlfing, F., Yamada, T. and Tsutsui, T. (1999). Electroabsorption spectroscopy on tris–(8–hydroxyquinoline) aluminum-based light emitting diodes, *J. Appl. Phys.*, 86, 4978, https://doi.org/10.1063/1.371469.
29. Yamada, T., Rohlfing F. and Tsutsui, T. (2000). *Jpn. J. Appl. Phys.*, 39, 1382, https://doi.org/10.1143/JJAP.39.1382.
30. Simmons, J. G. (1963). Generalized formula for the electric tunnel effect between Similar electrodes separated by a thin insulating film, *J. Appl. Phys.*, 34, 1793–1803, https://doi.org/10.1063/1.1702682.
31. Kubota, T., Wada, M., Iwamoto, M., Noshiro, H. and Sekine, M. (1992). Electrical transport properties of Josephson junctions with a Nb/Au/PI/(Pb–Bi) structure, *Thin Solid Films*, 210–211, 277–279, https://doi.org/10.1016/0040-6090(92)90233-2.
32. Noguchi, Y., Majima, Y. and Iwamoto, M. Analysis of space charge effect on the current–voltage characteristics with a step structure in metal/polyimide/rhodamine–dendrimer/polyimide/metal junction, *J. Appl. Phys.*, 90, 1368–1375, 2001; Y. Noguchi, M. Iwamoto, T. Kubota, S. Mashiko, Addendum: space charge effect and the step voltages in metal/polyimide/rhodamine-dendrimer/polyimide/metal junctions, *J. Appl. Phys.*, 92, 1174–1176, 2002, https://doi.org/10.1063/1.1487896.
33. Nishi, S. (2015). *Denkai Yuuki Hikari Dainiji Kouchuoha Hasseihou wo Motiita Yuuki Handoutai/Zetsuentai 2 Sou Kouzou Sosi no Denki Dendou Kikou no Kaiseki*, Master Thesis, Tokyo Institute of Technology (in Japanese).
34. Nishi, S., Taguchi, D., Manaka, T. and Iwamoto, M. (2015). Analysis of current–voltage characteristics of Au/pentacene/fluorine polymer/indium

zinc oxide diodes by electric-field-induced optical second-harmonic generation, *J. Appl. Phys.*, 117, 245502, https://doi.org/10.1063/1.4923058.
35. Miyazawa, R., Taguchi, D., Manaka, T. and Iwamoto, M. (2011). Probing of Electric Field Distribution in ITO/PI/P3HT/Au Using Electric Field Induced Second Harmonic Generation, *IEICE Trans. Electron.*, E94–C, 185–186, https://doi.org/10.1587/transele.E94.C.185.
36. Lim, E., Taguchi, D., Iwamoto, M. (2013). Study of rectifying property of ITO/PI/TIPS–pentacene/Au diodes by electric field induced optical second harmonic generation, *Org. Electron.*, 14, 1903–1908, https://doi.org/10.1016/j.orgel.2013.04.021.
37. Nishi, S., Taguchi, D., Manaka, T. and Iwamoto, M. (2016). Study of electrical conduction mechanism of organic double-layer diode using electric field induced optical second harmonic generation measurement, *J. Nanosci. Nanotech*, 16, 3364–3367, https://doi.org/10.1166/jnn.2016.12305.
38. Lim, E., Taguchi, D. and Iwamoto, M. (2014). Study of stress biasing effect observed in C–V characteristics of ITO/PI/TIPS–pentacene/Au diodes by electric field induced optical second harmonic generation, *Org. Electron.*, 15, 3590–3594, https://doi.org/10.1016/j.orgel.2014.09.044.
39. Lim, E., Taguchi, D. and Iwamoto, M. (2013). Probing electric-field distribution change in ITO/PI/TIPS-pentacene/Au diodes by time-resolved optical second harmonic generation measurement, *Chem. Phys. Lett.*, 561–562, 97–100, https://doi.org/10.1016/j.cplett.2013.01.034.
40. Lim, E., Taguchi, D. and Iwamoto, M. (2012). Probing carrier behavior in organic semiconductor device by electric field induced optical second harmonic generation measurement, *Org. Electron.*, 13, 2489–2493, https://doi.org/10.1016/j.orgel.2012.06.031.
41. Zhang, L., Taguchi, D., Li, J., Manaka, T. and Iwamoto, M. (2011). Transport limited interfacial carrier relaxation in a double-layer device investigated by time-resolved second harmonic generation and impedance spectroscopy, *Appl. Phys. Lett.*, 98, 092109, https://doi.org/10.1063/1.3560054.
42. Zhang, L., Taguchi, D., Manaka, T. and Iwamoto, M. (2011). Bulk–trap modulated Maxwell–Wagner type interfacial carrier relaxation process in a fullerene/polyimide double-layer device investigated by time-resolved second- harmonic generation, *J. Appl. Phys.*, 110, 033715, https://doi.org/10.1063/1.3610952.
43. Zhang, L., Taguchi, D., Li, J., Manaka, T. and Iwamoto, M. (2011). Investigation of interfacial voltage establishment and relaxation processes in double-layer device by time resolved optical second harmonic generation, *Jpn. J. Appl. Phys.*, 50, 04DK13, https://doi.org/10.7567/JJAP.50.04DK13.
44. Noma, T., Taguchi, D., Manaka, T., Lin, H. and Iwamoto, M. (2017). Determination of carrier mobility of semiconductor layer in organic metal–insulator–semiconductor diodes by displacement current and electric-field-induced optical second-harmonic generation measurements, *Orga. Electron.*, 43, 70–76, https://doi.org/10.1016/j.orgel.2017.01.004.

45. Li, X., Sunaga, M., Taguchi, D., Manaka, T., Lin, H. and Iwamoto, M. (2017). Study of carrier-mobility of organic thin film by dark-injection time-of-flight and electric-field-induced optical second-harmonic generation measurements, *Chem. Phys. Lett.*, 677, 50–54, https://doi.org/10.1016/j.cplett.2017.03.072.
46. Zhang, L., Taguchi, D., Li, J., Manaka, T. and Iwamoto, M. (2010). Probing of interfacial charging and discharging in double-layer device with a polyimide blocking-layer by time-resolved optical second harmonic generation, *J. Appl. Phys.*, 108, 093707, https://doi.org/10.1063/1.3483929.
47. Miyazawa, R., Taguchi, D., Manaka, T. and Iwamoto, M. (2011). Probing of Transient Electric Field Distribution in ITO/PI/P3HT/Au by Time-Resolved Optical Second Harmonic Generation Measurement, *Physics Procedia*, 14, 67–72, https://doi.org/10.1016/j.phpro.2011.05.015.
48. Taguchi, D., Manaka, T., Iwamoto, M., Sakuma, K., Watariguchi, K. and Kawahara, M. (2019). Probing internal electric field in organic photoconductors by using electric-field-induced optical second-harmonic generation, *IEICE Trans. Electron.*, E102–C, 113–118, https://doi.org/10.1587/transele.2018OMP0002.
49. Jacob, M. V., Bazaka, K., Taguchi, D., Manaka, T. and Iwamoto, M. (2012). Electron-blocking hole-transport polyterpenol thin films, *Chem. Phys. Lett.*, 528, 26–28, https://doi.org/10.1016/j.cplett.2012.01.031.
50. Bazaka, K., Jacob, M. V., Taguchi, D., Manaka, T. and Iwamoto, M. (2011). Investigation of interfacial charging and discharging in double-layer pentacene–based metal–insulator–metal device with polyterpenol blocking layer using electric field induced second harmonic generation, *Chem. Phys. Lett.*, 503, 105–111, https://doi.org/10.1016/j.cplett.2010.12.072.
51. Bazaka, K., Jacob, M. V., Taguchi, D., Manaka, T., Iwamoto, M. (2011). Effect of organic gate dielectric material properties on interfacial charging and discharging of pentacene MIM device, *Physics Procedia*, 14, 62–66, https://doi.org/10.1016/j.phpro.2011.05.014.
52. Taguchi, D., Manaka, T., Iwamoto, M., Anderson, L. J., Jacob, M. V. (2014). Study of carrier blocking property of poly–linalyl acetate thin layer by electric-field-induced optical second-harmonic generation, *Chem. Phys. Lett.*, 593, 69–71, https://doi.org/10.1016/j.cplett.2013.12.070.
53. Taguchi, D., Manaka, T., Iwamoto, M., Bazaka, K., Jacob, M. V. (2013). Analyzing hysteresis behavior of capacitance–voltage characteristics of IZO/C60/pentacene/Au diodes with a hole-transport electron-blocking polyterpenol layer by electric-field-induced optical second-harmonic generation measurement, *Chem. Phys. Lett.*, 572, 150–153, https://doi.org/10.1016/j.cplett.2013.04.030.
54. Taguchi, D., Weis, M., Manaka, T. and Iwamoto, M. (2009). Probing of carrier behavior in organic electroluminescent diode using electric field induced optical second-harmonic generation measurement, *Appl. Phys. Lett.*, 95, 263310, https://doi.org/10.1063/1.3277155.
55. Hosokawa, C., Tokailin, H., Higashi, H. and Kusumoto, T. (1992). *Appl. Phys. Lett.*, 60, 1220, https://doi.org/10.1063/1.107411.

56. Taguchi, D., Inoue, S., Zhang, L., Li, J., Weis, M., Manaka, T. and Iwamoto, M. (2010). Analysis of organic light-emitting diode as a Maxwell–Wagner effect element by time-resolved optical second harmonic generation measurement, *J. Phys. Chem. Lett.*, 1, 803–807, https://doi.org/10.1021/jz1000215.
57. Taguchi, D., Zhang, L., Li, J., Weis, M., Manaka, T. and Iwamoto, M. (2010). Analysis of Carrier Transients in Double-Layer Organic Light-Emitting Diodes by Electric-Field-Induced Second-Harmonic Generation, *J. Phys. Chem. C*, 114, 15136–15140, https://doi.org/10.1021/jp104712m.
58. Taguchi, D., Zhang, L., Li, J., Weis, M., Manaka, T. and Iwamoto, M. (2011). Direct Probing of Carrier Behavior in Electroluminescence IZO/α–NPD/Alq3/LiF/Al Diode by Time-Resolved Optical Second-Harmonic Generation, *Jpn. J. Appl. Phys.*, 50, 04DK08, https://doi.org/10.7567/JJAP.50.04DK08.
59. Sadakata, A., Osada, K., Taguchi, D., Yamamoto, T., Fukuzawa, M., Manaka, T., Iwamoto, M. (2012). Probing interfacial charge accumulation in ITO/α–NPD/Alq3/Al diodes under two electroluminescence operational modes by electric-field induced optical second-harmonic generation, *J. Appl. Phys.*, 112, 083723, https://doi.org/10.1063/1.4762014.
60. Sadakata, A., Osada, K., Taguchi, D., Manaka, T. and Iwamoto, M. (2014). Probing space charge effect on electroluminescence of indium tin oxide (ITO)/N,N′–di–[(1–naphthyl)–N,N′–diphenyl]–(1,1′–biphenyl)–4,4′–diamine (α–NPD)/tris(8–hydroxyquinolinato) aluminum (III) (Alq3)/Al diodes by time-resolved electric-field-induced optical second-harmonic generation measurement, *Thin Solid Films*, 554, 110–113, https://doi.org/10.1016/j.tsf.2013.08.015.
61. Sadakata, A., Ohshima, Y., Taguchi, D., Fukuzawa, M., Manaka, T. and Iwamoto, M. (2011). Electroluminescence enhanced from electrode interface in ITO/tetracene/Al diodes, *Mol. Cryst. Liq. Cryst.*, 538, 112–117, https://doi.org/10.1080/15421406.2011.563671.
62. Sadakata, A., Taguchi, D., Yamamoto, T., Fukuzawa, M., Manaka, T., Iwamoto, M. (2011). Analyzing Two Electroluminescence Mode of ITO/α–NPD/Alq3/Al Diodes by Using Large A. C. Square Voltages, *J. Appl. Phys.*, 110, 103707, https://doi.org/10.1063/1.3662149.
63. Sadakata, A., Oda, Y., Taguchi, D., Fukuzawa, M., Manaka, T. and Iwamoto, M. (2013). Study of hole-injection assisted layer in double-layer organic light-emitting diodes by electric-field-induced optical second-harmonic generation measurement, *Jpn. J. Appl. Phys.*, 52, 05DC03, https://doi.org/10.7567/JJAP.52.05DC03.
64. Oda, Y., Sadakata, A., Taguchi, D., Manaka, T. and Iwamoto, M. (2016). Study of impact of pentacene interface layer in ITO/α–NPD/Alq3/Al OLEDs by Time-Resolved Electric-Field-Induced Optical Second-Harmonic Generation Measurement, *J. Nanosci. Nanotech.*, 16, 3188–3193, https://doi.org/10.1166/jnn.2016.12290.
65. Weis, M., Otsuka, T., Taguchi, D., Manaka, T. and Iwamoto, M. (2015). Charge injection and accumulation in organic light-emitting diode with

PEDOT:PSS anode, *J. Appl. Phys.*, 117, 155503, https://doi.org/10.1063/1.4918556.
66. Kawai, H. (1969). The Piezoelectricity of Poly (vinylidene Fluoride), 8, 975, https://doi.org/10.1143/JJAP.8.975.
67. Koga, K. and Ohigashi, H. (1986)Piezoelectricity and related properties of vinylidene fluoride and trifluoroethylene copolymers, *Journal of Applied Physics*, 59, 2142–2150, https://doi.org/10.1063/1.336351.
68. Tajitsu, Y., Ogura, H., Chiba, A., Furukawa, T. (1987). Investigation of Switching Characteristics of Vinylidene Fluoride/Trifluoroethylene Copolymers in Relation to Their Structures, *Jpn. J. Appl. Phys.*, 26, 554–560, https://doi.org/10.1143/JJAP.26.554.
69. Yoshita, S., Tamura, R., Taguchi, D., Weis, M., Lim, E., Manaka, T. and Iwamoto, M. (2009) Displacement current analysis of carrier behavior in pentacene field effect transistor with poly(vinylidene fluoride and tetrafluoroethylene) gate insulator, *J. Appl. Phys.*, 106, 024505, https://doi.org/10.1063/1.3176487.
70. Tamura, R., Lim, E., Yoshita, S., Manaka, T. and Iwamoto, M. (2008). Analysis of threshold voltage shift of pentacene field effect transistor with ferroelectric gate insulator as a Maxwell–Wagner effect, *Thin Solid Films*, 516, 2753–2757, https://doi.org/10.1016/j.tsf.2007.04.106.
71. Ou-Yang, W., Weis, M., Taguchi, D., Chen, X., Manaka, T. and Iwamoto, M. (2010). Modeling of Threshold Voltage in Pentacene Organic Field-Effect Transistors, *J. Appl. Phys.*, 107, 124506, https://doi.org/10.1063/1.3449078.
72. Chen, X., Ou-yang, W., Weis, M., Taguchi, D., Manaka, T., Iwamoto, M. (2010). Reduction of Hysteresis in Organic Field-Effect Transistor by Ferroelectric Gate Dielectric, *Jpn. J. Appl. Phys.*, 49, 021601, https://doi.org/10.1143/JJAP.49.021601.
73. Yuan, Y., Reece, T. J., Sharma, P., Poddar, S., Ducharme, S., Gruverman, A., Yang, Y. and Huang, J. (2011.). Efficiency enhancement in organic solar cells with ferroelectric polymers, *Nat. Mater.*, 10, 296–302, https://doi.org/10.1038/nmat2951.
74. Weis, M., Li, J., Taguchi, D., Manaka, T. and Iwamoto, M. (2011). Effect of Photogenerated Carriers on Ferroelectric Polarization Reversal, *Appl. Phys. Express*, 4, 121601, https://doi.org/10.1143/APEX.4.121601.
75. Eguchi, Mototarô (1919). On Dielectric Polarisation, Nippon Sugaku-Buturigakkwai Kizi Dai 3 Ki, 1, pp. 326–331; Eguchi, M. (1919). Variation of Electrical Conductivity of Oils and Waxes, Nippon Sugaku-Buturigakkwai Kizi Dai 3 Ki, 1, pp. 320–326, https://doi.org/10.11429/ppmsj1919.1.10-11_320.
76. Li, J., Taguchi, D., OuYang, W., Manaka, T. and Iwamoto, M. (2011). Interaction of interfacial charge and ferroelectric polarization in a pentacene/poly(vinylidene fluoride–trifluoroethylene) double-layer device, *Appl. Phys. Lett.*, 99, 063302, https://doi.org/10.1063/1.3624477.

77. Li, J., Weis, M., Taguchi, D., Manaka, T. and Iwamoto, M. (2012). Analyzing a two-step polarization process in a pentacene/poly(vinylidenefluoride–trifluoroethylene) double-layer device using Maxwell–Wagner model, *J. Appl. Phys.*, 111, 023706, https://doi.org/10.1063/1.3677943.
78. Cui, X., Taguchi, D., Manaka, T. and Iwamoto, M. (2013). Analysis of carrier behavior in C60/P(VDF–TrFE) double-layer capacitor by using electric-field-induced second-harmonic generation measurement, *J. Appl. Phys.*, 114, 234504, https://doi.org/10.1063/1.4853035.
79. Cui, X., Taguchi, D., Manaka, T. and Iwamoto, M. (2014). Analysis of carrier behavior in α–NPD/P(VDF–TrFE) double-layer capacitor by using electric-field-induced optical second-harmonic generation, *Jpn. J. Appl. Phys.*, 53, 02BB05, https://doi.org/10.7567/JJAP.53.02BB05.
80. Cui, X., Taguchi, D., Manaka, T., Pan, W. and Iwamoto, M. (2013). Study of effect of inserted pentacene layer in ITO/P(VDF–TrFE)/α–NPD/Au capacitor using electric-field-induced optical second-harmonic generation and displacement current, *Org. Electron*, 15, 537–542, https://doi.org/10.1016/j.orgel.2013.12.008.
81. Shi, Z., Taguchi, D., Manaka, T. and Iwamoto, M. (2016). Observation of turnover of spontaneous polarization in ferroelectric layer of pentacene/poly–(vinylidene–trifluoroethylene) double-layer capacitor under photo illumination by optical second-harmonic generation measurement, *J. Appl. Phys.*, 119, 165502, https://doi.org/10.1063/1.4947584.
82. Lee, K., Weis, M., Lin, J., Taguchi, D., Majkova, E., Manaka, T., Iwamoto, M. (2011). Trapping effect of metal nanoparticle mono- and multilayer in the organic field-effect transistor, *J. Appl. Phys.*, 109, 064512, https://doi.org/10.1063/1.3567916.
83. Jacob, M. V., Taguchi, D., Iwamoto, M., Bazaka, K., Rawat, R. S. (2016). Resistive switching in graphene-organic device: Charge transport properties of graphene-organic device through electric field induced optical second harmonic generation and charge modulation spectroscopy, *Carbon*, 112, 111–116, https://doi.org/10.1016/j.carbon.2016.11.005.
84. Tang, C. W. (1986). *Appl. Phys. Lett.*, 48, 183, https://doi.org/10.1063/1.96937.
85. Hiramoto, M., Fujiwara H. and Yokoyama, M. (1991). *Appl. Phys. Lett.*, 58, 1062, https://doi.org/10.1063/1.104423.
86. For example, Kaji, T., Zhang, M., Nakao, S., Iketaki, K., Yokoyama,K., Tang, C. W. and Hiramoto, M. (2011) *Adv. Mater.*, 23, 3320–3325, https://doi.org/10.1002/adma.201101305.
87. Taguchi, D., Shino, T., Chen, X., Zhang, L., Li, J., Weis, M., Manaka, T. and Iwamoto, M. (2011). Analyzing photovoltaic effect of double-layer organic solar cells as a Maxwell–Wagner effect system by optical electric-field-induced second-harmonic generation measurement, *J. Appl. Phys.*, 110, 103717, https://doi.org/10.1063/1.3662914.
88. Taguchi, D., Shino, T., Zhang, L., Li, J., Weis, M., Manaka, T. and Iwamoto, M. (2011). Direct Probing of Photovoltaic Effect Generated in

Double-Layer Organic Solar Cell by Electric-Field-Induced Optical Second-Harmonic Generation, *Appl. Phys. Express*, 4, 021602, https://doi.org/10.1143/APEX.4.021602.

89. Taguchi, D., Shino, T., Chen, X., Zhang, L., Li, J., Weis, M., Manaka, T. and Iwamoto, M. (2011). Analyzing Carrier Lifetime of Double-Layer Organic Solar Cells by Using Optical Electric-Field-Induced Second-Harmonic Generation Measurement, *Appl. Phys. Lett.*, 98, 133507, https://doi.org/10.1063/1.3574002.
90. Chen, X., Taguchi, D., Manaka, T. and Iwamoto, M. (2013). Interfacial charging of cupper phthalocyanine/C60 double-layer organic solar cells induced by photoillumination: Effect of photoconductivity change, *Thin Solid Films*, 554, 158–161, https://doi.org/10.1016/j.tsf.2013.04.094.
91. Chen, X., Taguchi, D., Shino, T., Manaka, T. and Iwamoto, M. (2011). Analysis of interface carrier accumulation and relaxation in pentacene/C60 double-layer organic solar cell by impedance spectroscopy and electric-field-induced optical second harmonic generation, *J. Appl. Phys.*, 110, 074509, https://doi.org/10.1063/1.3642964.
92. Chen, X., Taguchi, D., Lee, K., Manaka, T. and Iwamoto, M. (2011). Analyzing interfacial carrier charging in pentacene/C60 double-layer organic solar cells by optical electric field induced second-harmonic generation measurement, *Chem. Phys. Lett.*, 511, 491–495, https://doi.org/10.1016/j.cplett.2011.06.082.
93. Chen, X., Taguchi, D., Manaka, T. and Iwamoto, M. (2012). Study of blocking effect of Cu–phthalocyanine layer in zinc oxide/pentacene/CuPc/C60/Al organic solar cells by electric field-induced optical second harmonic generation measurement, *Org. Electron.*, 14, 320–325, https://doi.org/10.1016/j.orgel.2012.11.014.
94. Chen, X., Taguchi, D., Weis, M., Manaka, T. and Iwamoto, M. (2012). Analyzing photo-induced internal electric field in pentacene/C60 double-layer organic solar cells under various external voltages by electric field induced optical second harmonic generation measurement, *Jpn. J. Appl. Phys.*, 51, 041605, https://doi.org/10.1143/JJAP.51.041605.
95. Chen, X., Taguchi, D., Lee, K., Manaka, T. and Iwamoto, M. (2013). Analysis of interfacial charging process in pentacene/C60/bathocuproine triple-layer organic solar cells using a Maxwell–Wagner model, *Jpn. J. Appl. Phys.*, 52, 04CR05, https://doi.org/10.7567/JJAP.52.04CR05.
96. Chen, X., Taguchi, D., Manaka, T. and Iwamoto, M. (2014). Study of multiple photovoltaic processes in stacked organic active layers, Org. *Electron*, 15, 2014–2020, https://doi.org/10.1016/j.orgel.2014.05.034.
97. Chen, X., Taguchi, D., Manaka, T. and Iwamoto, M. (2013). Interfacial charging originated from the conductivity decrease of C60 layer in IZO/pentacene/C60/Al organic double-layer solar cells, *Org. Electron.*, 15, 162–168, https://doi.org/10.1016/j.orgel.2013.11.005.
98. Chen, X., Taguchi, D., Lee, K., Manaka, T. and Iwamoto, M. (2012). Analysis of Anomalous Discharging Processes in Pentacene/C60 Double-Layer Organic Solar Cell, *Jpn. J. Appl. Phys.*, 51, 02BK01, https://doi.org/10.7567/JJAP.51.02BK01.

99. Chauhan, A. K., Gusain, A., Jha, P., Veerender, P., Koiry, S. P., Sridevi, C., Aswal, D. K., Gupta, S. K., Taguchi, D., Manaka, T. and Iwamoto, M. (2016). Interfacial charge trapping in the polymer solar cells and its elimination by solvent annealing, *AIP Advances*, 6, 095012, https://doi.org/10.1063/1.4963014.
100. Ahmad, Z., Abdullah, S. M., Taguchi, D., Sulaiman, K., Iwamoto, M. (2015). A way for studying the impact of PEDOT:PSS interface layer on carrier transport in PCDTBT:PC71BM bulk hetero junction solar cells by electric field induced optical second harmonic generation measurement, *J. Appl. Phys.*, 117, 163101, https://doi.org/10.1063/1.4917462.
101. Alrougy, I. M., Taguchi, D. and Manaka, T. (2019). Spectroscopic study of electric field induced optical second harmonic generation from PCPDTBT and PC71BM thin films, *IEICE Trans. Electron.*, E102–C, 119–124, https://doi.org/10.1587/transele.2018OMP0007.
102. Chen, X., Taguchi, D., Manaka, T. and Iwamoto, M. (2014). Selective observation of photo-induced electric fields inside different material components in bulk-heterojunction organic solar cells, *Appl. Phys. Lett.*, 104, 013306, https://doi.org/10.1063/1.4861620.
103. Lei Lei Yin Win, Taguchi, D. and Manaka, T. (2019). Transient carrier visualization of organic–inorganic hybrid perovskite thin films by using time-resolved microscopic second-harmonic generation (TRM–SHG), *Org. Electron.*, 75, 105416, https://doi.org/10.1016/j.orgel.2019.105416.
104. Ahmad, Z., Noma, T., Paek, S., Cho, K. T., Taguchi, D., Iwamoto, M., Manaka, T., Nazeeruddin, M. K., Touati, F. and Al-Muhtaseb, S. A. (2019). Stability in 3D and 2D/3D hybrid perovskite solar cells studied by EFISHG and IS techniques under light and heat soaking, *Org. Electron.*, 66, 7–12, https://doi.org/10.1016/j.orgel.2018.12.009.
105. Ahmad, Z., Taguchi, D., Paek, S., Mishra, A., Bhadra, J., Iwamoto, M., Manaka, T., Nazeeruddin, M. K. (2020). Detection of voltage pulse width effect on charge accumulation in PSCs using EFISHG measurement, *Res. Phys.*, 17, 103063, https://doi.org/10.1016/j.rinp.2020.103063.
106. Ahmad, Z., Taguchi, D., Paek, S., Mishra, A., Bhadra, J., Iwamoto, M., Manaka, T. and Nazeeruddin, M. K. (2020). Detection of voltage pulse width effect on charge accumulation in PSCs using EFISHG measurement, *Res. Phys.*, 17, 103063, https://doi.org/10.1016/j.rinp.2020.103063.
107. Taguchi, D., Sumiyoshi, R., Chen, X., Manaka, T. and Iwamoto, M. (2014). Study of interface layer effect in organic solar cells by electric-field-induced optical second-harmonic generation measurement, *Thin Solid Films*, 554, 51–53, https://doi.org/10.1016/j.tsf.2013.05.056.
108. Suzue, Y., Manaka, T. and Iwamoto, M. (2005). Current–voltage characteristics of pentacene films: Effect of UV/ozone treatment on Au electrodes, *Jpn. J. Appl. Phys.*, 44, 561–565, https://doi.org/10.1143/JJAP.44.561.
109. Li, C.-Q., Manaka, T. and Iwamoto, M. (2004). Electric-field-induced second harmonic generation from vacuum-evaporated metallophthalocyanine film/metal electrode interface, *Jpn. J. Appl. Phys.*, 43, 2330–2334, https://doi.org/10.1143/JJAP.43.2330.

Chapter 8

Application of MDC and SHG for Modelling and Analysis of Organic Devices: Three-electrode System

MDC and SHG provide a powerful way for modelling and analysing organic devices. Gauss's law, in Maxwell's electromagnetic field theory, suggests us an insightful idea for probing the carrier motion in organic devices. That is, by probing "*the electric field diverging from moving electrons and holes in devices*, the carrier motion can be directly monitored on-time if the polarization induced in organic material surrounding the moving electrons and holes is probed. Time-resolved microscopic optical second-harmonic generation (TRM-SHG) measurement is a technique that is useful to probe propagation of the dielectric polarization induced by electrons and holes which are moving across the organic materials. In TRM-SHG, nonlinear polarization induced in solids due to the coupling with incident electromagnetic waves, *e.g.*, laser beam, and d.c. static electric field generated from moving carriers is visualized on-time. SHG image on a charge-coupled device (CCD) camera directly captures motion of carriers in organic devices. This Chap. 8 is a companion to Chap. 7. Analysis and modelling of three-electrode system such as organic field-effect transistors (OFET) are described based on an idea that was established by using TRM-SHG measurement. Section 8.1 and Sec. 8.2 are devoted to describe steady state and transient state carrier processes, respectively. Determination of carrier mobility will be described in Sec. 8.2. Section 8.3 describes visualization of anisotropic carrier motion.

8.1 Probing and modelling of steady state by EFISHG

8.1.1 *Monopolar OFET*

OFET is one of the most important organic electronics devices. OFET is basically three-electrode system, and functions as a switching element that changes on-state and off-state alternatingly, according to the input Gate voltage. That is, Drain–Source current flows at on-state, whereas it does not flow at off-state.

To analyse the I–V characteristics of OFET using organic semiconductors, it is quite important to get information on the charge distribution along the OFET channel, and this will be of help for modelling the carrier behaviors. Conventionally, the analysis of OFET characteristics starts on the basis of device physics, in particular, on the basis of semiconductor device physics [1]. However, it should be noted that the Maxwell–Wagner model can be well utilized for analysing the I–V characteristics of OFETs, even in the case the active layer of OFETs shows dielectric behavior, rather than semiconductor behavior. The key concept of the "transistor" is to transfer the accumulated charge Q_s at the interface between the gate insulator/organic semiconductor layer along the interface by applying the external electric field between source and drain electrodes. Charge accumulation happens at any kinds of two material interface, due to the Maxwell–Wagner effect (see Chap. 4). This is the basic concept of the transistor proposed by Shockley [2].

According to this idea, we can derive the I–V characteristics of the three-electrode system such as OFET illustrated in Fig. 8.1, in the way as [3]

$$I_{ds} = When\ v = Wh\frac{Q_s}{h}v = WQ_s v \quad \text{with} \quad Q_s = C_g\bar{V} \quad \text{and}$$

$$v = \mu\bar{E} = \mu\frac{V_{ds}}{L}, \tag{8.1}$$

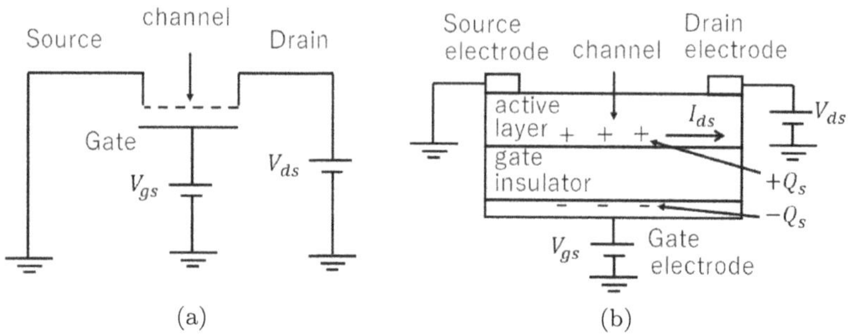

Figure 8.1. (a) A symbol of FET. (b) An OFET structure corresponding to the symbol.

where W, L, and h, respectively, are channel width, length and height. $\bar{V}$ is the average voltage across the gate insulator, Q_s is average charge density in channel, and C_g is capacitance of gate insulator. Generally, the electrical resistance of insulator is high in comparing with the resistance of the active layer, and $\bar{V}$ is approximately given as a function of V_{gs} (gate–source voltage), V_{ds} (drain–source voltage) and V_{th} (threshold voltage). Noteworthy the above equation is derived without assuming the semiconductor property of the organic semiconductor. Assuming that interfacial charge Q_s is mobile and available as carrier transporter in the OFET channel, under the linear approximation of the potential along the channel (gradual channel model), drain–source current I_{ds}, which flows along the interface is derived in the way as

$$I_{ds} = WQ_s\mu\frac{V_{ds}}{L} \quad \text{with} \quad V_{th} = 0 \quad \text{and} \quad Q_s = C_g\left(V_{gs} - \frac{1}{2}V_{ds}\right). \tag{8.2}$$

On the other hand, assuming that Q_s is partially available for conduction of I_{ds}, owing to the V_{th} shift, Eq. (8.2) is modified accordingly.

$$I_{ds} = WQ_g\mu\frac{V_{ds}}{L} \text{ with nonzero } V_{th} \quad \text{and}$$
$$Q_g = C_g\left(V_{gs} - V_{th} - \frac{1}{2}V_{ds}\right). \tag{8.3}$$

Q_g represents the mobile charge accumulated linearly along the channel at the interface between gate insulator and semiconductor, and they are conveyed by the constant electric field $\bar{E} = V_{ds}/L$. The threshold voltage V_{th} is the external voltage at which mobile carrier begins to accumulate. Noteworthy, this Eq. (8.3) is the same with the I_{ds}–V_{ds} and I_{ds}–V_{gs} equation which can be derived based on the semiconductor device physics, in the limit where the accumulated charges are confined at the interface. The OFET characteristics analysed using Eq. (8.3) well agree with the experimental data, as shown in Fig. 8.2 [3], with relation:

$$I_{ds} = WQ'_g\mu\frac{V_{ds}}{L} \quad \text{with} \quad V_{th} = 0 \quad \text{and} \quad {Q'}_g = \theta Q_s$$
$$\text{with } 0 < \theta < 1. \tag{8.4}$$

θ is a fitting parameter, and it represents a ratio of mobile charge with reference to accumulated charge Q_s.

Noteworthy that Eq. (8.4) can be replaced with the following Eq. (8.5), assuming that carrier mobility changes in the way as $\mu' = \theta\mu$ with $0 < \theta < 1$.

$$I_{ds} = WQ_g\mu'\frac{V_{ds}}{L} \quad \text{with } V_{th} = 0. \tag{8.5}$$

The I_{ds}–V_{ds} characteristics changes in proportion to θ in the same way for Eq. (8.4) and Eq. (8.5), indicating that we can fit the experimental results by employing this fitting parameter θ, appropriately, but it is difficult to distinguish the contribution to θ merely based on the I–V measurement. In other words, it is necessary to directly probe the fundamental carrier process along the channel.

TRM-SHG measurement [4] is available for this purpose as follows: At first, for deeply analysing the transistor characteristics, it is important to get information on the potential distribution along the interface from the edge of Source electrode to the Drain electrode, because we can directly determine carrier distribution along the channel. By using the TRM-SHG measurement, the electric field distribution along the channel of pentacene-OFETs can be explored. The electric field distribution in the channel of pentacene-OFETs

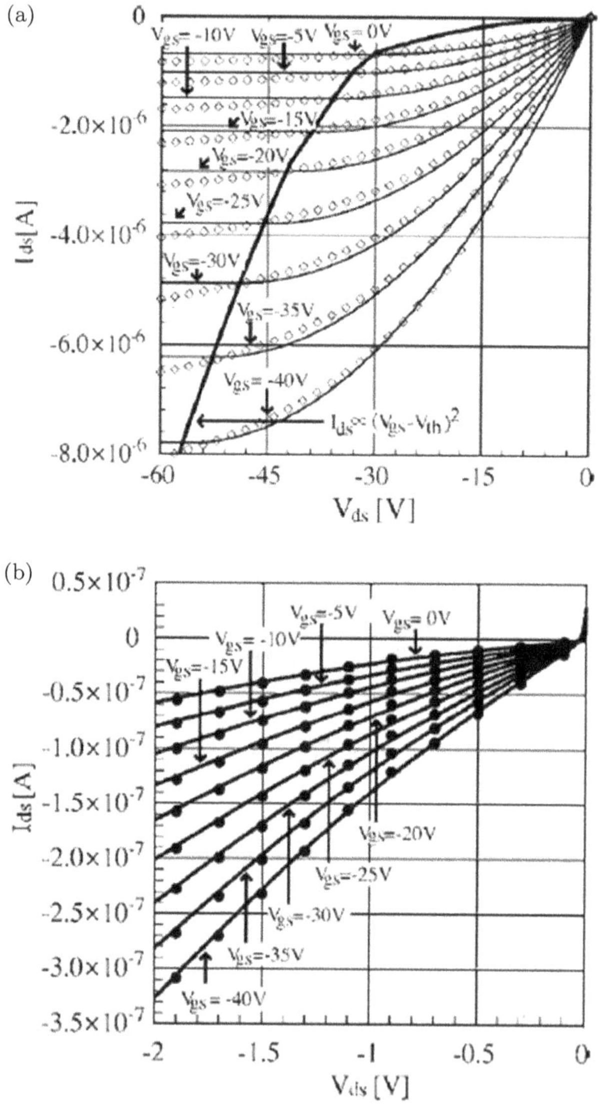

Figure 8.2. Typical I_{ds}–V_{ds} characteristics of pentacene FET, (a) in the middle electric and high electric field regions and (b) in the low electric field region. Reprinted from R. Tamura, E. Lim, T. Manaka, M. Iwamoto, J. Appl. Phys., Vol. 100, 114515, 2006, with the permission of AIP Publishing [3].

before and after hole injection into the channel were well explored by making use well of the delay time in the TRM-SHG measurement [5, 6], and the electric field distribution before hole injection is shown to be in good agreement with the distribution analyzed using the rigorous solution of the Laplace field of the electrode system of OFETs [7, 8]. That is, the observed field distribution $\vec{E}(0)$ is well described by Gauss law $\nabla \cdot \vec{E}(0) = \frac{\rho}{\varepsilon_s \varepsilon_0}$ with $\rho = 0$. Furthermore, the electric field distribution due to injected carriers has been determined from the TRM-SHG, and it has been shown that the charge distribution along the OFET channel is basically determined on the basis of a Maxwell–Wagner model that accounts for carrier accumulation at the interface between the gate-insulator and active organic layer [7, 8].

Figure 8.3 shows a typical example of the microscopic SHG measurement carried out for top-contact pentacene OFETs with Au source and drain electrodes. Under the condition of no carrier injection of the OFETs, EFISHG is shown to be activated to satisfy the electric field distribution formed in a three-electrode system without charge injection, as shown in Fig. 8.3(a). On the other hand, under the condition of hole-injection, the intensity of EFISHG decreases due to the space charge field formed along the channel by injected holes, as shown in Fig. 8.3(b). Using Gauss law and the relation

$$I(2\omega) \propto |\vec{P}_{NL}(2\omega)|^2 = |\varepsilon_0 \chi^{(3)}(-2\omega; 0, \omega, \omega) \vdots \vec{E}(0)\vec{E}_i(\omega)\vec{E}_j(\omega)|^2, \tag{8.6}$$

the carrier distribution along the channel is determined. These results show that carrier accumulation along the channel is governed by the Maxwell–Wagner effect, and the distribution is basically analyzed based on a Maxwell–Wagner model [7].

The most intuitive idea would be the charge distribution that follows a gradual channel approximation model, where the potential distribution is built linearly along the channel with a constant slope from the edge of the Drain electrode to the edge of the Source electrode. However, the microscopic SHG study of pentacene OFET shows that the carrier distribution along the channel in the steady-state follows a solution of a classical differential equation of current

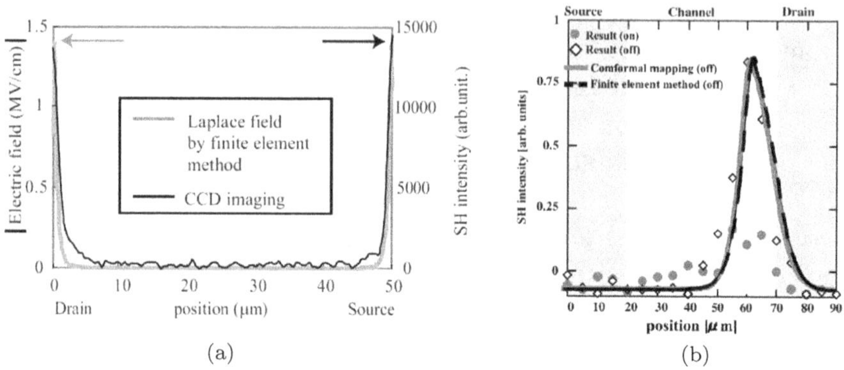

Figure 8.3. (a) The electrostatic field (Laplace field) formed in the pentacene field-effect transistor. The black curve is the electric field distribution which was directly visualized by TRM-SHG measurement. The thick gray curve is a numerical calculation by a finite element method. Reprinted from D. Yamada, T. Manaka, E. Lim, R. Tamura, M. Weis, M. Iwamoto, J. Appl. Phys., Vol. 104, 074502, 2008, with the permission of AIP Publishing [7]. (b) The electric field distribution at the drain electrode edge of the pentacene field-effect transistor. The filled circles and open squares plot electric field distributions with and without hole injection, under d.c. voltage application to the gate electrode with reference to the Au Source electrode ($V_{gs} < 0$). Reprinted from D. Yamada, T. Manaka, E. Lim, R. Tamura, M. Iwamoto, J. Appl. Phys., Vol. 103, 084118, 2008, with the permission of AIP Publishing [8].

that describes the continuity of carrier transport along the channel. That is, classical transmission line model based on an extension of the Maxwell–Wagner model [9, 10] well accounts for steady state charge distribution in pentacene OFET.

Analytical results show that the electrostatic potential distribution along the FET channel is not linear, but it changes on satisfying the square root dependence along the channel, in a manner as

$$\begin{aligned} V(x) &= V_{ds} - V_{gs} + V_0\sqrt{1 - \frac{x}{L}} \quad \text{with } V_0 = -V_{ds}, \\ V(0) &= -V_{gs} \quad \text{and} \quad V(L) = V_{ds} - V_{gs}, \end{aligned} \tag{8.7}$$

where $x = 0$ and $x = L$ represents the edge of the source and the drain electrodes, respectively. $V(x)$ is potential at OFET channel in reference to gate electrode potential. Accordingly, the electric field along with the channel length direction is given as

$$E(x) = -\frac{dV(x)}{dx} = \frac{V_0}{2\sqrt{L(L-x)}} \tag{8.8}$$

EFISHG probes this electric field distribution well, and generates the signals that follow Eq. (8.6) with Eq. (8.8), where $E(0)$ of Eq. (8.6) is replaced by the $E(x)$ of Eq. (8.8). To further clarify the details, Fig. 8.4(a) shows the plots of calculation (solid line) and experimental results (symbols) of the normalized electric field of the conduction carriers in the channel region of OFET in the on-state [10]. The calculation is based on the rigorous solution of the differential equation, but a small difference is identified in experimental electric intensity and the calculated result of the electric field of the charged layer (indicated by solid curves), possibly owing to the presence of trapped carriers that contribute to the additional electric field.

Figure 8.4(b) shows the electric field distribution obtained from SHG experiment plotted in the log–log scale, and the line represents the linear fit, suggesting the square root dependence of electric field along the channel. As described above, it is clear that EFISHG is available for probing the steady-state carrier distribution along the channel of OFETs, and for actual OFET transistors, the potential distribution along the channel changes on satisfying the square root dependence given by Eq. (8.7) based on a transmission line model. Therefore, the equation of the drain–source current I_{ds} given by Eq. (8.2) under the gradual approximation model should be slightly modified. After a simply calculation, we can reach a conclusion that the coefficient 1/2, representing a linear potential profile, changes into 1/3, in a manner as Eq. (8.10).

For gradual channel model

$$Q_s = \frac{1}{L}\int_0^L C_g\left\{V_{ds} - V_{gs} + V_0\left(1 - \frac{x}{L}\right)\right\}dx = -C_g\left(V_{gs} - \frac{1}{2}V_{ds}\right)$$

$$I_{ds} = WC_g\left(V_{gs} - V_{th} - \frac{1}{2}V_{ds}\right)\mu\frac{V_{ds}}{L}. \tag{8.9}$$

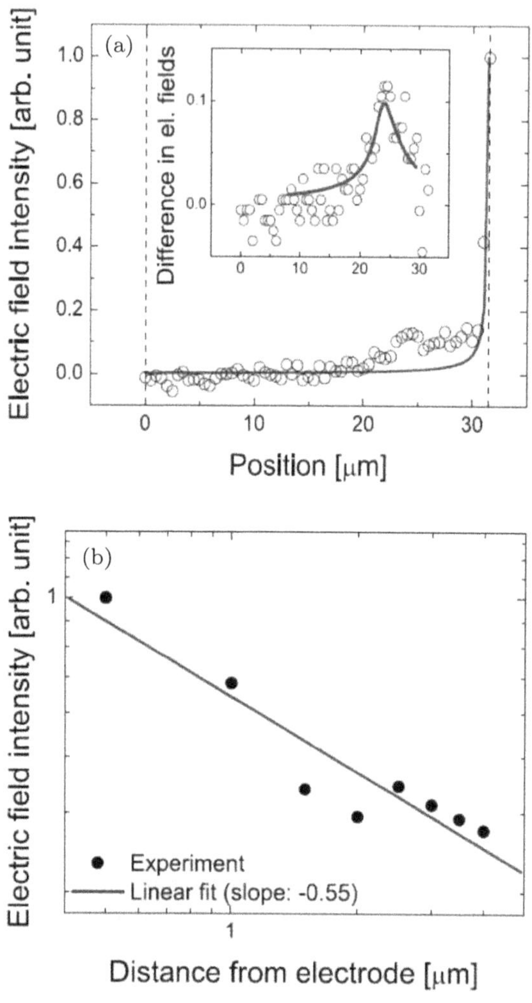

Figure 8.4. (a) Electric field $E(x)$ in the channel region of pentacene OFET in the on-state. Source and Drain electrodes are positioned at 0 and 30 μm, respectively. The solid curve is electric field calculated based on transmission line model (TLM). (b) Log–log plot of $E(x)$ as a function of distance from Drain electrode edge. A good fit between calculated curve and SHG measurement result indicates carriers in the pentacene OFET follow TLM rather than gradual channel model. Reprinted from M. Weis, T. Manaka, M. Iwamoto, J. Appl. Phys., Vol. 105, 024505, 2009, with the permission of AIP Publishing [10].

For transmission line model

$$
\begin{aligned}
Q_s &= \frac{1}{L}\int_0^L C_g\left\{V_{ds} - V_{gs} + V_0\sqrt{1-\frac{x}{L}}\right\}dx \\
&= -C_g\left(V_{gs} - \frac{1}{3}V_{ds}\right) \quad \text{with zero } V_{th} \\
Q_s &= -C_g\left(V_{gs} - V_{th} - \frac{1}{3}V_{ds}\right) \quad \text{with nonzero } V_{th} \\
I_{ds} &= WC_g\left(V_{gs} - V_{th} - \frac{1}{3}V_{ds}\right)\mu\frac{V_{ds}}{L},
\end{aligned}
\tag{8.10}
$$

when the square root dependent electric field is formed along the channel. This means that the I–V characteristics of the transistor slightly deviates from Eq. (8.9), but the deviation is not large. This is the reason why Eq. (8.9) and Eq. (8.10) reproduce well the experimental data, as shown in Fig. 8.2.

As mentioned above, probing the electric field distribution along the channel by EFISHG is very important to model I–V of OFET. It is instructive here to note that there are carrier traps in organic semiconductor layer, owing to defects, impurities, low crystallinities, *etc.* In this situation the number of carriers that can transport across FET channel might be restricted. One way to take into consideration of this effect is to introduce the parameter θ in the derived equation in Eq. (8.4), where θ describes the thermodynamic equilibrium condition between trapped carriers and free carriers. This idea is adopted in the SCLC analysis of carrier transport of electrical insulators [11]. The potential distribution well reflects the electrical property of materials used in OFET. Pentacene-OFET shows square-root dependence following the transmission line model, TIPS–pentacene OFET shows linear dependence following the gradual channel approximation model, and so on, indicating that interfacial charge distribution is depending on electronic structure of organic semiconducting materials used in OFET [12]. These results clearly demonstrate that probing of potential distribution along the channel is very effective to model I–V characteristics of OFET.

8.1.2 *Ambipolar OFET*

Some organic materials show ambipolar behavior, which means that electrons and holes are allowed to enter materials from the facing electrodes, respectively, and they are serving as carriers of devices. In the field of electrical insulation engineering, this carrier behavior is called double-injection [13]. Consequently, probing electron and hole injection, and their distribution in organic devices is very important. For example, for organic light emitting device application, double carrier injection, *i.e.*, electron and hole injection, is a key for bright light emission. Using ambipolar organic light emitting transistors (OLETs) with an active layer of the green-light-emitting polymer of poly(9,9–dioctylfluorene–*alt*–benzothiadiazole) (F8BT) [14, 15], which gives a photo luminescent peak at a wavelength of 560 nm, the EFISHG measurement has been employed using an incident laser beam with a wavelength of 840 nm (SHG wavelength 420 nm) [16, 17]. In this EFISHG experiment, an extra band pass filter was used so that we can selectively probe EFISHG that is generated at a wavelength of 420 nm only and luminescence from F8BT is totally blocked.

Figure 8.5(a) shows the structure of OLETs, where hole and electron transfer into the active layer (F8BT) is essential for light emission. Figure 8.5(b) shows an image of electroluminescence (EL) from the F8BT OLETs, which is locally generated from the channel region and it is visualized as a bright band by using a CCD camera. This emission band corresponds to the meeting position of electrons and holes, followed by recombination for EL, but this image does not directly exhibit the distribution of holes and electrons along the channel. An EFISHG image shown in Fig. 8.5(c) uncovers the distribution of holes and electrons, because EFISHG is generated in proportion to the intensity of electric field arising from electrons and holes, as given by Eq. (8.6) [17]. Therefore, we can map the carrier distribution along the channel region experimentally by making use well of the Gauss law.

The EFISHG experiments showed that there is actually a zero-potential position in the channel, which corresponds to the meeting point of electrons and holes for the electroluminescence [17]. The electric field in the gate insulator is pointing from positive charge in the

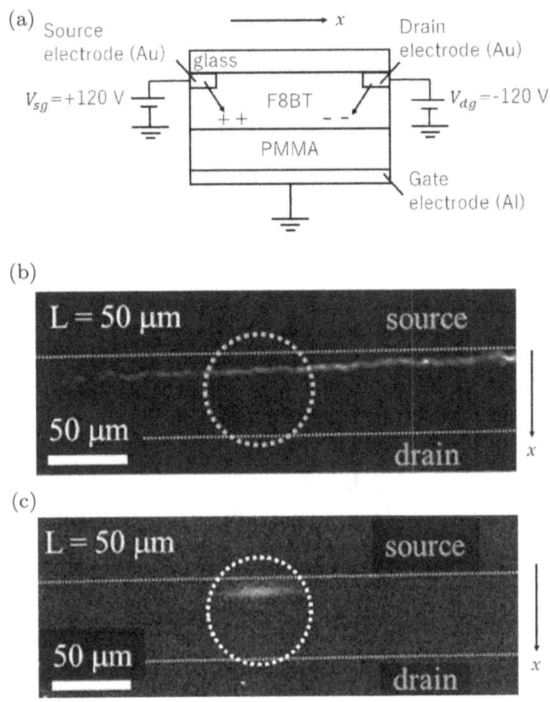

Figure 8.5. (a) Device structure of OLET. F8BT and PMMA thickness are 80 nm and 1 μm, respectively. Channel length is 50 μm. (b) EL emission (wavelength 560 nm) in steady state. (c) EFISHG (wavelength 420 nm) of F8BT in steady state. Dotted lines are edge of source and drain electrodes. Dotted circle is a laser spot area for EFISHG measurement [17].

channel toward negative charges induced on gate electrode surface, in the region from source electrode edge to the zero-potential position. The electric field changes direction from the zero-potential position to drain electrode edge in gate insulator, the field is pointing from positive induced charge on gate electrode toward negative charges in the channel. Figure 8.6 illustrates this distribution of the intensity of EL and evaluated lateral electric field (in x-direction according to Fig. 8.5(a)) in OLET with a channel length of 50 μm.

Interestingly, the EL reaches maximum at the position of 9 μm (from the source electrode), which is almost the same as a position of the maximum electric field (at the position of 8.5 μm). The difference between the peak positions is on the level of the EFISHG

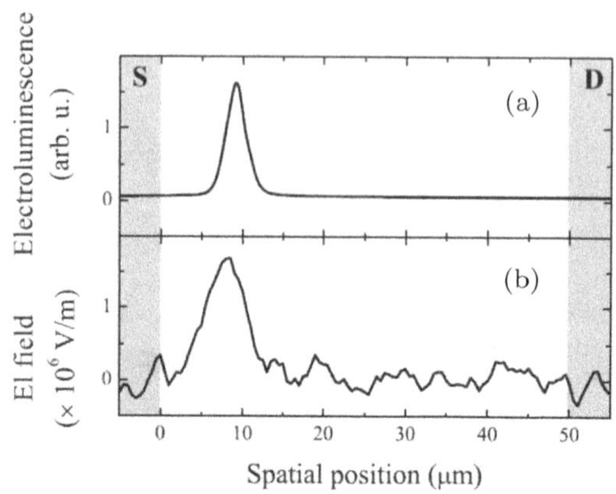

Figure 8.6. The spatial distribution of (a) emitted electroluminescent (EL) light and (b) lateral electric field (El field) in OLET, evaluated by using EFISHG experiment. Gray areas represent the source (S) and drain (D) electrodes. Reprinted from Y. Mashiko, D. Taguchi, M. Weis, T. Manaka, M. Iwamoto, Appl. Phys. Lett., Vol. 101, 243302, 2012, with the permission of AIP Publishing [17].

experiment resolution ($0.5\,\mu$m) and thus can be neglected. In other words, the maximum peak of EL intensity is recorded at the position of the strongest electric field. The result indicates that the hole–electron recombination process is governed by Coulomb attraction between holes and electrons at the meeting point (zero-potential position). This situation is totally different from inorganic light-emitting diodes, where diffusion of minority carrier over p–n junction and succeeding recombination is a major part of carrier process. We need to model carrier behavior based on this experimental fact.

To understand these experimental results, we need to pay attention to the potential profile across the channel region again. In the case of the ambipolar transport, the charge is accumulated on the organic semiconductor–gate insulator interface in accordance with the interface potential. It should be noted here that the ambipolar transport appears in the saturated region of both kinds of charge carriers, electrons and holes. Hence, the spatial limitation of the accumulated charge distribution must be present. As a result, a part of the channel is charged by holes, while beyond the *zero-potential* position x_0, the interface potential polarity changes and electrons are accumulated.

Creation of the space-charge field by both kinds of carriers, *i.e.*, electrons and holes, leads to simple relation for electrostatic potential at FET channel on satisfying the square root dependence, from the edges of Drain and of Source, respectively, in a manner as the relation given by Eq. (8.11).

$$V(x) = V_{sg}\sqrt{1 - \frac{x}{x_0}}, \quad (0 < x < x_0)$$
$$V(x) = V_{dg}\sqrt{1 - \frac{L-x}{L-x_0}}, \quad (x_0 < x < L). \tag{8.11}$$

On the other hand, the electric field distribution can be probed by using the EFISHG measurements, and it can be converted to the potential profile. Fitting result of experimental data by Eq. (8.11) is depicted in Fig. 8.7. Apparently, there is a good agreement between the experimental data and the dielectric model prediction. In addition, the dielectric approach does not require any assumption on the voltage or temperature dependent mobility, percolation, and overlap or localized states. Hence, the approach taking account of space-charge field of accumulated interface charges (the Maxwell–Wagner

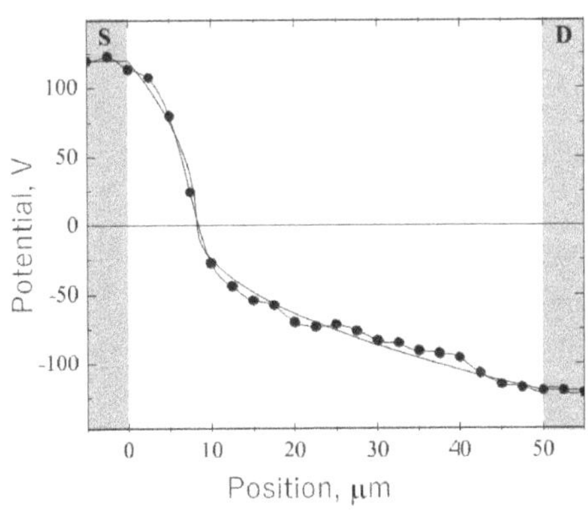

Figure 8.7. The potential profile reconstructed from the EFISHG experiment (black circles) and calculation based on the transmission line model for OLET (solid curve). Gray areas represent the source (S) and drain (D) electrodes. Reprinted from Y. Mashiko, D. Taguchi, M. Weis, T. Manaka, M. Iwamoto, Appl. Phys. Lett., Vol. 101, 243302, 2012, with the permission of AIP Publishing [17].

model) fits experiment without requirements on electrical parameters dependencies. The potential profiles lead to hole and electron steady current, I_h and I_e, in a way similarly to Eq. (8.10) as

$$I_h = \frac{2}{3} C_g V_{sg} \mu_h \frac{V_{sg}}{x_0}$$
$$I_e = \frac{2}{3} C_g V_{dg} \mu_e \frac{V_{dg}}{L - x_0}. \tag{8.12}$$

With non-zero threshold voltage for mobile charge generation for holes $V^h_{th,g}$ and for electrons $V^e_{th,g}$ (the subscript g indicates the value is in reference to gate electrode potential), Eq. (8.12) is modified as

$$I_h = \frac{2}{3} C_g (V_{sg} - V^h_{th,g}) \mu_h \frac{V_{sg}}{x_0}$$
$$I_e = \frac{2}{3} C_g (V_{dg} - V^e_{th,g}) \mu_e \frac{V_{dg}}{L - x_0}. \tag{8.13}$$

The coefficient of 2/3 is coming from square-root dependence of potential profile observed in SHG measurement. If the observed profile is linearly changing, appropriate conduction current model is

$$I_h = \frac{1}{2} C_g V_{sg} \mu_h \frac{V_{sg}}{x_0}$$
$$I_e = \frac{1}{2} C_g V_{dg} \mu_e \frac{V_{dg}}{L - x_0}. \tag{8.14}$$

In Eq. (8.14), the coefficient 1/2 appears reflecting linear potential profiles.

The zero-potential position (called pinch-off in unipolar transport) in the ambipolar transport region can be estimated from the equality of currents I_{ds} due to electrons and holes (balance between the electron and the hole current) as follows:

$$x_0 = \frac{\mu_h (V_{gs} - V^h_{th,s})^2}{\mu_h (V_{gs} - V^h_{th,s})^2 + \mu_e (V_{gd} - V^e_{th,s})^2} L \tag{8.15}$$

$V^h_{th,s}$ and $V^e_{th,s}$ indicates non-zero threshold voltage in reference to Source electrode. Assuming applied voltages in the experiment and threshold voltages and mobilities for electrons and holes estimated from the transfer characteristic, $x_0 = 7.6\,\mu\text{m}$ is obtained. This value supports the experimental observation ($8.5\,\mu\text{m}$) and the deviation dwells in the imperfect estimation of electrical parameters due to additional effects, such as charge trapping. As mentioned above, the EL and potential distribution in the channel region of the ambipolar OFET was observed and it was analysed to model carrier behavior of OLET. It was found good agreement between the electrostatic potential profiles determined by experiment and the potential profile predicted by using the transmission line model. This analysis of OFET potential profiles in ambipolar transport modes points out the importance of the discussion based on Maxwell–Wagner effect charge distributions and the calculation based on an appropriate model for further development of OFET and OLET device technologies. EFISHG measurement provides a powerful way to directly probe potential profile along OFET and OLET for modelling a device operation.

Noteworthy that I–V of OLET based on transmission line model (Eq. (8.12)) and gradual channel model (Eq. (8.14)), location of zero potential position is the same expression in Eq. (8.15). Accordingly, merely based on the identification of zero potential position resulting from current balance relationship, we cannot judge whether gradual channel model or transmission line model is better for modelling of a device. By using EFISHG measurement, the valid model is determined based on the visualized charge distribution in OLET channel. As mentioned above, steady state carrier distribution is well probed by using EFISHG, and well accounts for the I–V characteristics of OLET. The interfacial charge Q_s gives also greatly effect on the transient carrier behaviors of double-injection device (see Sec. 7.7). For example, EL emission governed by interfacial charge Q_s is found in tetracene OLET [18]. In other words, EFISHG measurement of tetracene OLET evidently showed that space charge field originating from Q_s directly gives effects on the EL [19].

8.1.3 *Memory OFET*

Ferroelectric polarization reversal and charge trapping are utilized to make memory OFETs. In analysis and modelling of memory

OFETs, effect of polarization reversal and charge trapping are freshly installed in addition to a model for ordinary OFET. Keeping this in mind, carrier injection [20], accumulation [21–23], transport processes [24], and other fundamental carrier processes, have been analyzed in terms of Maxwell–Wagner effect. By assuming that electrostatic fields from spontaneous polarization P_s and trapped charge ρ_t are falling onto gate electrode, electrostatic potential at FET channel $V(x)$ of OFET (see Eq. (8.7)), in reference to gate electrode potential, is shifted [25]: for OFET with upward $+P_s$ and downward $-P_s$ polarization in gate insulator,

$$\Delta V = \pm \frac{P_s}{C_0}, \tag{8.16}$$

and, for OFET with trapped charge $\rho_t(z)$ in gate insulator:

$$\Delta V = \frac{Q_t'}{C_0} \quad \text{with} \quad Q_t' = \frac{\bar{x}}{d} Q_t. \tag{8.17}$$

C_0 $(=\varepsilon/d)$ in Eq. (8.16) and Eq. (8.17) is a capacitance of gate insulator as a parallel capacitor with thickness d and dielectric constant ε. Q_t is total trapped charge $Q_t = \int_0^d \rho_t(z)dz$ and charge centroid $\bar{x}$ is as defined in Eq. (3.24). The electrostatic potential ΔV is one of the candidates that govern memory behavior. Probing of mobile charge Q_s in the channel of FET, in relation to the state of P_s and Q_t' is key to analyse and model memory FETs.

As an example of device structure of memory OFET, pentacene–P(VDF–TrFE) OFET with ferroelectric polarization is illustrated in Fig. 8.8. In the following, memory behavior of this OFET is analyzed. Figure 8.9 shows I_{ds}–V_{gs} and I_{gs}–V_{gs} characteristics of the memory OFET. In Fig. 8.9(a), a hysteresis loop of drain–source current I_{ds} is seen, reflecting upward and downward polarization of ferroelectric gate insulating layer. For example, polarization and carrier distribution change as illustrated in Fig. 8.9(c), and give the hysteresis loop in I_{ds}–V_{gs} characteristics. At $V_{gs} = +40V$ (see panel 4 of Fig. 8.9(c)), in a part of FET channel nearby source and drain electrodes, polarization is upward ($P_s > 0$) and electrons accumulate ($Q_s < 0$) only there. The electrons path is broken in the middle of channel and I_{ds} is zero. By sweeping gate

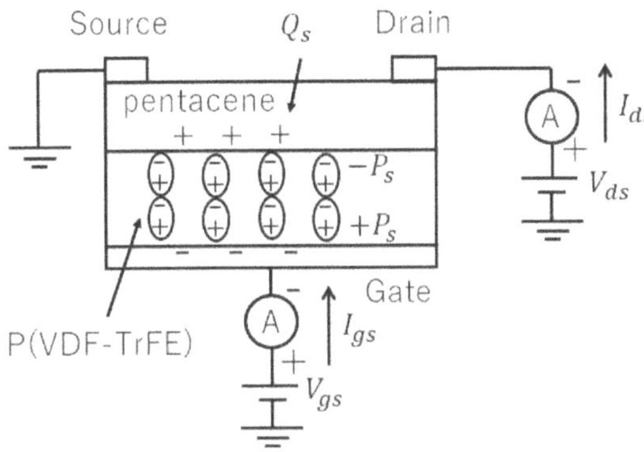

Figure 8.8. Device structure of Al (Gate)/P(VDF–TrFE)/pentacene/Au (Source and Drain) memory OFET. P(VDF–TrFE) is ferroelectric gate insulating layer. P(VDF–TrFE) and pentacene thickness are 200 nm and 50 nm, respectively. Channel length and width are 50 μm and 2 mm.

voltage from +40 V to −40 V, upward polarization turns-over to downward polarization at $V_{gs} = V_{th1} = -16\,\text{V}$, and non-zero I_{ds} begins to flow with spreading of holes onto downward polarization. At $V_{gs} = -40\,\text{V}$, polarization is downward ($P_s < 0$) over entire region of FET channel and holes distribute over entire region ($Q_s > 0$) as expected from transmission line model. By sweeping V_{gs} from −40 V to +40 V, polarization reversal occurs when the electrostatic field in the ferroelectric gate layer increases beyond the coercive field E_c (=0.85 MV/cm) of ferroelectric layer. At $V_{gs} = V_{th2} = +14\,\text{V}$, polarization reversal from downward to upward occurs and drain–source current returns to zero $I_{ds} = 0$ (see transition from panel 3 to 4 in Fig. 8.9(c)). Owing to the different state of spontaneous polarization, threshold voltage depends on the sweeping direction of gate voltage $V_{th1} \neq, V_{th2}$, and the hysteresis loop is seen. The results suggest that spontaneous polarization governs the formation and removal of conductive FET channel, but an actual situation, for example validity of transmission line approximation, is unknown. EFISHG measurement is available to directly probe turn-over of polarization P_s and charge behavior Q_s by visualizing electrostatic field originating from them. An example of EFISHG measurement of memory devices in two-electrode system is described in Sec. 7.8.

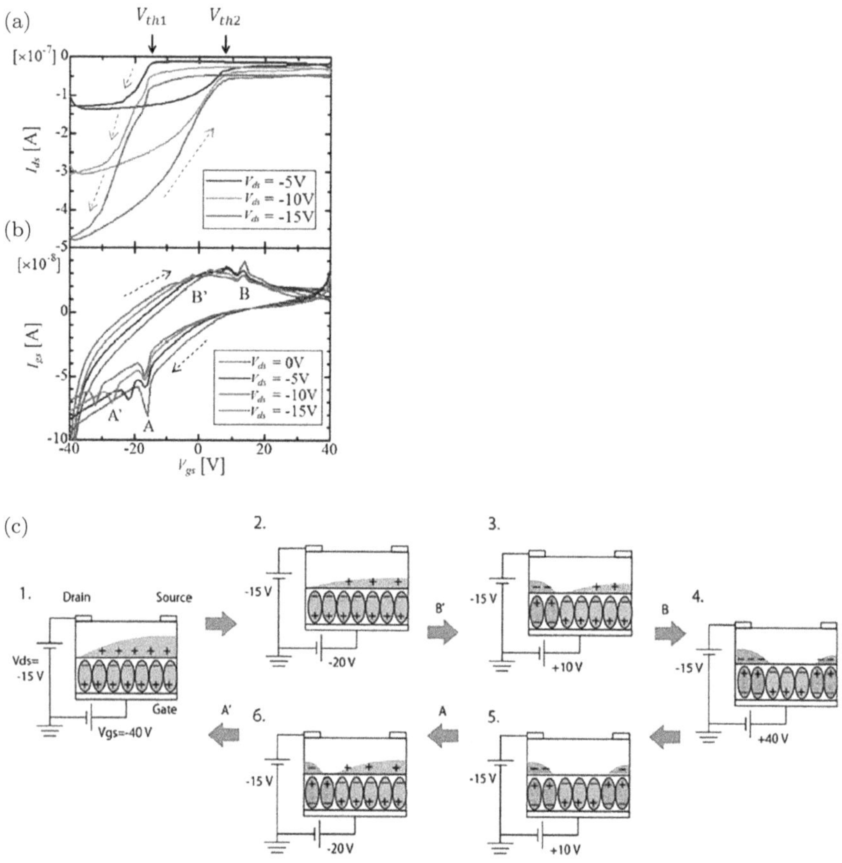

Figure 8.9. (a) I_{ds}–V_{gs} characteristics of memory OFET (Al/P(VDF–TrFE)/pentacene/Au) where two different source–drain current levels at V_{gs} = 0 V are switched on account of polarization direction of ferroelectric gate insulating layer. (b) DCM (I_{gs}–V_{gs}). (c) Electron and hole distribution and polarization distribution in a course of I–V measurement. Reprinted from T. Otsuka, D. Taguchi, T. Manaka, M. Iwamoto, J. Appl. Phys., Vol. 122, 185501, 2017, with the permission of AIP Publishing [26].

Figure 8.10 shows hole and electron distribution visualized by using charge modulation reflectometry (CMR) [26]. Accumulated holes and electrons are illustrated in dark in the figure. Figures 8.10(a) and (b) is an image of hole distribution in the channel at $V_{gs} = -40$ V ($V_{ds} = 0$ V). Figures 8.10(c) and (d) is an image of

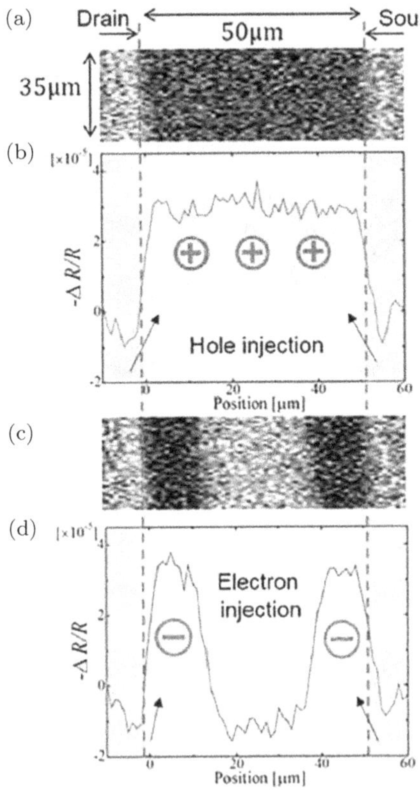

Figure 8.10. Hole and electron distribution in Memory OFET (Al/P(VDF–TrFE)/pentacene/Au). (a) Hole distribution visualized as a darkness of CMR image ($V_{gs} = -40\,\text{V}$). (b) Hole distribution along channel direction obtained from the CMR image. (c) Electron distribution visualized by using CMR image ($V_{gs} = +40\,\text{V}$). (d) Electron distribution along channel direction obtained from the CMR image. The source and drain electrodes are short-circuited ($V_{ds} = 0\,\text{V}$). Reprinted from T. Otsuka, D. Taguchi, T. Manaka, M. Iwamoto, J. Appl. Phys., Vol. 122, 185501, 2017, with the permission of AIP Publishing [26].

electrons in the channel at $V_{gs} = +40\,\text{V}$ ($V_{ds} = 0\,\text{V}$). Different hole and electron distributions are clearly seen. In Figs. 8.10(a) and (b), CMR image is dark in the entire region between source and drain electrode located at both side of the image. This image shows that holes are occupying whole region of the FET channel. On the other hand, as Figs. 8.10(c) and (d) show, electrons are only accumulating

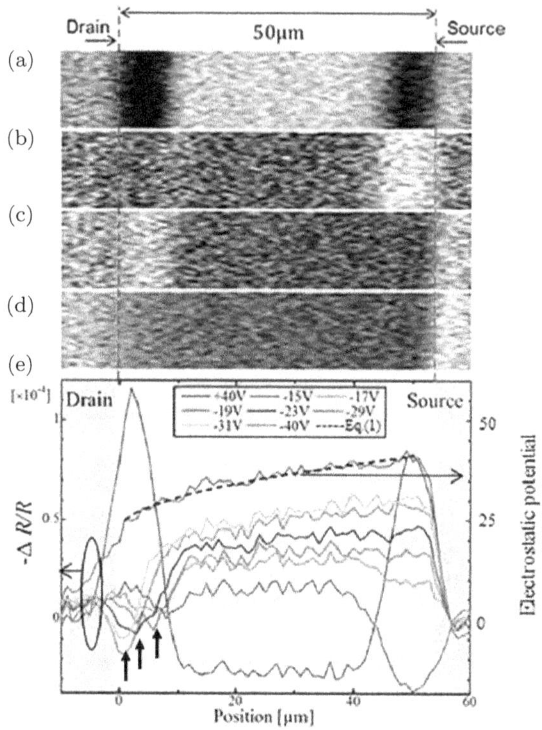

Figure 8.11. Hole and electron distribution in memory OFET (Al/P(VDF–TrFE)/pentacene/Au) visualized as a darkness of CMR image at $V_{ds} = -15\,\text{V}$. (a) $V_{gs} = +40\,\text{V}$, (b) $V_{gs} = -15$ V, (c) $V_{gs} = -23\,\text{V}$, and (d) $V_{gs} = -40\,\text{V}$. (e) Charge distribution across the channel region. Dotted curve is hole distribution fitted by assuming transmission line model. A series of vertical arrows indicates the location of frontier of polarization turn-over. Reprinted from T. Otsuka, D. Taguchi, T. Manaka, M. Iwamoto, J. Appl. Phys., Vol. 122, 185501, 2017, with the permission of AIP Publishing [26].

in the region with 10 μm from the source and drain electrode edges. These results indicate that spontaneous polarization $-P_s$ in the entire region of channel is supporting holes while this polarization turn-over to $+P_s$ only in the region nearby source and drain electrodes. Figure 8.11 is charge distribution by CMR with non-zero drain–source voltage $V_{ds} = -15\,\text{V}$ is applied where I_{ds} is allowed to flow across the FET channel. At $V_{gs} = -40\,\text{V}$, holes distribute in entire region of FET channel, which are well fitted by a potential distribution expected by using transmission line model, *i.e.*, Eq. (8.7) (dotted curve in Fig. 8.11(e)). This result indicates that the transmission line model is valid for memory OFET with ferroelectric layer.

The hole distribution from $V_{gs} = -23\,\mathrm{V}$ to $-40\,\mathrm{V}$ shows the frontier of polarization reversal penetrates toward drain electrode as indicated by a series of vertical arrows in Fig. 8.11(e). This voltage region is just corresponding to $V_{gs} < V_{th1}$ where non-zero drain–source current I_{ds} begins to flow. Assuming that the holes in the channel distribute in accordance with transmission line model, the location x of frontier of polarization reversal is calculated by using CMR images. That is, the location x where electrostatic potential $V(x)$ in Eq. (8.7) satisfies the relation $V(x) = E_c d$ (E_c: coercive field of ferroelectric layer, d: thickness of ferroelectric layer), is calculated. Figure 8.12 shows the calculation (curve) and experimental observation (circles) of the location x. The good agreement between the calculation and experimental results indicates that, in Al/P(VDF–TrFE)/pentacene/Au memory OFET, polarization reversal from $+P_s$ to $-P_s$ goes, on accompanying holes spreading on $-P_s$, and governs flow of drain–source current I_{ds} in voltage

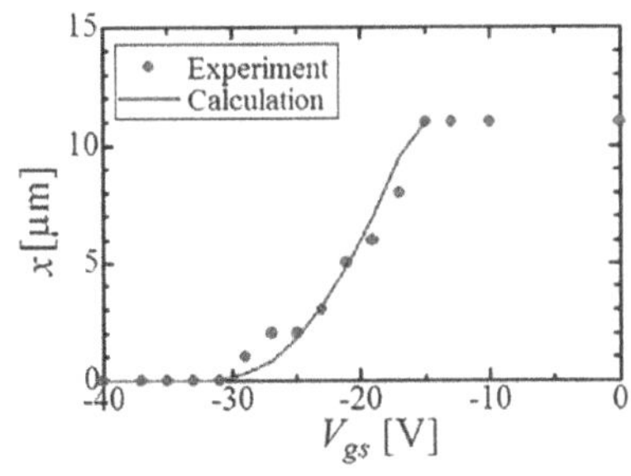

Figure 8.12. The position of polarization reversal frontier in the channel x ($x = 0$ at Source electrode edge) of memory OFET (Al/P(VDF–TrFE)/pentacene/Au). The solid curve is the calculation based on transmission line model (Eq. (8.7)). The circles represent experimental results from CMR imaging. The gate voltage is swept from $V_{gs} = -10\,\mathrm{V}$ to $-40\,\mathrm{V}$. In a course of voltage sweep, polarization reversal region penetrates toward Source electrode edge, and governs I_{ds}–V_{gs} characteristics of memory OFET. Reprinted from T. Otsuka, D. Taguchi, T. Manaka, M. Iwamoto, J. Appl. Phys., Vol. 122, 185501, 2017, with the permission of AIP Publishing [26].

region $V_{gs} < V_{th1}$. Connection between polarization reversal and carrier distribution in other voltage region is also well analysed, and is very useful to model the device operation of memory OFET. It is instructive to note that both holes and electrons are candidates for conductive carrier in memory OFET, similarly to ambipolar OFET such as OLET (Sec. 8.1.2). In OLET, hole and electron distributions are governed by a balance between hole current and electron current. By contrast, in memory OFET, distribution of spontaneous polarization dominates hole and electron distributions.

It is also worthy to note that TRM-SHG measurement is useful to analyse the relationship between spontaneous polarization and transient carrier behaviour in memory OFET. The details of TRM-SHG measurement is described in the next section, Sec. 8.2. Briefly, the transient behavior of hole injection and propagation in memory OFET is described [24]. The device structure is Al (Gate)/poly(vinylidene fluoride–tetrafluoroethylene) (P(VDF–TeFE))/pentacene/Au (Source and Drain) that is similar to a memory OFET shown in Fig. 8.8. P(VDF–TeFE) is a ferroelectric layer.

The two polarization conditions were set in prior to TRM-SHG measurement: case (i), d.c. gate voltage $V_{gs} = -80\,\text{V}$ was applied and spontaneous polarization is oriented downward, *i.e.*, negative charge $-P_s$ is on FET channel and positive charge $+P_s$ is on gate electrode: case (ii), d.c. gate voltage $V_{gs} = +80\,\text{V}$ was applied and spontaneous polarization is oriented upward, *i.e.*, positive charge $+P_s$ is on FET channel and negative charge $-P_s$ is on gate electrode. After setting the polarization condition, positive voltage pulse $V_s = +20\,\text{V}$ is applied to a source electrode for hole injection. TRM-SHG measurement was carried out to visualize propagation of the injected holes in the FET channel. Results showed that the hole propagation followed $x \sim \sqrt{t}$ relation (see Sec. 8.2) in both case (i) and case (ii), suggesting that the transmission line model is valid in the transient carrier motion in memory OFET. But the carrier mobility ($\mu_{h+} = 5.60 \times 10^{-4}\,\text{cm}^2/\text{Vs}$) of holes on downward polarization and the mobility ($\mu_{h-} = 0.93 \times 10^{-4}\,\text{cm}^2/\text{Vs}$) on upward polarization were different. This result indicates that hole motion on $-P_s$ is smooth but the motion on $+P_s$ is limited by speed of polarization reversal propagation, electrostatic repulsion between holes $Q_s > 0$ and spontaneous polarization $P_s > 0$, and others.

8.2 Probing and modelling of transient carrier behavior by EFISHG

8.2.1 *Determination of carrier mobility by TRM-SHG*

Using the TOF measurement, we can determine the carrier mobility of organic semiconductor layers in OFETs [27]. The basic idea is similar to the case of TOF measurement for organic layer sandwiched between two electrodes, with a structure of MIM, where carriers generated around one electrode by laser radiation (Laser-TOF) [28] or by carrier injection (e-TOF) [29, 30] are crossing along the layer at constant velocity under external d.c. electric field formed between the two electrodes. Noteworthy that basic concept of e-TOF is the same with that of laser-TOF, and we can discuss the carrier transfer in the same way (Chap. 2). That is, carriers generated around the electrode move by the external electric field, V/L. Here V is applied d.c. voltage and L is the sample thickness.

Keeping in mind the concept of TOF measurement, we can carry similar experiments by applying external d.c. voltage V_{ds} between Source and Drain electrodes. By applying the external step voltage V_{ds}, we can assume that carriers are injected from the Source electrode and they are conveyed along the electric field formed between the Source and Drain electrodes, $E_{ds} = V_{ds}/L$. Here, L is the distance between the Source and Drain electrode edge. We speculate that injected carriers move along the channel at a constant velocity $v = \mu E_{ds}$, and carrier mobility can be estimated by using the transit time $t_r = L/v$ recorded in the trace of transient current. However, these are merely speculation. We do not know, for example, carriers are moving at the constant speed, *etc.* If we can directly observe carrier transport across the FET channel, we can get information on carrier transport mechanism and can determine carrier mobility from a visualized image of the carrier transfer.

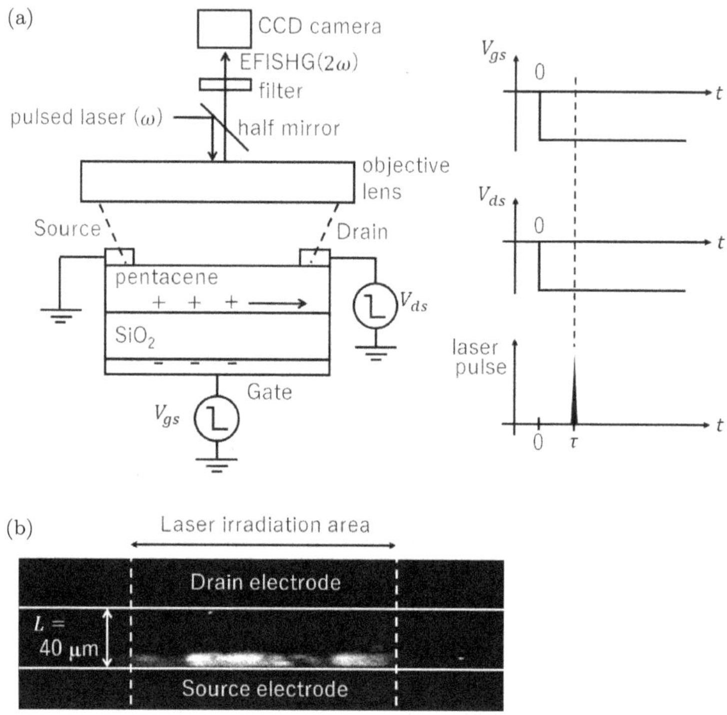

Figure 8.13. (a) TRM-SHG experimental setup and an example of voltage waveform used in the measurement. (b) Example of visualized electric field image in the channel region of pentacene OFET. Holes injected from Source electrode is visualized as a white band of EFISHG emission [4].

It is possible to probe carrier motion by means of the time-resolved microscopic SHG (TRM-SHG) measurement. Using the electrode system illustrated in Fig. 8.13(a), *i.e.*, top-contact pentacene FETs with Au Source and Drain electrodes, TRM-SHG measurement has been carried out using OFETs with a channel length of 40 μm. Here the pentacene was deposited at room temperature, *i.e.*, around 20°C.

Figure 8.13(b) shows the TRM-SHG image in the channel region of the pentacene FET, where the SHG image was obtained along with carrier motions after a step negative voltage was applied. That is, a negative voltage −140 V was applied to the gate and drain electrodes with respect to the Source electrode electrically grounded, *i.e.*, $V_{ds} = V_{gs} = -140\,\text{V}$. At $\tau = 0\,\text{ns}$, the laser pulse coincides with the falling edge of the voltage pulse, and SHG signals are found near the edge of the Source electrode, indicating that the charges are only induced on Source electrode. That is, merely the electric field $\vec{E}_0$ originating from positive charges induced on Source electrode to negative charges induced on gate electrode is formed in the FET. Consequently, the field component of $\vec{E}_0$ that is parallel to the channel is formed only around the Source electrode.

Figure 8.14 shows more clearly the transport of carriers, corresponding to the penetration of white band in SHG images. The emission band of the SHG gradually moves in the channel from the Source to Drain electrode. The visualized emission band motion, which starts from the Source electrode, evidently indicates hole-injection from the Au Source electrode. That is, the pentacene FET shows a p-type semiconductor behavior, where the majority of carriers are holes injected from the Source electrode and the carrier mobility is $0.10\,\text{cm}^2\,\text{V}^{-1}\,\text{s}^{-1}$. Note that the SHG image appears only in the region within a radius of around $150\,\mu$m, corresponding to the spot size of the laser (the area between two vertical dotted lines in Fig. 8.13(b)).

To further clarify the hole transport mechanism across the channel, we plotted the frontier position of SHG band with respect to delay time τ, as shown in Fig. 8.14(b) [31, 32]. From Figs. 8.14(a) and (b), it is clear that holes travel in the direction from the source Au electrode to the drain Au electrode, on satisfying the square root time dependence $x \sim \sqrt{t}$ [31, 33], *i.e.*, not linear time dependence $x \sim t$. It looks like a diffusion process, but not diffusion process itself. The front of the SHG emission band moves but it does not blur. We need to reconsider the idea that carriers move under the external electric field $E_{ds} = V_{ds}/L$. Indeed, for the carrier propagation along the channel interface, space charge field caused from interface charges traveling along the channel makes a significant contribution to drive the interface charges [34, 35]. This time dependence reflects the interface charge propagation process, and well supports a Maxwell–Wagner effect model that is employed for analyzing transient carrier behaviors along the OFET channel [31, 32].

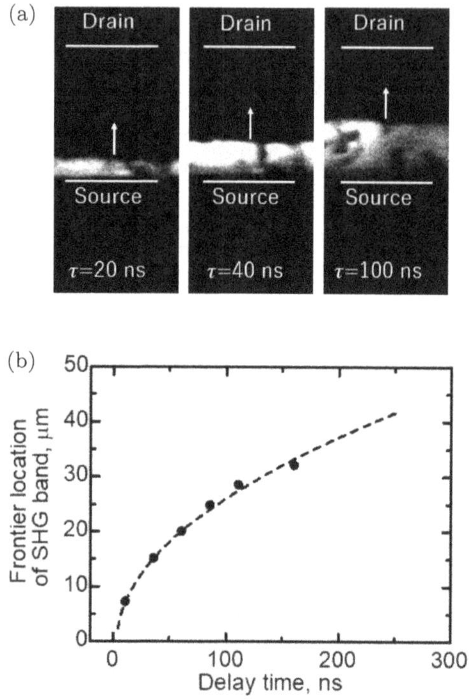

Figure 8.14. TRM-SHG of holes in pentacene-OFET. (a) Hole motion in pentacene OFET after voltage pulse $V_{gs} = V_{ds} = -140\,\text{V}$ is applied to gate and drain electrodes in reference to Source electrode grounded. (b) Plot of a location of frontier of SHG band as a function of delay time τ [4].

This is also well supported by a transmission line model that can be derived as an extension of the Maxwell–Wagner model [9, 27, 36]. In physics, carriers propagate along the interface between the gate insulator and active layer, and this situation is roughly described in the way as that the interface accumulated charge $Q_s = WC_g(V_{gs} - V_{th} - \frac{1}{2}V_{ds})$ travels with a velocity of $v = \mu\frac{\bar{V}}{L}$, using a RC ladder circuit [9]. Here, C_g is the gate capacitor and W is FET channel width. V_{gs} is the applied gate–source voltage, V_{th} is the threshold voltage, and V_{ds} is the applied drain–source voltage. μ is carrier mobility, and L is channel length. $\bar{V}$ ($\approx V_{gs} - V_{th} - \frac{1}{2}V_{ds}$) is the average driving voltage for carrier drift in the FET channel region. In this model, the transit time $t_r = L/v$ is given by

$$t_r \approx \frac{L^2}{\mu(V_{gs} - V_{th} - \frac{1}{2}V_{ds})} \tag{8.18}$$

Equation (8.18) suggests that the carrier migrates into the channel region on satisfying the square root time dependence, *i.e.*, $L \propto \sqrt{t_r}$. The computer simulation based on a classical drift–diffusion equation supports this prediction [5, 31, 37]. Noteworthy that Eq. (8.18) suggests that this square root time dependence appears even when $V_{ds} = 0$.

$$t_r \approx \frac{L^2}{\mu(V_{gs} - V_{th})} \tag{8.19}$$

This means that the space charge field formed by injected holes is playing an important role for the carrier propagation along the interface. Actually, we can observe the charge propagation into the channel region from the source and drain electrodes when $V_{ds} = 0$ under the application of a negative step voltage $V_{gs} = -100\,\mathrm{V}$ (see Fig. 8.15) [38]. On the other hand, under the application of a positive step voltage $V_{gs} > 0$, the position of the SHG peak does not move, and the SHG signals are only concentrated around the edge of the Source electrode during the measurement. Results suggest that electron injection is restricted very strongly and this process was negligible. Note that these experiments have been carried out in air. In vacuum, electron injection to pentacene FET is observed by charge modulation spectroscopy and displacement current measurement [39].

Interestingly, Eq. (8.18) implies that hole propagation is possible even with $V_{gs} = V_{ds} = 0$ if V_{th} is positive $V_{th} > 0$. That is, for OFET like depression type transfer characteristics, transit time in Eq. (8.18) is,

$$t_r \approx \frac{L^2}{\mu V_{th}}. \tag{8.20}$$

In pentacene FET with SiO_2 insulator coated with a poly-4-vinylphenol (PVP), hole injection and transit are well visualized by TRM-SHG after switching V_{gs} and V_{ds} from $+140\,\mathrm{V}$ to zero [35].

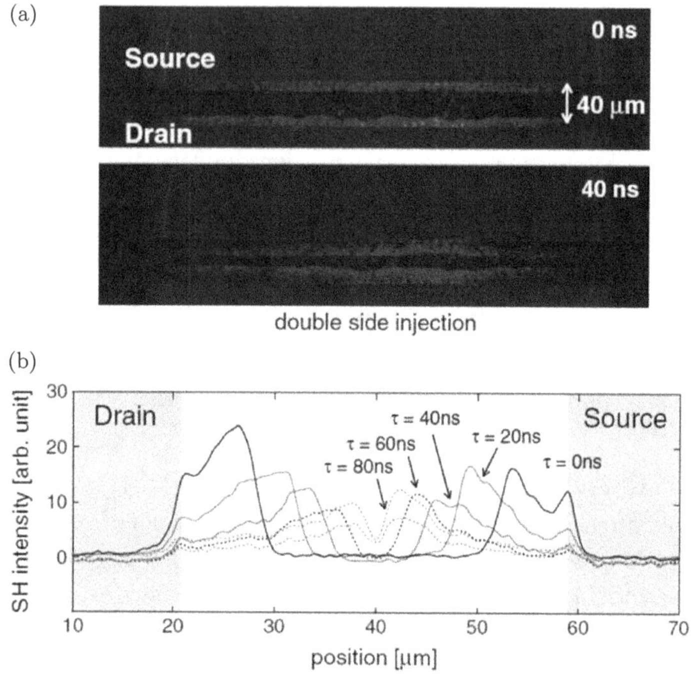

Figure 8.15. Hole injection from source and drain electrodes with $V_{gs} = -100\,\text{V}$ and $V_{ds} = 0\,\text{V}$. (a) SHG image, and (b) SHG profiles which are SHG intensity along the channel direction from drain to source electrode. Reprinted from Thin Solid Films, Vol. 519, M. Iwamoto, T. Manaka, Probing and modeling of carrier motion in organic devices by optical second harmonic generation, p. 961, Copyright (2010), with permission from Elsevier [38].

It is instructive here to note that carrier injection process is greatly influenced by electrode materials, where the energy barrier height for hole injection should be a very important parameter. As TRM-SHG can visualize the carrier injection and succeeding carrier transport process, we can study carrier injection process and transport process one-by-one, experimentally. Figure 8.16(a)(b) shows the TRM-SHG images carried for pentacene FET with Au and Ag electrodes, respectively [33]. The work function of Au is larger than Ag, and it is anticipated that hole injection into the HOMO level of pentacene from the Au source electrode is smooth in comparison with hole injection from the Ag electrode. As a result, the SHG image

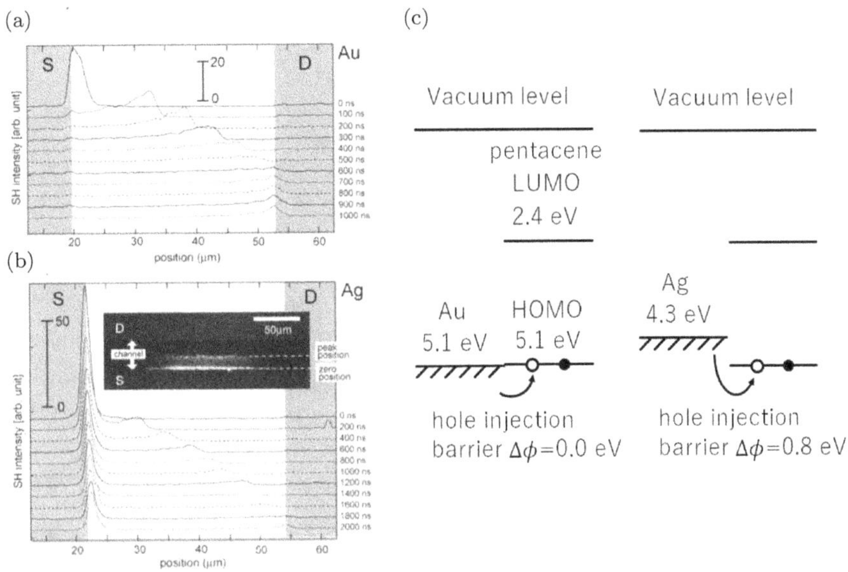

Figure 8.16. TRM-SHG of hole injection to pentacene OFET. (a) Au–pentacene OFET and (b) Ag–pentacene OFET. (c) Energy diagram of Au, Ag, and pentacene. Reprinted with permission from T. Manaka, F. Liu, M. Weis, M. Iwamoto, J. Phys. Chem. C, Vol. 113, 10279, 2009. Copyright (2009) American Chemical Society [33].

around the Ag source electrode remains bright for long time, in comparing with the image around the Au source electrode, after applying a negative voltage $V_{gs} = V_{ds} = -100\,\text{V}$ to gate and drain electrodes in reference to the source electrode. This result indicates that, in pentacene OFET with Ag electrodes, a potential drop remains for so long time owing to the difficulty of hole injection [40]. That is, as injected holes contribute to form space charge field around the Source electrode, and the electric field around the electrode is relaxed in proportion to the accumulated holes at the interface between pentacene and gate insulator, the remaining bright SHG image suggests that the potential drop V_{drop} caused by the application of the positive pulse voltage remains for long time [41]. In other words, the degree of the hardness of the hole injection owes to the contact resistance around the source electrode [42].

In the presence of non-zero potential drop V_{drop} at source electrode edge with reference to OFET channel, Eq. (8.18) is modified as

$$t_r \approx \frac{L^2}{\mu(V_{gs} - V_{th} - \frac{1}{2}V_{ds} - V_{drop})}. \tag{8.21}$$

Interestingly, if we consider this potential drop in the analysis of carrier transport along the FET channel, we can get the same value of carrier mobility for pentacene FETs with Ag and Au electrodes [40, 43], and we can observe the SHG image propagation in the same way if we apply additional voltage on the Ag electrode that correspond to the potential drop. As mentioned above, probing carrier transit by TRM-SHG is very effective to study the carrier transport mechanism, and to determine the carrier mobility of active layer in organic FETs.

8.2.2 *Probing of carrier transits caused by double carrier injection*

As described in previous section, carrier injection is governed by the conditions of electrode materials, *i.e.*, effective barrier height for electrons and holes is a key parameter. Similarly, the film preparation condition also contributes greatly. We therefore can prepare pentacene layers that exhibit ambipolar behavior by controlling the substrate temperature during the deposition of the pentacene layers. Figure 8.17(a) shows the I_{ds}–V_{gs} characteristics of pentacene FET with a 500 nm thick SiO_2 gate insulator, where the pentacene film was deposited at a substrate temperature of 70°C and the channel length is 50 μm. An ambipolar behavior is seen in the I_{ds}–V_{gs} characteristics. That is, both electrons and holes serve as carriers, depending on the applied V_{gs}.

By employing the TRM-SHG experiment, we can observe the ambipolar behavior and get the relationship between the position of the holes and electrons front with elapsed time by applying the step voltage V_{gs} (Figs. 8.17(b) and (c)). The results show that the hole mobility is much higher than the electron mobility, *i.e.*, about 100 times higher [44]. With consideration of these results,

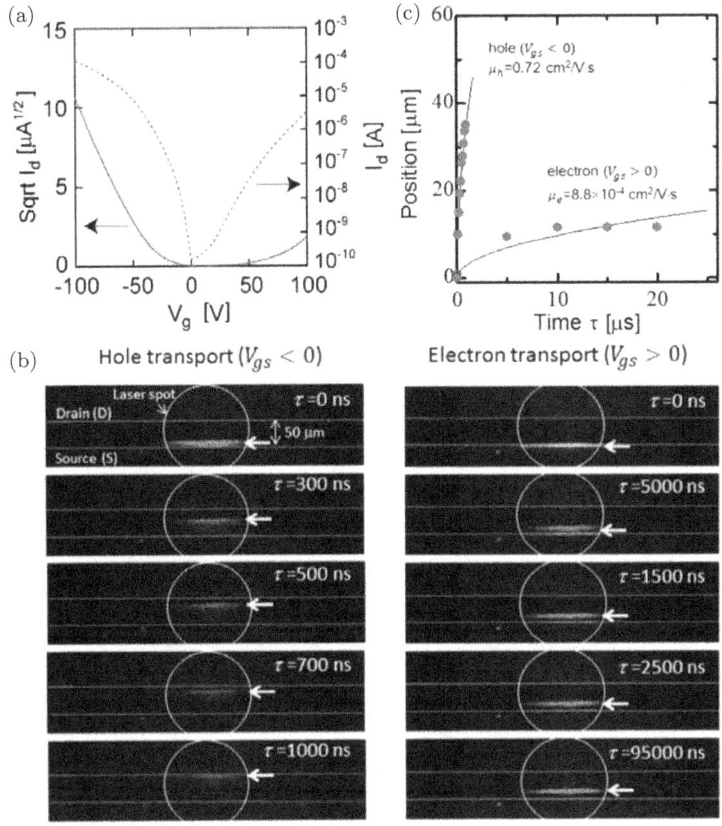

Figure 8.17. (a) I_d–V_g curve of ambipolar pentacene OFET. Reprinted from Physics Procedia, Vol. 14, T. Manaka, S. Kawashima, M. Iwamoto, Charge modulation spectroscopy for probing ambipolar carrier injection into pentacene field effect transistors, p. 209, Copyright (2011), with permission from Elsevier [39]. (b) TRM-SHG images directly visualizing hole (left) and electron (right) transport in the ambipolar pentacene OFET. Holes and electrons are injected from the Source electrode by the negative and positive voltage application $V_{gs} < 0$ and $V_{gs} > 0$, respectively. The frontier of carrier propagation, indicated by arrows, shifts from the source electrode edge to drain electrode, as the time τ elaspsed after voltage application. (c) Time dependence of the frontier of carrier propagation, for holes and electrons moving in the pentacene OFET channel. Republished with permission of IOP Publishing, Ltd, from Modeling and visualization of carrier motion in organic films by optical second harmonic generation and Maxwell-displacement current, M. Iwamoto, T. Manaka, D. Taguchi, Journal of Physics D: Applied Physics, Vol. 48, 373001, 2015; permission conveyed through Copyright Clearance Center, Inc. [45].

we can visualize the double carrier injection process that leads to recombination between electrons and holes injected from the source and drain electrodes [45]. Figure 8.18 shows the TRM-SHG experiment, where a negative step voltage of $V_{dg} = -100\,\text{V}$ was applied on the drain electrode with respect to the gate, and a positive step voltage of $V_{sg} = +100\,\text{V}$ was applied on the source electrode. As can be seen in the images of Fig. 8.18, double carrier injection followed by the electron–hole recombination is clearly demonstrated. At the time corresponding to carrier recombination, a bright band is seen due to the formation of a high static electric field between the electrons and holes injected into the channel.

As shown in Fig. 8.18, the TRM-SHG measurements allow us to visualize the double carrier transport process, *i.e.*, double carrier injection and recombination. Consequently, this technique is available for studying the electro-light emitting process that starts from double carrier injection leading to hole–electron recombination (Sec. 7.7), and also for visualizing the process that leads to the pre-electrical breakdown in devices (Sec. 9.1).

Further we should note that carrier propagation is dependent on the interfacial film condition of the channel, *e.g.*, carrier traps, morphology, *etc.* Equation (8.6) suggests that TRM-SHG is available for monitoring the contribution of the surface electronic states of the channel where electrostatic field $\vec{E}(0) = \vec{E}_t$ originating from trapped charge contributes to EFISHG. To confirm the potentiality of the TRM-SHG measurements, experiments were carried out for pentacene-FETs with a 100 nm thick poly–methylmethacrylate (PMMA) layer on the surface of the gate SiO_2 insulator [5, 34]. The SHG intensity is concentrated in the front region of the propagating SHG profile [5]. Results indicate that the space-charge field distribution is being formed along the channel due to the trapping of holes. Analysis based on a carrier trap model accounts well for the result of TRM-SHG experiments. That is, TRM-SHG is effective to probe the interface electronic states of the channel region.

As we discussed here, the results show that EFISHG measurement is very helpful for studying the carrier transport in pentacene FETs. Further, we should note that we can detect carrier motions in

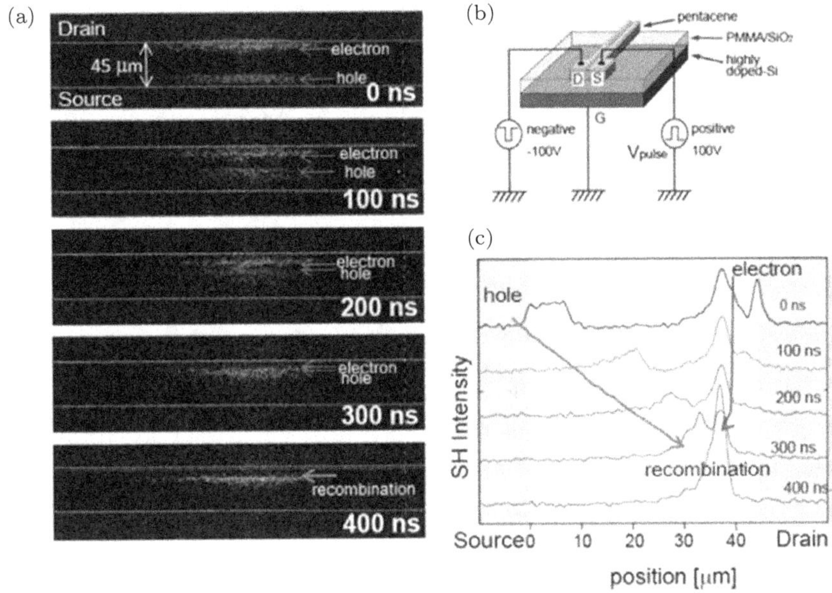

Figure 8.18. TRM-SHG measurement of ambipolar pentacene OFET. (a) SHG image visualizing carrier motion of both holes and electrons. (b) Experimental setup for pulse voltage application to OFETs. Electrons were injected from the drain electrode prior to hole injection from the source electrode. (c) SHG intensity profile across the OFET channel from the source electrode edge ($0\,\mu$m) to the drain electrode edge ($45\,\mu$m). The SHG peak corresponding to the frontier where propagating holes meet electrons at around $33\,\mu$m. The enhanced local electric field due to the hole–electron recombination process is clearly observed in the bottom profile (400 ns). Republished with permission of IOP Publishing, Ltd, from Modeling and visualization of carrier motion in organic films by optical second harmonic generation and Maxwell-displacement current, M. Iwamoto, T. Manaka, D. Taguchi, Journal of Physics D: Applied Physics, Vol. 48, 373001, 2015; permission conveyed through Copyright Clearance Center, Inc. [45]

OFETs with an active layer of double-layer system, *e.g.*, OFETs with a C60/pentacene active layer, *etc.* [46–48], where the hole transport in the pentacene layer and electron transport in C60 were selectively probed by choosing appropriate wavelengths of incident laser, which activate EFISHG from pentacene at 1120 nm and C60 at 1000 nm, respectively. That is, we can directly probe carrier motions in one of the multilayers by choosing an appropriate incident laser wavelength. The EFISHG measurement for pentacene/C60 OFETs showed that electrons are actually transported in the C60 layer, *etc.* Some examples were demonstrated in Chap. 7, on focusing carrier behavior in sandwich-type devices. We can study carrier propagation process in active layer in OFETs by using TRM-SHGs. Many organic semiconductor layers have been developed for the use of organic devices, and TRM-SHG provided a powerful way to visualize carrier behavior in them [49–52].

8.3 Anisotropic carrier behavior

Carrier transport is strongly dependent on the film preparation condition, crystallinity of deposited films, and so forth. There are many film preparation methods, such as dipping, spin-coating, evaporation, and others. Accordingly, anisotropic property is induced in deposited films, and the organic film shows anisotropic carrier transport property; carrier transport along the long-molecular axis direction is different from transport perpendicular to the long-molecular axis direction [53, 54]. Evaluating the anisotropic carrier behavior is very important for device technology, and also for understanding the carrier transport mechanism.

By using the TRM-SHG measurement, we can visualize the anisotropic carrier transport behavior in organic thin films by using a round shape electrode system [55–58]. Noteworthy that Eq. (8.18) in previous section and observed SHG images suggest us that we do not need to apply external electric field along the channel. In other words, spreading of injected charges from the Source electrode can start under the condition $V_{ds} = 0$. This fact suggests us to use one electrode system for the study of anisotropic carrier transport, as illustrated in Fig. 8.19. Note that it is assumed that carriers leaving from the electrode can enter organic thin films in all of the directions.

Figure 8.19(a) shows the visualized SHG images by holes injected into poly[(9,9–dioctylfluorenyl–2,7–diyl)–*co*–bithiophene]) (F8T2) films from the edge of the round shape Au electrode [57]. The F8T2 films were prepared using the floating film transfer method (FTM) [59–61]. As we can see in the set of figures (see Figs. 8.19(a–1)–(a–3)), the spreading speed of holes from the edge of the round

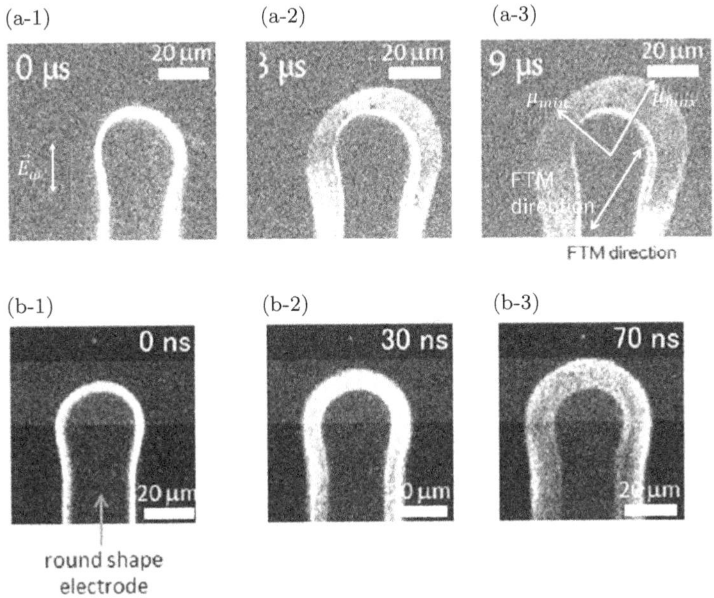

Figure 8.19. (a–1)–(a–3) Anisotropic hole transport in F8T2 film prepared by the FTM method. The holes are injected from the round-shaped single electrode, and transported in the film where maximum carrier mobility (μ_{max}) is mostly coincident with FTM direction as indicated in (a–3). $\vec{E}_\omega$ in (a–1) is polarizing direction of probing laser beam. (b–1)–(b–3) Isotropic hole transport in pentacene film prepared by conventional vacuum evaporation. TRM-SHG visualized the isotropic carrier transport of the pentacene film, reflecting isotropic morphology of the pentacene film by vacuum evaporation. Republished with permission of IOP Publishing, Ltd, from Modeling and visualization of carrier motion in organic films by optical second harmonic generation and Maxwell-displacement current, M. Iwamoto, T. Manaka, D. Taguchi, Journal of Physics D: Applied Physics, Vol. 48, 373001, 2015; permission conveyed through Copyright Clearance Center, Inc. [45].

shape electrode is totally directional dependent on the film. As we can see in the distorted SHG image of injected holes around the edge of Au electrode, carrier spreading is seen along the FTM direction, *i.e.*, parallel to the molecular long-axis direction of F8T2. From the trace of anisotropic carrier transport, it is estimated that the anisotropy of carrier mobility along the molecular long axis $\mu_{\|}$ with respect to its short axis $\mu_{\perp}$ is $\mu_{\|}/\mu_{\perp} = 4.5$. On the other hand, the visualized carrier behavior of holes in pentacene films deposited by thermal evaporation is isotropic, and carriers spread identically toward all outer directions, as shown in Figs. 8.19(b–1)–(b–3). In this way, we can evaluate the anisotropic carrier behaviors by using the TRM-SHG microscopic images.

8.4 Summary

Analysis and modelling of three-electrode system was described based on the TRM-SHG measurement. As examples of three-electrode organic devices, OFET, OLET, and memory OFET were discussed. Steady and transient carrier motion were well visualized by using TRM-SHG. Based on the visualized TRM-SHG images, it was shown that transmission line model well accounts for carrier behavior in pentacene-FET. Visualizing anisotropic carrier transport was also described. It is anticipated that EFISHG is broadly available for analysis and modelling of carrier behaviors in electronic devices. It is instructive here to note that by coupling MDC and EFISHG measurements with other methods, we can carry more powerful measurements. For example, the coupling with charge modulation spectroscopy and others will be effective [26, 62–70]. Finally, we should note that this method has a potentiality to study carrier behaviors in inorganic devices, such as amorphous silicon (a-Si) solar cells [71], AlGaN/GaN high electron mobility transistors (HEMT) [72, 73], and organic–inorganic perovskite OFET [74].

References

1. Sze, S. M. (1981) *Physics of semiconductor devices*, 2nd ed. (John Wiley & Sons, New York Chichester Brisbane Toronto Singapore).
2. Shockley, W. and Pearson, G. L. (1948). Modulation of conductance of thin films of semi-conductors by surface charges, *Phys. Rev.*, 74, 232–233, https://doi.org/10.1103/PhysRev.74.232.
3. Tamura, R., Lim, E., Manaka, T. and Iwamoto, M. (2006) Analysis of pentacene field effect transistor as a Maxwell–Wagner effect element, *J. Appl. Phys.*, 100, 114515, https://doi.org/10.1063/1.2372433.
4. Manaka, T., Lim, E., Tamura, R. and Iwamoto, M. (2007). *Nat. Photon.*, 1, 581, https://doi.org/10.1038/nphoton.2007.172.
5. Manaka, T., Liu, F., Weis, M. and Iwamoto, M. (2010). Influence of traps on transient electric field and mobility evaluation in organic field-effect transistors, *J. Appl. Phys.*, 107, 043712, https://doi.org/10.1063/1.3285503.

6. Liu, F., Lin, J., Manaka, T. and Iwamoto, M. (2011). Modeling carrier transport and electric field evolution in Gaussian disordered organic field-effect transistors, *J. Appl. Phys.*, 109, 104512, https://doi.org/10.1063/1.3590154.
7. Yamada, D., Manaka, T., Lim, E., Tamura, R., Weis, M. and Iwamoto, M. (2008). Injected carrier distribution in a pentacene field effect transistor probed using optical second harmonic generation, *J. Appl. Phys.*, 104, 074502, https://doi.org/10.1063/1.2988137.
8. Yamada, D., Manaka, T., Lim, E., Tamura, R. and Iwamoto, M. (2008). Probing of electric field in pentacene using microscopic optical second harmonic generation, *J. Appl. Phys.*, 103, 084118, https://doi.org/10.1063/1.2907965.
9. Weis, M., Lin, J., Taguchi, D., Manaka, T. and Iwamoto, M. (2010). The charge transport in organic field-effect transistor as an interface charge propagation: The Maxwell–Wagner effect model and transmission line approximation, *Jpn. J. Appl. Phys.*, 49, 071603, https://doi.org/10.1143/JJAP.49.071603.
10. Weis, M., Manaka, T. and Iwamoto, M. (2009). Origin of electric field distribution in organic field-effect transistor: Experiment and analysis, *J. Appl. Phys.*, 105, 024505, https://doi.org/10.1063/1.3062601.
11. Lampert, M. A. and Mark, P. (1970). *Current injection in solids* (Academic, New York and London) Chapter 2.4 Shallow traps.
12. Matsubara, K., Manaka, T. and Iwamoto, M. (2015). Band-like transport observed in TIPS–pentacene thin film by time-resolved microscopic optical second-harmonic generation imaging, *Appl. Phys. Express*, 8, 041601, https://doi.org/10.7567/APEX.8.041601.
13. Lampert, M. A. and Mark, P. (1970) *Current injection in solids* (Academic, New York and London).
14. Kajii, H., Koiwai, K., Hirose, Y. and Ohmori, Y. (2010). Top-gate-type ambipolar organic field-effect transistors with indium–time oxide drain/source electrodes using polyfluorene derivatives, *Org. Electron.*, 11, 509–513, https://doi.org/10.1016/j.orgel.2009.12.004.
15. Kajii, H., Tanaka, H., Kusumoto, Y., Ohtomo, T. and Ohmori, Y. (2015). In-plane light emission of organic light-emitting transistors with bilayer structure using ambipolar semiconducting polymers, *Org. Electron.*, 16, 26–33, https://doi.org/10.1016/j.orgel.2014.10.032.
16. Ohshima, Y., Lim, E., Manaka, T., Iwamoto, M. and Sirringhaus, H. (2011). Observation of electron behavior in ambipolar polymer-based light-emitting transistor by optical second harmonic generation, *J. Appl. Phys.*, 110, 013715, https://doi.org/10.1063/1.3606423.
17. Mashiko, Y., Taguchi, D., Weis, M., Manaka, T. and Iwamoto, M. (2012). The Maxwell–Wagner model for charge transport in ambipolar organic field-effect transistors: The role of zero-potential position, *Appl. Phys. Lett.*, 101, 243302, https://doi.org/10.1063/1.4771989.
18. Ohshima, Y., Kohn, H., Manaka, T. and Iwamoto, M. (2009). Space charge field effect on light emitting from tetracene field-effect transistor under AC electric field, *Thin Solid Films*, 518, 583–587, https://doi.org/10.1016/j.tsf.2009.07.022.

19. Ohshima, Y., Satou, H., Hirako, N., Kohn, H., Manaka, T. and Iwamoto, M. (2011). Direct observation of carrier behavior leading to electroluminescence in tetracene field-effect transistor, *Jpn. J. Appl. Phys.*, 50, 04DK14, https://doi.org/10.7567/JJAP.50.04DK14.
20. Yoshita, S., Tamura, R., Taguchi, D., Weis, M., Lim, E., Manaka, T. and Iwamoto, M. (2009). Displacement current analysis of carrier behavior in pentacene field effect transistor with poly(vinylidene fluoride and tetrafluoroethylene) gate insulator, *J. Appl. Phys.*, 106, 024505, https://doi.org/10.1063/1.3176487.
21. Tamura, R., Lim, E., Yoshita, S., Manaka, T. and Iwamoto, M. (2008). Analysis of threshold voltage shift of pentacene field effect transistor with ferroelectric gate insulator as a Maxwell–Wagner effect, *Thin Solid Films*, 516, 2753–2757, https://doi.org/10.1016/j.tsf.2007.04.106.
22. Tamura, R., Yoshita, S., Lim, E., Manaka, T. and Iwamoto, M. (2008). Analysis of charge accumulation in pentacene field effect transistor with ferroelectric gate insulator using Maxwell–Wagner model, *Jpn. J. Appl. Phys.*, 47, 3170–3173, https://doi.org/10.1143/JJAP.47.3170.
23. Tamura, R., Yoshita, S., Lim, E., Manaka, T. and Iwamoto, M. (2008). Pentacene field-effect transistor with ferroelectric gate insulator as Maxwell–Wagner effect element, *Jpn. J. Appl. Phys.*, 47, 476–479, https://doi.org/10.1143/JJAP.47.476.
24. Tamura, R. (2009). Yuuden Bunkyoku Genshou Ni Chakumoku Shita Maxwell–Wagner Kouka Soshi Toshiteno pentacene FET No Tokusei Kaiseki To Seigyo, Ph. D. Thesis (Tokyo Institute of Technology, Tokyo) (in Japanese).
25. Tamura, R., Yoshita, S., Lim, E., Manaka, T. and Iwamoto, M. (2009). Analysis of pentacene field-effect transistor with a ferroelectric P(VDF–TeFE) gate insulator as an element of Maxwell–Wagner effect system, *Jpn. J. Appl. Phys.*, 48, 021501, https://doi.org/10.1143/JJAP.48.021501.
26. Otsuka, T., Taguchi, D., Manaka, T. and Iwamoto, M. (2017). Direct visualization of polarization reversal of organic ferroelectric memory transistor by using charge modulated reflectance imaging, *J. Appl. Phys.*, 122, 185501, https://doi.org/10.1063/1.5004002.
27. Basu, D. and Dodabalapur, A. (2010). *Drift Velocity and Drift Mobility Measurement in Organic Semiconductors Using Pulse Voltage Organic Electronics*, Advances in Polymer Science, Vol. 223, G. Meller, T. Grasser eds. (Springer, Heidelberg, Dordrecht, London, New York), https://doi.org/10.1007/12_2009_4.
28. Kitamura, M., Imada, T., Kato, S. and Arakawa, Y. (2004). Time-of-flight measurement of lateral carrier mobility in organic thin films, *Jpn. J. Appl. Phys.*, 43, 2326–2329, https://doi.org/10.1143/JJAP.43.2326.
29. Weis, M., Lin, J., Taguchi, D., Manaka, T. and Iwamoto, M. (2009). Analysis of transient currents in organic field effect transistor: The time-of-flight method, *J. Phys. Chem. C*, 113, 18459–18461, https://doi.org/10.1021/jp908381b.

30. Lin, J., Weis, M., Taguchi, D., Manaka, T. and Iwamoto, M. (2011). Carrier propagation dependence on Applied potentials in petancene organic field effect transistors investigated by impedance spectroscopy and electrical time-of-flight techniques, *Jpn. J. Appl. Phys.*, 50, 04DK01, https://doi.org/10.7567/JJAP.50.04DK01.
31. Manaka, T., Liu, F., Weis, M. and Iwamoto, M. (2008). Diffusionlike electric-field migration in the channel of organic field-effect transistors, *Phys. Rev. B*, 78, 121302(R), https://doi.org/10.1103/PhysRevB.78.121302.
32. Manaka, T., Liu, F., Weis, M. and Iwamoto, M. (2010). Influence of traps on transient electric field and mobility evaluation in organic field-effect transistors, *J. Appl. Phys.*, 107, 043712, https://doi.org/10.1063/1.3285503.
33. Manaka, T., Liu, F., Weis, M., and Iwamoto, M. (2009). Studying transient carrier behaviors in pentacene field effect transistors using visualized electric field migration, *J. Phys. Chem. C*, 113, 10279–10284, https://doi.org/10.1021/jp900779d.
34. Tanaka, Y., Manaka, T. and Iwamoto, M. (2011). Study of trap-filling effect on transient carrier transport in pentacene field effect transistors by time-resolved optical second harmonic generation, *Chem. Phys. Lett.*, 507, 195–198, https://doi.org/10.1016/j.cplett.2011.03.068.
35. Taguchi, D., Masada, H., Manaka, T., Iwamoto, M., Nishiura, T., Iizuka, T. and Takemori, T. (2012). Probing a two-step channel formation process in injection-type pentacene field-effect transistors by time-resolved electric-field-induced optical second-harmonic generation measurement, *Org. Electron.*, 13, 2801–2806, https://doi.org/10.1016/j.orgel.2012.08.030.
36. Weis, M., Taguchi, D., Manaka, T. and Iwamoto, M. (2010). Organic electronics: Relaxation time controlled devices, *Jpn. J. Appl. Phys.*, 49, 04DK15, https://doi.org/10.1143/JJAP.49.04DK15.
37. Liu, F., Manaka, T. and Iwamoto, M. (2013). Master equation model for Gaussian disordered organic field-effect transistors, *J. Appl. Phys.*, 114, 074502, https://doi.org/10.1063/1.4818497.
38. Iwamoto, M. and Manaka, T. (2010). Probing and modeling of carrier motion in organic devices by optical second harmonic generation, *Thin Solid Films*, 519, 961–963, https://doi.org/10.1016/j.tsf.2010.08.022.
39. Manaka, T., Kawashima, S. and Iwamoto, M. (2011). Charge modulation spectroscopy for probing ambipolar carrier injection into pentacene field effect transistors, *Physics Procedia*, 14, 209–212, https://doi.org/10.1016/j.phpro.2011.05.042.
40. Nakao, M., Manaka, T., Weis, M., Lim, E. and Iwamoto, M. (2009). Probing carrier injection into pentacene field effect transistor by time-resolved microscopic optical second harmonic generation measurement, *J. Appl. Phys.*, 106, 014511, https://doi.org/10.1063/1.3168434.
41. Manaka, T., Nakao, M., Weis, M., Liu, F. and Iwamoto, M. (2009). Transient charge accumulation in pentacene field effect transistor with silver electrode, *Thin Solid Films*, 518, 485–488, https://doi.org/10.1016/j.tsf.2009.07.028.
42. Weis, M., Lin, J., Taguchi, D., Manaka, T. and Iwamoto, M. (2010). Insight into the contact resistance problem by direct probing of the potential drop

in organic field-effect transistors, *Appl. Phys. Lett.*, 97, 263304, https://doi.org/10.1063/1.3533020.

43. Manaka, T., Liu, F., Weis, M. and Iwamoto, M. (2009). Mobility measurement based on visualized electric field migration in organic field-effect transistors, *Appl. Phys. Express*, 2, 061501, https://doi.org/10.1143/APEX.2.061501.
44. Manaka, T., Nakao, M., Lim, E. and Iwamoto, M. (2009). Ambipolar injection into pentacene field effect transistor observed by time-resolved optical second harmonic generation imaging, Materials Science—Poland, Vol. 27, No. 3., pp. 709–717.
45. Iwamoto, M., Manaka, T. and Taguchi, D. (2015). Modeling and visualization of carrier motion in organic films by optical second harmonic generation and Maxwell–displacement current, *J. Phys. D*, 48, 373001, https://doi.org/10.1088/0022-3727/48/37/373001.
46. Shibata, Y., Nakao, M., Manaka, T., Lim, E. and Iwamoto, M. (2009). Probing electric field distribution in underlayer of an organic double-layer system by optical second-harmonic generation measurement, *Jpn. J. Appl. Phys.*, 48, 021504, https://doi.org/10.1143/JJAP.48.021504.
47. Taguchi, D., Manaka, T. and Iwamoto, M. (2015). Analysis of carrier behaviors in organic semiconductors by electric field induced optical second harmonic generation, Kogaku, 44, pp. 111–118. Fig. 4 courtesy of Dr. L. Zhang.
48. Zhang, L., Taguchi, D., Manaka, T. and Iwamoto, M. (2012). Direct probing of selective electron and hole accumulation processes along the channel of an ambipolar double-layer field-effect transistor by optical modulation spectroscopy, *Appl. Phys. Lett.*, 100, 103301, https://doi.org/10.1063/1.3692581.
49. Abe, Y., Taguchi, D., Manaka, T. and Iwamoto, M. (2013). Study of carrier transport in flexible organic field-effect transistors: Analysis of bending effect and microscopic observation using electric-field-induced optical second-harmonic generation, *Thin Solid Films*, 554, 166–169, https://doi.org/10.1016/j.tsf.2013.05.160.
50. Chauhan, A. K., Gupta, S. K., Taguchi, D., Manaka, T., Jha, P., Veerender, P., Sridevi, C., Koiry, S. P., Gadkari, S. C. and Iwamoto, M. (2017). Enhancement of the carrier mobility of conducting polymers by formation of their graphene composites, *RSC Advances*, 7, 11913–11920, https://doi.org/10.1039/C6RA26195G.
51. Ahmad, Z., Abdllah, S. M., Taguchi, D., Sulaiman, K., Manaka, T. and Iwamoto, M. (2014). Investigation of carrier transit motion in PCDTBT by optical SHG technique, *Laser Physics*, 24, 105701, https://doi.org/10.1088/1054-660X/24/10/105701.
52. Mir, I. A., Ahuja, T., Kumar, D., Taguchi, D., Manaka, T. and Iwamoto, M. (2010). Carrier injection from polypyrrole coated gold electrode in pentacene field effect transistors, *Synth. Met.*, 160, 2116–2120, https://doi.org/10.1016/j.synthmet.2010.07.041.
53. Sakai, M., Iizuka, M., Nakamura, M. and Kudo, K. (2004). *Jpn. J. Appl. Phys.*, 43, 2362, https://doi.org/10.1143/JJAP.43.2362.

54. Onoda, M., Okamoto, T., Tada K. and Nakayama, H. (1999). *Jpn. J. Appl. Phys.*, 38, L1070–L1072, https://doi.org/10.1143/JJAP.38.L1070.
55. Manaka, T., Matsubara, K., Abe, K. and Iwamoto, M. (2013). Direct observation of anisotropic carrier transport in organic semiconductor by time-resolved microscopic optical second-harmonic imaging, *Appl. Phys. Express*, 6, 101601, https://doi.org/10.7567/APEX.6.101601.
56. Yamamichi, R., Manaka, T., Taguchi, D. and Iwamoto, M. (2017). Mobility control of TIPS–pentacene thin films prepared by blade-coating method, *IEICE Trans. Electron.*, E100–C, 130–132, https://doi.org/10.1587/transele.E100.C.130.
57. Abe, K., Manaka, T. and Iwamoto, M. (2017). Direct evaluation of anisotropic carrier mobility in uniaxially aligned polymer semiconductor film by time-resolved microscopic optical second-harmonic generation measurement, *J. Phys. D*, 50, 015103, https://doi.org/10.1088/1361-6463/50/1/015103.
58. Matsubara, K., Abe, K., Manaka, T. and Iwamoto, M. (2016). Carrier transport mechanism in single crystalline organic semiconductor thin film elucidated by visualized carrier motion, *J. Nanosci. Nanotech.*, 16, 3388–3393, https://doi.org/10.1166/jnn.2016.12321.
59. Morita, T., Singh, V., Nagamatsu, S., Oku, S., Takashima, W. and Kaneto, K. (2009). Enhancement of transport characteristics in Poly(3–hexylthiophene) films deposited with floating film transfer method, *Appl. Phys. Express*, 2, 111502, https://doi.org/10.1143/APEX.2.111502.
60. Dauendorffer, A., Nagamatsu, S., Takashima, W. and Kaneto, K. (2012). Optical and transport anisotropy in Poly(9,9′–dioctyl–fluorene–*alt*–bithiophene) films prepared by floating film transfer method, *Jpn. J. Appl. Phys.*, 51, 055802, https://doi.org/10.1143/JJAP.51.055802.
61. Tripathi, A. S. M., Pandey, M., Sadakata, S., Nagamatsu, S., Takashima, W., Hayase, S. and Pandey, S. S. (2018). Anisotropic charge transport in highly oriented films of semiconducting polymer prepared by ribbon-shaped floating film, *Appl. Phys. Lett.*, 112, 123301, https://doi.org/10.1063/1.5000566.
62. Manaka, T., Kawashima, S. and Iwamoto, M. (2010). Charge modulated reflectance topography for probing in-plane carrier distribution in pentacene field-effect transistors, *Appl. Phys. Lett.*, 97, 113302, https://doi.org/10.1063/1.3490716.
63. Noma, T., Taguchi, D., Manaka, T. and Iwamoto, M. (2018). Modeling and analysis of I–V hysteresis behaviors caused by defects in tin perovskite thin films, *J. Appl. Phys.*, 124, 175501, https://doi.org/10.1063/1.5050557.
64. Otsuka, T., Taguchi, D., Manaka, T. and Iwamoto, M. (2017). Study of carrier energetics in ITO/P(VDF–TrFE)/pentacene/Au diode by using electric-field-induced optical second harmonic generation measurement and charge modulation spectroscopy, *J. Appl. Phys.*, 121, 065501, https://doi.org/10.1063/1.4975484.
65. Lim, E., Taguchi, D. and Iwamoto, M. (2014). Analysis of carrier transport and carrier trapping in organic diodes with polyimide–6,13–Bis(triisopropylsilylethynyl)pentacene double-layer by charge modulation

spectroscopy and optical second harmonic generation measurement, *Appl. Phys. Lett.*, 105, 073301, https://doi.org/10.1063/1.4893760.

66. Zhang, L., Taguchi, D., Manaka, T. and Iwamoto, M. (2013). Dependence of the carrier transport characteristics on the buried layer thickness in ambipolar double-layer organic field-effect transistors investigated by electrical and optical measurements, *Jpn. J. Appl. Phys*, 52, 05DC01, https://doi.org/10.7567/JJAP.52.05DC01.
67. Zhang, L., Taguchi, D., Manaka, T. and Iwamoto, M. (2012). Direct probing of selective electron and hole accumulation processes along the channel of an ambipolar double-layer field-effect transistor by optical modulation spectroscopy, *Appl. Phys. Lett.*, 100, 103301, https://doi.org/10.1063/1.3692581.
68. Miyazawa, R., Taguchi, D., Weis, M., Manaka, T. and Iwamoto, M. (2010). Study of organic field-effect transistors using charge modulation spectroscopy: Behavior of injected carriers, *Jpn. J. Appl. Phys.*, 49, 04DK07, https://doi.org/10.1143/JJAP.49.04DK07.
69. Lei, L. Y. W., Taguchi, D. and Manaka, T. (2019). Direct observation of carrier transport in organic–inorganic hybrid perovskite thin film by transient photoluminescence imaging measurement, *Jpn. J. Appl. Phys.*, 58, SBBG18, https://doi.org/10.7567/1347-4065/ab027c.
70. Babajanyan, A., Lee, K., Lim, E., Manaka, T., Iwamoto, M. and Friedman, B. (2007). *Appl. Phys. Lett.*, 90, 182104, https://doi.org/10.1063/1.2734500.
71. Ou-Yang, W., Manaka, T., Naitou, S., Kunitomo, K. and Iwamoto, M. (2012). Optical second-harmonic generation in hydrogenated amorphous silicon single- and double-junction solar cells, *Jpn. J. Appl. Phys.*, 51, 070209, https://doi.org/10.1143/JJAP.51.070209.
72. Katsuno, T., Manaka, T., Ishikawa, T., Ueda, H., Uesugi, T. and Iwamoto, M. (2014). Current collapse imaging of Schottky gate AlGaN/GaN high electron mobility transistors by electric field-induced optical second-harmonic generation measurement, *Appl. Phys. Lett.*, 104, 252112, https://doi.org/10.1063/1.4885838.
73. Katsuno, T., Manaka, T., Ishikawa, T., Ueda, H., Uesugi, T. and Iwamoto, M. (2014). Degradation analysis and current collapse imaging of AlGaN/GaN HEMTs by measurement of electric field-induced optical second-harmonic generation, *Microelectron. Reliab.*, 54, 2227–2231, https://doi.org/10.1016/j.microrel.2014.07.025.
74. Lei, L. Y. W., Taguchi, D. and Manaka, T. (2019). Transient carrier visualization of organic-inorganic hybrid perovskite thin films by using time-resolved microscopic second-harmonic generation (TRM–SHG), *Org. Electron.*, 75, 105416, https://doi.org/10.1016/j.orgel.2019.105416.

Chapter 9

EFISHG as a Tool for Analyzing Electronic Phenomena in Dielectric Materials

This chapter describes a new way for analyzing electronic phenomena in dielectric materials. Since the discovery of "electricity" in the 17th century, electronic phenomena such as electrification and electrical breakdown have been well recognized, and many electrical methods have been developed to study these electrical phenomena. However, these methods are still insufficient, owing to the difficulty of distinguishing origins of electrostatic phenomena and of detecting electric field distribution formed in dielectric materials. It is necessary to develop other techniques for further deep understanding. By coupling the conventional electrical method with Electric-field-induced optical second-harmonic generation (EFISHG) measurement, we can find a new way for exploring the dielectric phenomena in solids. This chapter describes the application of EFISHG measurement to the classical research field related to electrical breakdown and tribo-electrification of solids. Finally, future scope is overviewed, on paying attention to the advantage of the detection of the electric field in solids by employing electric-field-induced optical second-harmonic generation methods.

9.1 Detecting pre-electrical breakdown phenomena by EFISHG

Electrical breakdown is an important research topic in electrical engineering. Sustainable electric power transmission and supply is key to continue our safe and comfortable daily-life, but electrical breakdown of power cables is a serious issue. The development of diagnostic technology of power cables in use is very important. Among the diagnostic techniques, a technique that is capable of detecting a sign of electrical breakdown is highly anticipated, because we are able to replace the cables in use before they suffer from actual electrical breakdown. If we can detect pre-electrical breakdown signals from power cables, it will be of great help for our sustainable daily life. In the field of electrical insulation engineering, electroluminescence (EL) is known as one of pre-electrical breakdown phenomena [1]. The recombination of positive and negative charges injected from opposite electrodes into insulating materials under high electric field leads to the generation of electroluminescence (EL), where it is believed that the generation of EL is caused by the recombination of electrons and holes. However, the intensity of this electroluminescence (EL) is very weak, and usually invisible by naked-eyes. A high-sensitive camera is necessary for the detection. Nevertheless, the appearance of very weak EL has been identified as pre-electrical breakdown phenomena of polymer insulators [2–4]. For example, the generation of EL from 1 cm-thick polymer insulator under 100 kV was successfully detected. Similarly, the emission of very weak EL has been identified from small molecules [5, 6], alkali halide crystals [7], and others, as a sign of pre-electrical breakdown phenomena.

On the other hand, electrical breakdown is common for all electrical and electronic devices, and it is not limited to the case of insulators used in power cables. Therefore, the development of diagnostic measurement that is capable of detecting pre-electrical breakdown of organic materials, including semiconductors, conductors, capacitors and other electronic components, will be very useful for further advancement of electronic device technology. That is, the impact of the development of diagnostic technique concerning pre-electrical breakdown is not confined in the field of electrical insulation engineering. Over the past several ten years, many functional materials have been paid much attention in electronics, and many new materials have been discovered and synthesized. Among them are conducting polymers, organic semiconductors, and others. They are utilized in organic light-emitting diodes, organic solar cells, organic transistors and other organic and molecular devices, for the realization of technologies in the 21st century. Accordingly, research and development (R&D) in the field of organic electronics and molecular electronics is very active, and many researchers in the world are working very hard. In order to realize novel device operation, very thin organic layers are installed in organic and molecular devices, where the thickness of thin layer is less than several 10 nms. Accordingly, the electric field formed in the organic layer is usually high, even though the applied voltage is less than 10 volts, *e.g.*, the electric field is 100 kV/cm for 100 nm thick layer under the applied voltage of 1 V. It will not be peculiar that electroluminescence is emerging from organic thin layer as a sign of pre-electrical breakdown whilst it is in use. The technique that can detect the source of electroluminescence leading to pre-electrical breakdown will be of help for further developing organic devices.

Among organic devices, organic light-emitting diodes (OLED) are attracting much attention as a new way of display technology in the 21st century [8, 9]. Since the pioneering work of OLED with a double-layer structure by Tang [10], many researchers are studying OLEDs with multi-layer structures, in which hole–electron recombination proceeds very efficiently with accompanying electroluminescence. The intensity of EL emitted from OLED is very high, and of course very bright and dazzling for naked-eyes. Life of OLED is dependent on the OLED materials, device structures, operation voltages, *etc.*, but they are always suffering from unpredictable electrical breakdown during operation because organic semiconductor layers in devices are being subjected to high electric field. There will be EL generated as a sign of pre-electrical breakdown phenomenon, separated from ordinal ELs from OLEDs. The fundamental process of pre-electrical breakdown will be different from the process of the ordinal EL emission. Direct recombination of electrons and holes that are injected into organic layer from facing two electrodes under the high field will cause EL related to pre-electrical breakdown. The behaviour of this EL is ruled by intermittent carrier injection process into multi-layers in OLEDs. That is, it is expected that this EL is generated in a discrete fashion, and totally different from ordinal

ELs generated continuously from OLEDs. Furthermore, the intensity of EL generated as a sign of pre-electrical breakdown phenomenon will be very weak, invisible, and the wavelength of EL might be different from that of ordinal bright EL from OLEDs. The mechanism of EL generated in a pre-breakdown process of OLEDs will be rather similar to the case of the EL that has been observed from insulators as a sign of pre-electrical breakdown phenomenon. Therefore, it is very reasonable to develop a technique that can detect charges involved in electrical breakdown from view-point of electrical insulation engineering. In other words, probing of electronic charge behaviors and related dielectric phenomena is key for the detection of pre-electrical breakdown of OLED. As that, electric-field-induced optical second-harmonic generation method (EFISHG method), which can directly probe the electric field emergent from moving carriers that are injected irregularly from electrodes, will be a promising candidate for the detection of pre-electrical breakdown. Noteworthy that EFISHG can probe the electric field arising from charges that contribute to the pre-electrical breakdown, before the EL related to pre-electrical breakdown is activated. This point is quite important for the EFISHG measurement, and it differs from the EL measurement used for the detection of pre-electrical breakdown. EL is always observed after the recombination of electrons and holes, but EFISHG can probe the charges related to the EL generated as pre-electrical breakdown signal before the EL is generated. In this section, detection of pre-electrical breakdown signal of OLED is discussed from view point of dielectric physics, on focussing the behaviors of electronic charges injected in a discrete fashion under external d.c. electric field.

Figure 9.1 shows a model of double-layer OLED. Hole transport layer and electron transport layer are sandwiched between two electrodes. The two layers are organic semiconductors. Holes and electrons, respectively, injected from facing two electrodes cross the hole transport and electron transport layers, respectively, and meet and recombine around the interface between the two layers to emit EL. The conductivity of organic semiconductor layer is rather low, and we can analyse the carrier behaviors by using a Maxwell–Wagner

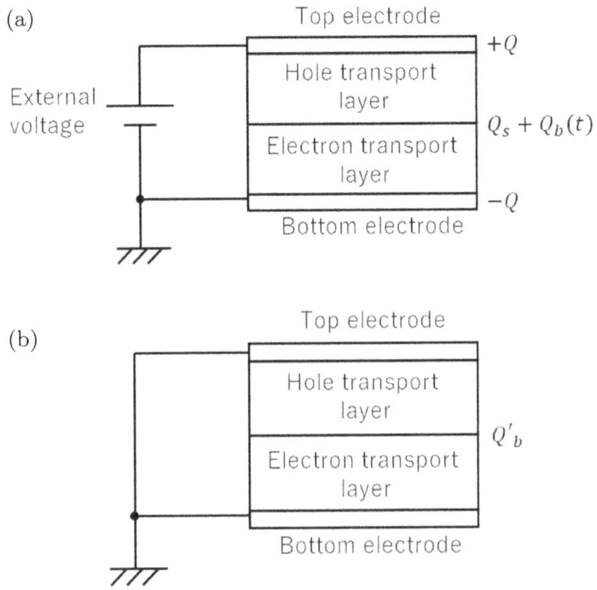

Figure 9.1. A double-layer OLED. Additional interfacial charge accumulation Q_b and Q_b' is a sign of pre-electrical breakdown phenomenon, (a) with non-zero external voltage application, and (b) with zero voltage application.

model, by assuming that the two layers exhibit dielectric property (see Sec. 7.3A).

Let us here assume that the external voltage is applied to top electrode while the bottom electrode is electrically grounded. With increasing positive voltage on the top electrode, hole and electron injection from the facing two electrodes is initiated and EL emission will be followed, depending on the amplitude of applied positive voltage. In low voltage region, carrier injection will be negligible and no current flows. As that, there will be no chance for EL to be activated. In other words, in this low voltage region, charges $+Q$ and $-Q$ are induced on top and bottom electrodes, respectively, but no carrier injection happens. Consequently, OLED is free from electrical breakdown. The electric field in hole transport layer (dielectric constant ε_1) is given as

$$E = \frac{Q}{\varepsilon_1} \tag{9.1}$$

The polarity of electric field E is defined as such that positive field is pointing from top electrode to bottom electrode. With further increasing external voltage, carrier injection will be initiated, and recombination of holes and electrons will happen around the interface, on accompanying the EL as follows: Under high external field, holes and electrons are allowed to be injected from the top electrode and bottom electrode, respectively. The injected holes are transported through the hole transport layer, and accumulate at the interface between the two transport layers, due to the Maxwell–Wagner effect. On the other hand, electrons enter the electron transport layer from the bottom electrode, and move to the interface. At the interface, accumulated holes and electrons recombine, to emit bright EL light. In ordinal OLED, electrostatic field in hole transport layer under operation is given as (see Sec. 7.7)

$$E = \frac{Q}{\varepsilon_1} - \frac{1}{d_1}\frac{Q_s}{C_1 + C_2} \tag{9.2}$$

$Q_s = Q_h + Q_e$ is accumulated space charge at the interface (Q_h (>0): holes, Q_e (<0): electrons). ε_1 and d_1 is dielectric constant and thickness of hole transport layer. C_1 and C_2 is capacitance of hole and electron transport layer, respectively. Noteworthy, the accumulation of Q_s is the result of Maxwell–Wagner effect (see Chap. 4 and Chap. 7), and the amount of charge Q_s is governed by the relaxation time difference between the hole transport and electron transport layers and the current density crossing the interface, and it does not change with time during EL emission in steady state.

This situation is broken if intermittent carrier injection process related to electrical breakdown happens. That is, owing to the injection of additional and excessive charge, the electric field E given by Eq. (9.2) will be unstable. This situation will be described by assuming that the time dependent charge $Q_b = Q_b(t)$ is effectively accumulated at the interface. That is, the electric field given by Eq. (9.2) in the transport layer will be changed to

$$E = \frac{Q}{\varepsilon_1} - \frac{1}{d_1}\frac{Q_s + Q_b}{C_1 + C_2}. \tag{9.3}$$

Noteworthy that the amount of charge Q_s can be estimated theoretically with taking into account the Maxwell–Wagner effect, but it is difficult to estimate the amount of Q_b theoretically, because we do not know the amount of injected charges due to the stochastic intermittent carrier injection related to the pre-electrical breakdown. In other words, we only can estimate the charge Q_b from the experiment that can probe the electric field E.

After the removal of external voltage to be zero, electrode charge Q and space charge Q_s will discharge to $Q = 0$ and $Q_s = 0$, according to the general discharging process of the double layer system (see Chap. 4 and Chap. 7). On the other hand, the charge Q_b will not follow the general discharging process and will remain, possibly because Q_b will be trapped at the interface states. Consequently, the electric field after the completion of discharging will be

$$E = -\frac{1}{d_1}\frac{Q'_b}{C_1 + C_2} \tag{9.4}$$

This situation is illustrated in Fig. 9.1(b). If this is a sign of pre-electrical breakdown in charge behaviors leading to weak EL emission, SHG gives a way to predict actual electrical breakdown by detecting these charge behaviors Q_b in Eq. (9.3) and Q'_b in Eq. (9.4).

9.1.1 *SHG–t measurement*

Figure 9.2 shows EFISHG measurement of pre-electrical breakdown phenomena of OLED. Top and bottom electrodes of OLED, respectively, are IZO and Al. α–NPD (CAS RN® 123847–85–8) is hole transport layer material, and Tris(8–quinolinolato) aluminum(III) (Alq3, CAS RN® 2085–33–8) is electron transport layer material. The total thickness of the double-layer of α–NPD and Alq3 is 100 nm. In SHG measurement, incident laser wavelength is set at 820 nm, while the emerging SHG wavelength is set at 410 nm. In this setup, electrostatic field in α–NPD layer E is selectively probed, where the SHG from Alq3 is negligible at this wavelength. IZO is a transparent electrode. The laser impinges on α–NPD layer through IZO electrode, and SHG transmits through the IZO electrode. The optical

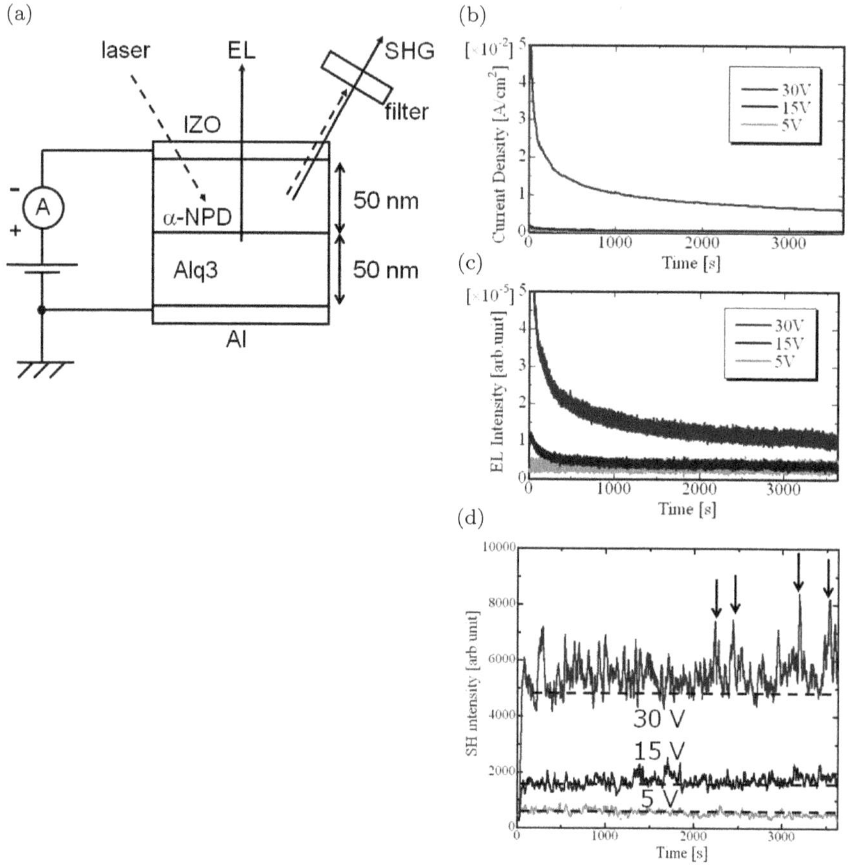

Figure 9.2. Detection of pre-electrical breakdown phenomena by SHG–t measurement. Spike-like change of SHG is detected as a signal of pre-electrical breakdown. (a) Experimental setup of SHG measurement of double-layer (α–NPD/Alq3) OLED, (b) current, (c) EL intensity, (d) SHG intensity, as a function of elapsed time t after applying d.c. voltage [11].

filter blocks reflected laser beam and allows only SHG to be transmitted for the detection. The SHG wavelength 410 nm is apart far from the EL spectrum of OLED. The ordinary EL of this OLED is mainly enhanced in the wavelength region longer than 500 nm. By applying d.c. external voltage, current (I–t), EL intensity (EL–t), and SHG intensity (SHG–t) are measured as a function of elapsed time t [11]. In other words, charge behavior as the pre-electrical breakdown, *i.e.*, $Q_b = Q_b(t)$, is detected as a function of time by means of SHG measurement. The experiment was carried out at three d.c. voltages, *i.e.*, at +5 V, +15 V, and +30 V. The d.c. voltage is applied to IZO electrode in reference to Al electrode grounded. At small voltage +5 V, EL is not emitted from OLED. At this voltage, hole and electron injection is not active, and OLED is free from electrical breakdown. EL intensity and current are zero after applying d.c. voltage. Electric field in α–NPD layer will be constant, according to Eq. (9.1). Actually, ordinal EFISHG is generated constant with time, indicating that the electric field in device is constant and it is free from electrical breakdown. At +15 V, bright EL is observed. Upon this d.c. voltage application, electrode charging proceeds, and charge is induced on the top and bottom electrodes. Subsequently, hole and electron injection start. Interfacial space charge accumulates at the double-layer interface in a short time, within less than 50 ms [12–14]. After that, this state continues in steady. Meanwhile, bright EL is coming out in stable during the hole–electron recombination process proceeds smoothly without electrical breakdown. That is, no sign of electrical breakdown is observed in EL. The observed EFISHG intensity is constant, supporting that the electric field in OLED is constant after electrode charging and space charge accumulation at the interface. The electric field is kept constant in a manner as described by Eq. (9.2).

By applying voltage at +30 V, the SHG signal is no longer constant with time t. A series of spike-like SHG signals are detected as indicated by an arrow in Fig. 9.2(d). According to Eq. (9.3), detection of spike-like SHG signals in a discrete fashion suggests that charge $Q_b(t)$ (<0) is frequently and non-periodically increased due to intermittent electron injection from the bottom electrode. Actually, upon application of +30 V, most of OLEDs are electrically broken in 1 hour. The train of spike-like SHG signals displayed in Fig. 9.2(d) is the evidence of additional carrier injection from electrodes, which leads to electrical breakdown. In other words, as a sign of pre-electrical breakdown phenomena, electron injection occurs irregularly in a discrete fashion, before actual electrical breakdown happens. This process is recorded as a series of spike-like SHG signals. It is noteworthy that EL intensity and current are stable, *i.e.*, smooth and continuous, until the actual breakdown is identified. That is, there is no sign of electrical breakdown in the general EL–t and I–t traces, though a series of spike-like signals are recorded in SHG–t. That is, pre-electrical breakdown is invisible and undetectable in the conventional EL–t and I–t traces, but clearly detectable in SHG–t. It is noteworthy that we can see the EL related to pre-electrical breakdown by means of the EFISHG, but cannot see its colour, *i.e.*, the wavelength of the EL to be emitted as a sign of pre-electrical breakdown phenomena, because what we see is the electric field coming out from the stochastic irregular charge injection that leads to the pre-electrical breakdown. In other words, EFISHG probes the charge accumulation process leading to EL of the pre-electrical breakdown, before this EL is emitted as a sign of pre-electrical breakdown.

9.1.2 *SHG–V measurement*

Figure 9.3 illustrates another example of SHG measurement of pre-electrical breakdown phenomena. The sample is a double-layer OLED with IZO/α–NPD/Alq3/Al structure. α–NPD layer is 50 nm in thickness and Alq3 layer is also 50 nm in thickness. By using laser wavelength 820 nm and SHG wavelength 410 nm, electric field formed in α–NPD layer is selectively probed. External voltage is applied to IZO electrode in reference to the Al electrode grounded. Carrier behavior is probed as a function of voltage. The external voltage is sequentially increased step by step as illustrated in Fig. 9.3(a). That is, d.c. voltage $V = 0\,\mathrm{V}, +1\,\mathrm{V}, 0\,\mathrm{V}, +2\,\mathrm{V}, 0\,\mathrm{V}, +3\,\mathrm{V}, 0\,\mathrm{V}, \ldots$ is applied for 30 s, sequentially. Figures 9.3(b), (c) and (d) are current, EL intensity and SHG intensity under the sequential voltage application. If there is a carrier behavior as an evidence of pre-electrical breakdown, Q_b in Eq. (9.3) under non-zero voltage and Q_b' in Eq. (9.4) at 0 V will be detected as a sign of pre-electrical breakdown phenomenon in the SHG intensity (Fig. 9.3(d)).

Figure 9.4(a) is a close look-up of Fig. 9.3(d). SHG intensity returns to zero at 0 V after non-zero voltage application, when the non-zero voltage is smaller than $V = +18\,\mathrm{V}$, indicating that charges accumulated inside OLED is totally discharged after switching voltage to zero. By contrast, non-zero SHG intensity remains

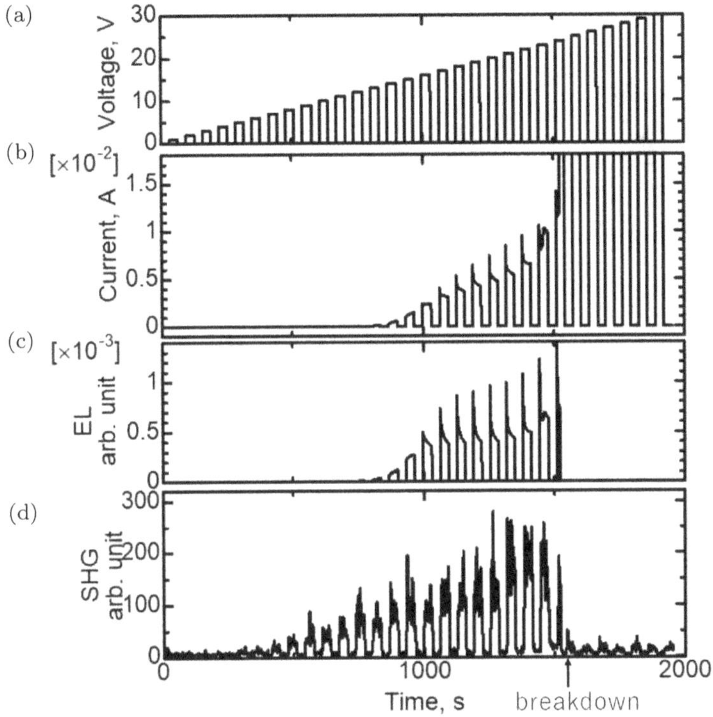

Figure 9.3. SHG–V measurement of pre-electrical breakdown phenomena of OLED. The OLED device structure is IZO/α–NPD/Alq3/Al. SHG probes electric field in α–NPD layer in this OLED with laser wavelength 820 nm and SHG wavelength 410 nm. Charge behavior as a function of increasing d.c. voltage is capable of predicting electrical breakdown. (a) D.c. voltage sequentially applied to OLED. (b) Current. (c) EL intensity. (d) SHG. At Time 1500 s indicated by arrow, OLED is electrically breakdown under d.c. voltage +24 V.

after voltage application larger than $V = +18\,\text{V}$. The non-zero SHG intensity is an evidence that charge Q'_b as a sign of pre-electrical breakdown remains for a time longer than 30 s in this voltage region. Figure 9.4(b) is the accumulated charge $Q_s + Q_b$ and Q'_b calculated from SHG in Fig. 9.4(a), by using Eq. (9.3) and Eq. (9.4). These results show that three voltage regions are identified in prior to electrical breakdown [15].

In the voltage region $0 < V < V_1$ ($V_1 = +12\,\text{V}$), hole and electron injections are minor (low voltage region in Fig. 9.4(b)). Current and EL intensity is zero (Fig. 9.3(b)(c)). In this region, carriers are only

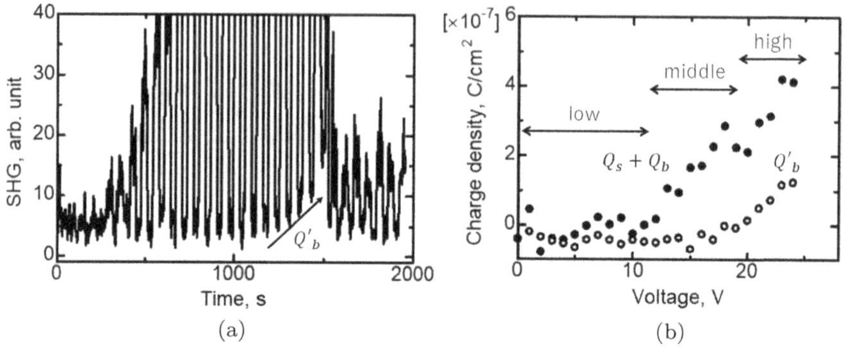

Figure 9.4. (a) SHG–V measurement of pre-electrical breakdown phenomena as a function of increasing d.c. voltage. Non-zero SHG at 0 V develops after applying voltage $V > +18$ V (Time after 1080 s), indicating that non-zero charge Q'_b remained for a time longer than 30 s as a pre-electrical breakdown phenomenon. (b) A charge $Q_s + Q_b$ (filled circles) and Q'_b (open circles) calculated from SHG measurement. Development of non-zero Q'_b is available as a sign of pre-electrical breakdown phenomenon.

induced on the top and bottom electrode under non-zero voltage application, and excess charges never accumulate at the interface, $Q_s = Q_b = Q'_b = 0$. Of course, no electrical breakdown occurs.

In the middle voltage region, $V_1 < V < V_2$ ($V_2 = +18$ V), holes are injected from the top electrode and electrons are coming from the bottom electrode. These carriers are accumulated at the interface between hole and electron transport layers, resulting in recombination of hole–electron pairs accompanied with EL emission. Consequently, current and EL intensity increases as the non-zero voltage increases. OLED is available for radiating bright EL light in the middle voltage region. In this voltage region, by increasing non-zero d.c. voltage, hole and electron injection Q_h and Q_e are enhanced, and EL intensity increases. The space charge $Q_s = Q_h - Q_e$ increases in proportion to the applied d.c. voltage, according to the Maxwell–Wagner effect (see Sec. 7.7). SHG measurement shows that $Q_s + Q_b$ smoothly increases as a function of voltage until OLED is electrically breakdown (Fig. 9.4(b)), indicating that merely Maxwell–Wagner charge Q_s is accumulated and is detected by means of SHG under non-zero voltage application. We cannot find a sign of pre-electrical breakdown phenomenon from $Q_s + Q_b$–V characteristics of this OLED. The result shows $Q_s \neq 0$ but $Q_b = 0$. By switching voltage from non-zero V to 0 V, the accumulated charge Q_s smoothly discharges in less than 50 ms [12–14]. The result shows $Q'_b = 0$. As expected, OLED in this voltage region is free from electrical breakdown.

In the high voltage region $V_2 < V < V_b$ ($V_b = +24$ V), carrier behavior changes and the OLED is electrically breakdown at $V = V_b$. The result shows that the space charge $Q_s + Q_b$ accumulated at the interface increases in the manner same as in the middle voltage region, *i.e.*, $Q_s \neq 0$ and $Q_b = 0$. But non-zero charges remain in OLED for long time ($t > 30$ s) after switching the voltage from V to 0 V. That is, $Q_b = 0$, but $Q'_b \neq 0$ gives a way for the prediction of electrical breakdown. The voltage dependence of accumulated charge provides one way to predict electrical breakdown. This way utilizes probing of charge $Q'_b = Q'_b(V)$ as a function of voltage.

9.1.3 *Spatial pattern of SHG image*

Figure 9.5 shows one more example of pre-electrical breakdown phenomena probed by using SHG measurement [16]. The sample is ITO/polyimide/C60/Al double-layer diode. The thickness of polyimide and C60 layer is 100 nm and 100 nm, respectively. Under constant d.c. voltage application, +50 V is applied to ITO electrode in reference to grounded Al electrode, electric field distribution in electrode area is visualized. In SHG measurement, laser wavelength 1000 nm and SHG wavelength 500 nm are used to visualize the electric field distribution in C60 layer. The visualized area is circular region in Fig. 9.5(a), which corresponds to a spot size of probe laser beam. The electric field distribution is tracked by taking an SHG image every 10 s after d.c. voltage +50 V is applied at time zero $t = 0$ s.

The evolution of visualized SHG pattern with time t until electrical breakdown occurs is as follows: For a fresh sample, the first pattern taken at $t = 10$ s, shows homogeneous distribution of electric field (Fig. 9.5(a)). Under the constant d.c. voltage application, the result indicates that the charge accumulated at C60/polyimide

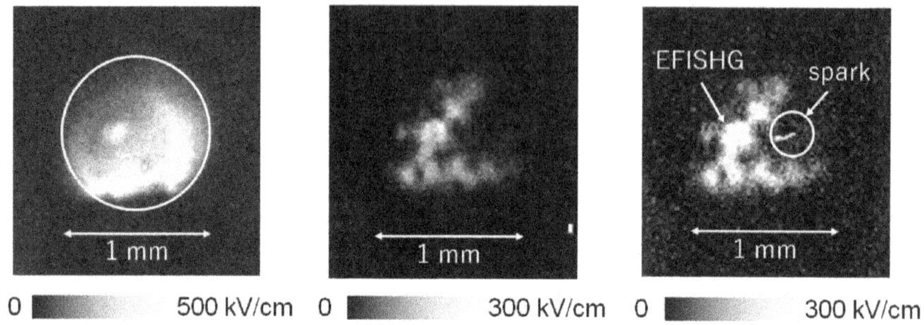

Figure 9.5. Development of pattern of SHG image as a pre-electrical breakdown phenomenon. In spatial pattern of SHG, dark region appears in prior to electrical breakdown. By setting laser wavelength 1000 nm and SHG wavelength 500 nm, electric field in C60 layer of ITO/polyimide/C60/Al double-layer diode is selectively visualized. Spatial pattern of SHG at (a) Time 10 s, (b) 3 hours 45 min, (c) 3 hours 45 min 10 s, after d.c. voltage application. In (a), just after applying voltage, whole area of laser irradiation indicated by a circle emits SHG and the device is free of electrical breakdown. But in (b), dark region is developing as a sign of pre-electrical breakdown phenomena. In (c), spark of actual electrical breakdown is recorded in this dark region.

interface is independent on location in the area probed by means of SHG. Afterwards, SHG pattern slowly develops and approaches the pattern shown in Fig. 9.5(b), after 3 hour and 45 min. Then 10 s after that, a discharge spark is recorded in the dark area in SHG pattern as Fig. 9.5(c).

The development of dark region in SHG pattern reflects the space charge accumulation is enhanced in these dark regions. In prior to the discharge, *i.e.*, electrical breakdown, local charge accumulation in dark region is enhanced as a pre-electrical breakdown phenomenon. This result suggests that the trace of spatial pattern of charge distribution $Q_b = Q_b(x, y)$ is available as a signal of possible electrical breakdown. It is noteworthy that location of discharge spark looks randomly distributed on electrode area but they are always located in the dark region in SHG image. Under the d.c. voltage application, negative charge Q_b accumulates at polyimide/C60 interface. Electric field in polyimide is increased while electric field in C60 layer is decreased. Accordingly, dark SHG region is a sign of local enhancement of electric field in polyimide layer, which is a cause of electrical breakdown in polyimide.

In this Sec. 9.1, SHG measurement as a tool to detect a pre-electrical breakdown phenomenon was described. Experimental results showed that detection of pre-electrical breakdown phenomena is possible by using electric-field-induced SHG (EFISHG) originating from electronic charges related to the pre-electrical breakdown. It is noteworthy that EFISHG detects charge accumulation, not EL emission itself, but we can predict electrical breakdown without seeing any EL signals related to pre-electrical breakdown.

9.2 Visualizing triboelectric charges by EFISHG

The discovery of "electricity" has attracted much attention from viewpoint of science and engineering [17]. The history of "electricity" is very long, in particular, in the field of "electrostatics". The charge-transfer at two-material interface has been well recognized with an idea of triboelectric series of materials. The development of Quantum Physics and Quantum Chemistry in the 20th has motivated researchers to study the charge transfer mechanism at two-material interface by employing an idea of "interfacial energy states", and many innovative instruments have been invented to clarify the details of the interfacial energy states and the electrostatic charge distribution in materials.

On the other hand, using "electricity" has also been motivated by possible applications in electronics, electrical engineering, and others. Among them is an idea of power generation by using electrostatic charge caused from tribo-electrification, where frictional force is controlled for the power generation. This idea itself is very old, and we can see many related ideas from the initial stage of the history of "electricity". Nevertheless, available power supplied by tribo-electrification was unstable, and its application was thus limited. However, this situation is now changing due to the discovery of new functional materials and the development of new processing technologies for the last few decades, along with the development of nano-technology. Further, low-energy consumption electronic circuit technologies are emerging and advancing. Therefore, we anticipate that electronics devices installing new electronic circuit technologies will show high performance by employing a novel high-power triboelectric generator [18]. For example, interactive books with electronic paper has been demonstrated, where triboelectric generator is used to power electronic paper [19].

Along with the emergence of new technology related to triboelectrification, high technology analysis tools are being applied to investigate the mechanism of triboelectric generation, where ultraviolet photoelectron spectroscopy (UPS) [20–22], Kelvin probe force microscope (KPFM) [23], energy-filtered transmission electron microscopy (EFTEM) [24], and others are utilized. In this section, SHG measurement is demonstrated as a novel tool to investigate electrostatic charges created in a new triboelectric generator [25–27].

It is instructive here to note that the following three-step is a basic electronic process that accounts for creation of triboelectric charges.

- Process 1: Electronic charges displace and dipoles rotate by rubbing.
- Process 2: The created electrostatic charges are isolated after the rubbing.
- Process 3: Electrical work is done in external circuit.

The novel triboelectric generator is innovative from two aspects, and totally different from the traditional triboelectric generator. One aspect is concerned with a main process of tribo-electrification, where mechanical energy is converted to electrical energy. In the traditional triboelectric generator, mechanical force is converted to electric energy in Process 2, whereas in a novel triboelectric generator, mechanical work is mainly converted to electricity in Process 1. This is evidently suggested by the fact that modification of material surface is effective for the improvement of triboelectric generation [18]. As that, advanced evaluation technique that can explore Process 1 is key.

The other aspect is concerned with the selection of material, based on the idea of "triboelectric series". Triboelectric series of materials has been listed-up in a book, mainly based on the experiment of charges remaining on a material surface after rubbing. Briefly, the concept of experiment is as follows: material A and B are rubbed together, on facing each other (Process 1). Then A is separated from B, and introduced into a Faraday cup (Process 2), in order to trace the amount of charge remained on A by measuring the induced charge on the Faraday cup by using electrometer (Process 3). This means that the triboelectric series is mainly determined by measuring net electronic charges that are transported between the two materials A and B.

On the other hand, as described in Chap. 2 and Chap. 3, two origins are postulated as the source of electric current flowing through external circuit. One is electronic charge displacement, and the other is dipolar rotation. These two origins are different microscopically, but we cannot distinguish the contribution of these origins to the electrical work consumed in an external circuit. Generation of MDC is the result of electrical energy transmission from the two origins to external circuit, *i.e.*, both charge displacement and dipole rotation are related to the electrical energy stored in materials and they are acting as a source of MDC. In other words, two routes are available to generate triboelectric power. One is to utilize the displacement of electronic charges between two materials by rubbing. The other one is to use rotation of dipoles in materials by rubbing. Making use well of these two contributions is important for designing high-power tribo-generator. That is, merely employing the concept of triboelectric series of materials, which mainly accounts for the tendency of electronic charge displacement at interfaces, is no longer sufficient.

As described in Chap. 6, SHG can probe charge displacement and dipole rotation, selectively and individually. In this section, probing of triboelectric generation by using SHG measurement is described from view point of generation of electricity, *i.e.*, electronic charge displacement and dipole rotation, on the basis of dielectric physics.

Figure 9.6 illustrates an idea of triboelectric generation, in which the two microscopic contributors are considered. One is the displacement of electronic charges created by rubbing, where we assume mechanical force f is merely acting to displace electronic charge with a distance of Δx in normal direction to the material surface. That is,

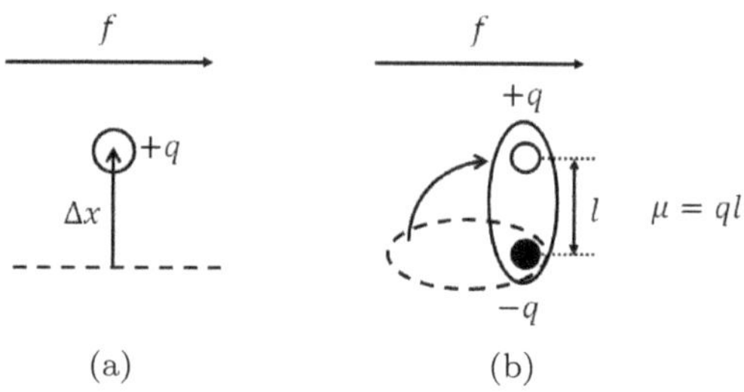

Figure 9.6. Idea of triboelectric generation from viewpoint of dielectric physics. Triboelectric generation by (a) charge displacement and (b) dipole rotation. Under mechanical force f, electronic charge $+q$ displaces by distance Δx and dipole with moment $\mu = ql$ (q: elementary charge, l: separation of positive and negative charge in a dipole) rotates, generating electrical power that can be taken out in external electronics circuit for use.

mechanical energy $W = f\Delta x$ is given to the material, and it is converted to electrical energy. During this rubbing process, positive and negative charges $+q$ and $-q$ are separated with a distance of Δx, whilst electronic energy $U = q^2/2C$ is stored. Here, C is effective capacitance created as the result of the charge displacement. Noteworthy that if this process proceeds with no heat flow during the rubbing, we may consider $W = U$ is satisfied. This energy U corresponds to the increase of internal energy, and it can be taken out in external circuit for use. That is, this electrical energy U is available in an electronic device by connecting to a triboelectric generator.

As mentioned above, if mechanical force f is merely acting to displace electronic charge in adiabatic process by rubbing, the electronic energy U is stored as internal energy in the generator. Consequently, the free energy ΔF available in the external circuit will be described as follows:

$$\Delta F = \Delta U(q\Delta x). \tag{9.5}$$

The $\Delta U(q\Delta x)$ expresses the increase of internal energy and it is given as a function of charge displacement $q\Delta x$. Noteworthy that $q\Delta x$ is created on contacting two material due to surface Fermi level alignment and others, and closely links to material structure in energy space [28].

The other origin of electromotive force is rotational motion of dipole by rubbing (Fig. 9.6(b)). During rotating a dipole $\mu = ql$ (l: separation of positive and negative charges in a dipole) by, for example, 90°, mechanical force f works with a distance l, producing mechanical work $W = fl$. Meanwhile dipoles align their direction and entropy decreases $\Delta S < 0$ owing to the orientational ordering. In an isothermal process, the energy corresponding to $Q' = -T\Delta S$ is dissipated to the external, $Q' = U - W$. As a result, free energy $\Delta F = \Delta U - T\Delta S$ ($\Delta S < 0$) is stored as an available electric energy. The increase of internal energy is given as $U = q^2/2C'$. Here C' is the effective capacitance in a course of dipolar alignment accompanying electronic polarization contribution. The energy $-T\Delta S(>0)$ is due to dipolar alignment.

As mentioned above, the free energy ΔF is available in an electrical device by connecting with this generator. That is, electrical work that is available in external circuit is given by [29]

$$\Delta F = \Delta U(q\Delta x) - T\Delta S(m). \tag{9.6}$$

The first term represents increase of internal energy due to electronic polarization, and is given as a function of charge displacement $q\Delta x$. The second term represents entropy contribution due to alignment of dipoles m. Noteworthy if there is no connection to electrical device, the process proceeds reversible in a quasi-static process of thermodynamic equilibrium. The orientational ordering returns to the initial state when the heat $Q' = -T\Delta S$ is coming in quasi-static process, and electrostatic energy U disappears meanwhile the mechanical position of external origin goes back to the original position. This process is similar to the case of pulling a rubber [30]. By pulling a rubber, positional entropy of polymer segments decreases, adding free energy by $-T\Delta S$ while an elastic deformation of polymer segments contributes to increase of free energy by ΔU. In a reversible process, free energy $\Delta F = \Delta U - T\Delta S$ goes back to zero. Meanwhile polymer segments re-distribute and make a mechanical work in a course of a shrink of rubber.

These two contributions to triboelectric charges, *i.e.*, charge displacement and dipole rotation, can be directly probed by SHG

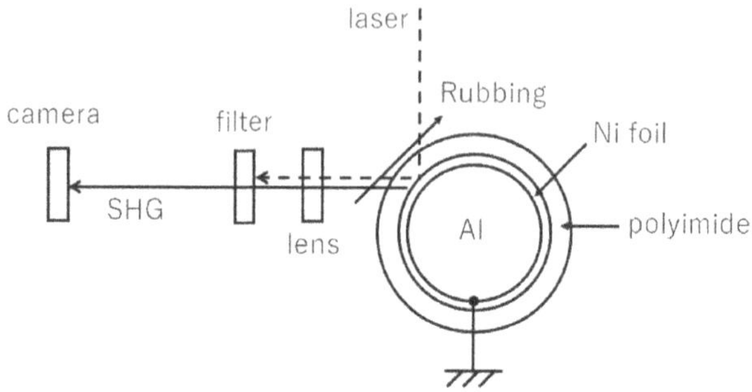

Figure 9.7. Optical setup for probing triboelectric charge generation by means of SHG and EFISHG. Polyimide layer coated on Ni foil is rubbed to generate electrical power. During a course of rubbing, charge displacement $+q\Delta x$ and dipole rotation $\mu = ql$ in Fig. 9.6 are identified and selectively visualized by EFISHG (*e.g.*, laser wavelength 1140 nm and SHG wavelength 570 nm) and SHG (laser wavelength 570 nm and SHG wavelength 285 nm), respectively.

measurement. The electromotive force caused by charge displacement $U(q\Delta x)$ in Eq. (9.5) and Eq. (9.6) is probed by using electric-field-induced SHG (EFISHG). The electromotive force by dipole corresponding to the heat $d'Q = -T\Delta S(m)$ in Eq. (9.6) is selectively probed by using SHG due to dipolar alignment.

Figure 9.7 illustrates an example of optical setup for probing triboelectric charge generation. A polyimide-coated Ni foil is attached to a grounded Al cylinder. A laser pulse impinges into polyimide layer, and SHG is radiated from the polyimide. *P*-polarized laser beam is used and *p*-polarized SHG is detected. Rubbing is carried out in the direction along *p*-polarization. By imaging SHG lights on a charge-coupled device (CCD) camera, electronic charge displacement and alignment of dipoles are visualized.

In the manner as described in Chap. 6, two SHG wavelengths are selected to visualize electronic charges and dipoles, individually. The selection of wavelengths was carried out to be able to identify microscopic origins [27]. The wavelength for electronic charge displacement was selected by using SHG spectrum and voltage dependence measurement on indium–zinc–oxide (IZO)/polyimide/Ag device. The device structure is illustrated in Fig. 9.8(a). The two-electrode sandwich structure is utilized to form external electrostatic field in a

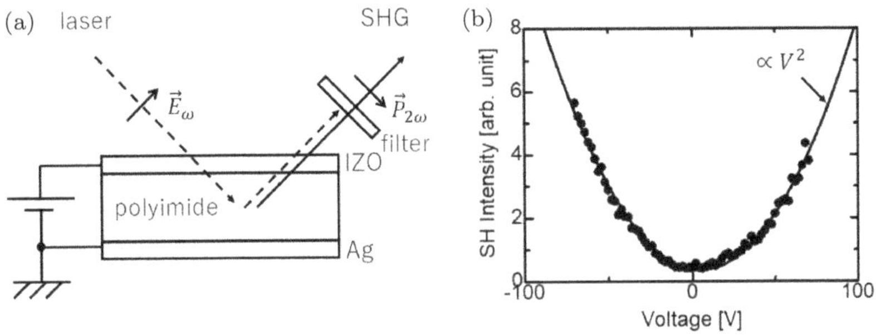

Figure 9.8. EFISHG response of polyimide layer in triboelectric generator. This EFISHG is available for visualization of electronic charge displacement in the two triboelectric generation mechanisms, *i.e.*, charge displacement and dipole rotation. (a) IZO/polyimide/Ag device used for identifying EFISHG signals from polyimide layer. External voltage was applied to IZO electrode in reference to Ag grounded electrode to form electrostatic field in polyimide layer. (b) Voltage dependence of SHG intensity. Laser wavelength is 900 nm, and SHG wavelength is 450 nm. The SHG intensity changes in proportion to the square of external voltage, identifying EFISHG process [27].

polyimide layer, to examine the response of SHG against electrostatic field. SHG spectrum (see Fig. 6.11) showed that the SHG intensity is enhanced by external electrostatic field at around SHG wavelength of 570 nm (laser wavelength 1140 nm). By installing the SHG and laser wavelength at 450 nm and 900 nm, respectively, SHG intensity dependence on external voltage is measured. Figure 9.8(b) is the SHG intensity dependence on external voltage. The SHG intensity increases in proportional to the square of external voltage. This result supports that the SHG at this wavelength well reflects electrostatic field originating from electronic charge displacement through EFISHG process.

The wavelength for probing dipole orientation is selected at the SHG wavelength 285 nm and laser wavelength 570 nm. Rubbing polyimide surface causes alignment of dipoles. By rubbing a polyimide/IZO device, SHG spectrum measurement was conducted. Two succeeding SHG spectra are shown in Fig. 9.9. The *p*-polarized laser was used and *p*-polarized SHG was detected. After rubbing polyimide surface, dipole aligns along the direction of rubbing. Consequently, after rubbing polyimide in the direction along

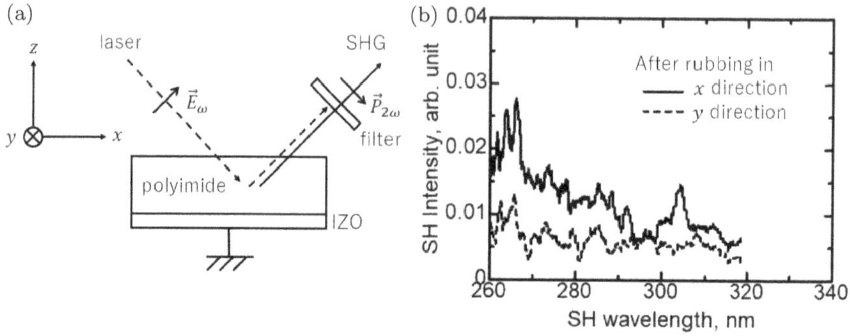

Figure 9.9. (a) Polyimide/IZO device used to select a laser and SHG wavelength that are available in measuring dipolar alignment of polyimide among the two triboelectric generation mechanisms, *i.e.*, charge displacement and dipole rotation. (b) SHG spectrum after rubbing polyimide/IZO device in x and y direction. P-polarized laser was used and p-polarized SHG was detected. The enhancement of SHG intensity after rubbing in x direction indicates that SHG in this wavelength region is available to visualize dipolar alignment [27].

p-polarization direction (x direction), p-polarized SHG is induced while s-polarized SHG is minor. After rubbing in the direction parallel to s-polarization direction (y direction), s-polarized SHG is active while p-polarized SHG is dark. The organic solvent application to the rubbed polyimide surface removes alignment of dipoles. The SHG is inactivated after the solvent application, identifying that the SHG at this wavelength is coming from dipolar alignment origin.

Figures 9.10 and 9.11 show SHG image of polyimide surface after rubbing in x direction. A picture of Fig. 9.10 is taken at a laser wavelength of 1140 nm (SHG wavelength 570 nm) and visualizes electronic charge distribution remained on polyimide after rubbing. A photograph of Fig. 9.11 is obtained by using a laser with a wavelength at 570 nm (SHG wavelength 285 nm) and image of dipolar alignment is obtained. Both images show that electronic charge and dipolar alignment induced on rubbing in triboelectric generation process. Note that, with no-rubbing, both images are dark. Clearly identifying these microscopic origins with electromotive force, SHG measurement is available as one of powerful tools to investigate the microscopic source, *i.e.*, charge displacement and dipole rotation, of triboelectric generation mechanisms.

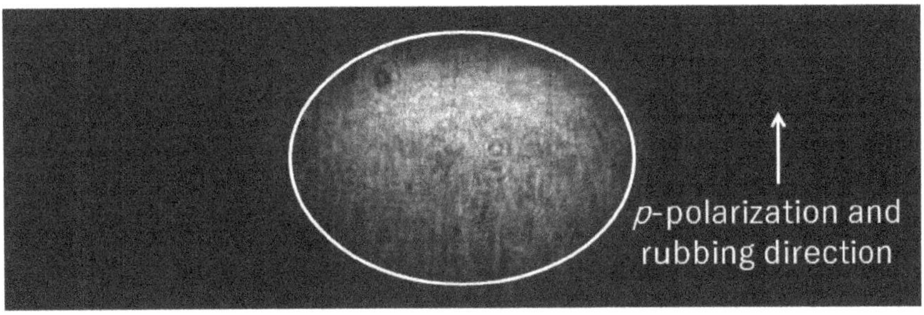

Figure 9.10. SHG image of electronic charge after rubbing polyimide/Ni foil. The electronic charges in the rubbed polyimide are visualized by focusing SHG at wavelength 570 nm and laser wavelength 1140 nm. A circle indicates laser area 2×1.4 mm [27].

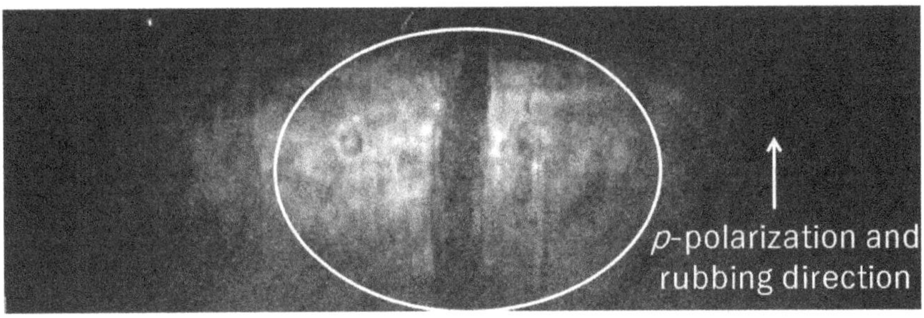

Figure 9.11. SHG image of dipolar alignment after rubbing polyimide/Ni foil. The dipolar alignment in the rubbed polyimide is visualized by focusing SHG at wavelength 285 nm and laser wavelength 570 nm. A circle indicates laser area 2×1.4 mm [27].

SHG is an optical method, meanwhile MDC is an electrical method. MDC gives a way to probe charge displacement and dipolar rotation by measuring electromotive force as an electrical current flowing through external circuit. By contrast, SHG is an optical way and gives an opportunity to get information on electrical phenomena in the optical aspect. Time-resolved measurement of triboelectric generation has been realized. Identifying the polarity of electronic charge, *i.e.*, positive and negative charge, has been reported [31]. A series of SHG images are recorded at 10 fps and dynamical motion during a course of charge generation is visualized. We can make well use of SHG measurement for investigating an impact of charge displacement and dipolar rotation during a course of triboelectric power generation [32].

Application of SHG to triboelectric charge generation has been described, but this is not all. Tribology has been investigated for long time from mechanical and physicochemical aspects [33]. Microscopic understanding of tribology is developing with a technology of scanning probe microscope. Nano scale tribology study is on-going step-by-step by using mechanical probe [34]. SHG uses a laser as a probe and can contribute to tribology by selectively visualizing electronic charge displacement and dipolar rotation.

9.3 Future scope

SHG measurement is available for detecting charge displacement and dipolar orientation in solids, where ideas based on dielectric physics are well-adopted. This measurement is anticipated to be promoted as a new tool for analysing electronic phenomena. In this section, future scope of this SHG measurement is overviewed.

9.3.1 *As evaluation technique*

In Chap. 6 and Appendix F, the details of SHG measurement is described, on focussing the principle, experimental setup, and validation of measurement system operation. In this section, the impact of SHG as an evaluation technique is briefly overviewed.

SHG is an optical method, but it can be used as a non-destructive technique for probing electrical and electronic phenomena. A laser beam is incident on solids and devices, and generated SHG is explored in terms of electrostatic phenomena. As described in Chap. 6, the generated SHG selectively probe electronic charge displacement and dipole alignment, where the incident laser wavelength is appropriately chosen. Two laser wavelengths λ_1 and λ_2 are selected to probe electronic charge displacement and dipole alignment, one-by-one. The size of an electron is 1 fm and of a dipole is 10 pm. It is a very hard task to identify these charges by the conventional electrical techniques, *e.g.*, scanning tunnelling microscope (STM). It is actually impossible to distinguish them. On the other hand, using SHG makes it possible by utilizing optical nonlinear polarization response originated from electrons and dipoles, due to the interaction between the electromagnetic waves of laser and electron clouds of molecules. This situation is similar to the case of X-ray diffraction (XRD) technique, where X-ray is used as a probe. X-ray is not focused on at atomic scale, but X-ray images atoms on a photographic plate due to the coupling of X-ray and atoms [35].

In Secs. 9.1 and 9.2, detection of pre-electrical breakdown phenomena and triboelectric generation by using SHG are described as examples of exploring electrostatic phenomena, which are induced in electrical and electronic components by d.c. biasing and rubbing, but these examples are not all. SHG will be available as an advanced evaluation method for clarifying physical, chemical, and biological phenomena induced in materials and devices by a variety of external stimuli such as heating, radiation, pressure, electrical biasing, and others, in terms of the microscopic origins, *i.e.*, charge displacement and dipolar alignment.

On the other hand, MDC is the electrical measurement and used to trace dynamical motion of charge displacement and dipolar orientation, but it cannot directly image carriers themselves. In Chap. 3, it is shown that dynamical carrier motions in solids, which appear in thermally stimulated process, are probed by TSC measurement. Therefore, it is anticipated that SHG measurement coupled with TSC can give more insightful information on the carrier dynamics in thermal relaxation process [36, 37]. That is, by additionally getting the SHG–T (T: temperature) traces, we can find a way to study carrier dynamics more directly from energetic point of view in the way as in the TSC measurement. For example, depth of carrier traps, activation energy of dipolar rotation will be evaluated.

Figure 9.12 shows an example of the SHG measurement coupled with TSC measurement, where the SHG measurement system equipped with a heating component is used. The interface trap distribution of double-layer OLED with IZO/α–NPD/Alq3/Al structure is analysed. In SHG measurement, incident laser wavelength is selected

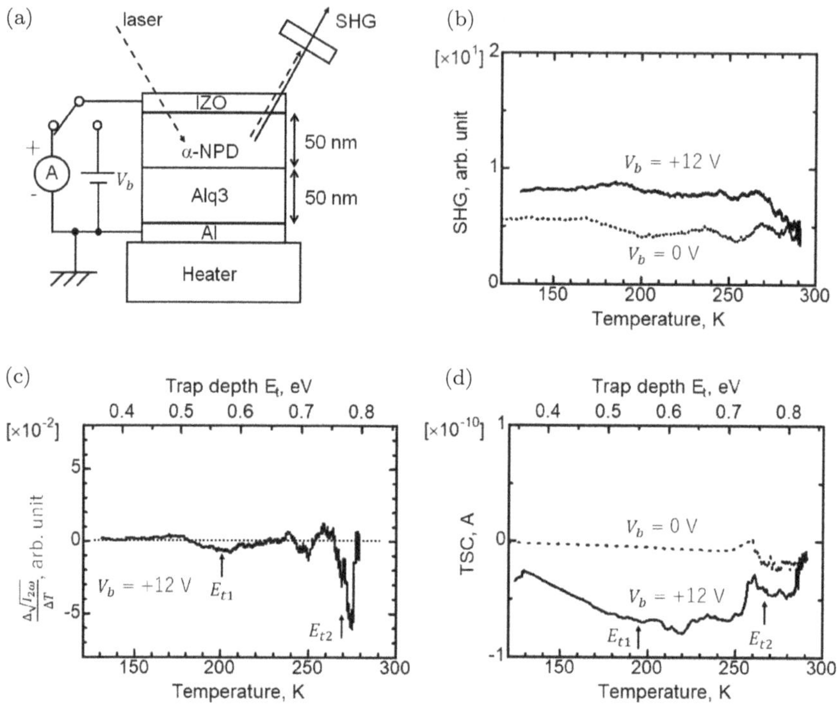

Figure 9.12. SHG measurement with thermal stimulation. Interfacial trap state of double-layer OLED with IZO/α–NPD/Alq3/Al structure is analyzed. (a) SHG system coupled with TSC for probing detrapping carrier processes. In SHG measurement, laser wavelength 820 nm and SHG wavelength 410 nm are selected to probe electric field in α–NPD layer originating from interfacial trapped charges. V_b is a d.c. battery for filling traps and an ammeter is equipped for TSC measurement. (b) SHG. (c) Rate of discharging calculated from SHG, *i.e.*, $\Delta\sqrt{I_{2\omega}}/\Delta T$ which is proportional to trap state density. (d) TSC traces trap state density. In (c) and (d), trap depth E_t is described on the top of plots. Analysis results by means of SHG and TSC are similar in that two trap depths E_{t1} and E_{t2} are resolved, but details are different. Trap states at E_{t2} dominates SHG analysis while E_{t1} looks main contributor in TSC analysis. The difference identifies that the E_{t1} is related to thermoelectric effect.

at 820 nm and SHG generated from the double-layer OLED is probed at 410 nm, by using experimental setup illustrated in Fig. 9.12(a). At this wavelength conditions, electric field in α–NPD layer originating from interfacial trap charges is selectively probed. After applying d.c. voltage +12 V to the IZO electrode, EL is brightly emitted from the

OLED. Upon keeping d.c. voltage application, the OLED is rapidly cooled to low temperature and trapped charges are frozen, and then IZO and Al electrodes are short-circuited following the basic way of TSC method (see Chap. 3 and Chap. 4). Afterwards, SHG intensity is recorded together with TSC during heating, from which the depth of carrier traps in OLED is analysed. TSC traces interfacial trap states N_{ss} as described in Sec. 4.5(c). On the other hand, as described in, Sec. 7.7, electric field in α–NPD layer changes in proportional to interfacial accumulated charge Q_s, and SHG intensity is proportional to the square of this electric field. That is, change of the square-root of SHG intensity $\Delta\sqrt{I_{2\omega}}$ is in proportion to change of interfacial trap charge density ΔQ_{ss} if we consider the accumulated charge Q_s is trapped in interfacial states and $Q_{ss} \approx Q_s$ at low temperature. Looking at Eq. (4.25b) and Eq. (4.25c), TSC is in proportion to the rate of changing of Q_{ss}, *i.e.*, $TSC \propto \Delta Q_{ss}/\Delta T$. Accordingly, the rate of SHG intensity change $\Delta\sqrt{I_{2\omega}}/\Delta T$ is in proportion to TSC which traces interfacial trap states.

Figure 9.12(b) is SHG intensity during a course of heating. After biasing at $V_b = +12\,\mathrm{V}$, SHG intensity is enhanced at low temperature, due to the presence of trapped charges Q_{ss}. These trapped charges are mainly thermally-released and SHG intensity returns to the level recorded with $V_b = 0\,\mathrm{V}$ at high temperature. Figures 9.12(c) and 9.12(d) show the results of SHG and TSC analyses, respectively. The temperature T during heating is scaled to trap energy depth by using the relation in Eq. (4.38), and is described on the top of the figures. Both SHG and TSC show that discharging occurs from traps of $E_{t1} = 0.55\,\mathrm{eV}$ and $E_{t2} = 0.75\,\mathrm{eV}$. A main peak appears at the temperature corresponding to E_{t2} in SHG, whereas E_{t1} governs the TSC peak. SHG coupled with TSC measurement suggests that the peak of E_{t1} is due to the thermoelectric effect leading to the conduction current. As mentioned above, the TSC measurement coupled with the SHG measurement is helpful for deep understanding of carrier dynamics in solids.

Similarly, by coupling SHG technique with TSC measurement, dipolar orientational motion in thermal relaxation process of optical nonlinear polymers has been evaluated to study the impact of disordering of polar molecules on the nonlinear optical property [38]. TSC is MDC as described in Chap. 3. As a result, both thermal relaxation process of charge detrapping and that of dipolar re-orientation is overlapped in TSCs. In nonlinear optical devices, external electric field is used to align dipoles of functional groups that are responsible for nonlinear optical property. This process is called poling. Dipolar relaxation of the functional groups generates TSC, and gives one way to characterize stability of nonlinear optical property. Meanwhile space charges migrate in the poling process, and make TSC analysis complex. SHG coupled with TSC is shown to be effective to identify microscopic origin, *i.e.*, dipolar alignment, as a cause of TSC peak, *etc.*

Photo-excitation is key in energy production process of biological systems such as retinal in eye, porphyrin in chloroplast, and others. By attaching the photo-excitation system to the SHG system, the energy transfer process in biological systems and others will be probed in terms of charge displacement and dipole rotation. In a similar way, by using SHG system equipped with chemical, electrochemical, and other stimulation systems, SHG measurement will be promoted to an advanced experimental technique that can clarify the origin of molecular and material functions in terms of charge displacement and dipolar rotation.

As an evaluation technique, SHG and EFISHG measurement is a powerful tool to investigate a variety of scientific and industrial issues. It is instructive here to point out that the evaluation technique is always associated with synthesis technique, and evaluation technique can develop synthesis technique. For example, scanning probe microscope (SPM) allows one to visualize sub-nano meter atoms and molecules, as an evaluation technique, where STM essentially utilizes nano-scale interaction such as quantum tunnelling effect. This is the reason why STM is powerful as an evaluation of nano scale. As an evaluation method, nano scale stimulation is transferred to cantilever, and detected as a signal to be used for analysis. In an inverse route, as a synthetic method, an external control signal is transferred to motion of cantilever, and nanoscale structures are constructed. This inverse route provides a new synthesis way of nano structure without using an idea of chemistry, and opens a new synthesis field that cannot be realized on the basis of chemistry [39–42].

The situation is similar in the case of SHG. SHG can selectively probe electronic charge displacement and dipolar alignment, as described in this book. This means that, nonlinear optical polarization provides an effective way to excite electrons in materials selectively for material synthesis. Pulsed laser at a wavelength λ_1 selectively excite an electron in symmetric π-electron cloud where EFISHG process is active. Pulsed laser at wavelength λ_2 which differs from λ_1 can identify an electron in polar group to be excited. That is, photo-synthesis route is selected by choosing λ_1 and λ_2. This way makes it possible to activate photo-chemical reaction depending on electronic charge displacement and dipolar alignment of molecules.

For example, diacetylene monomer is polymerized to polydiacetylene (PDA) by using three-photon process [41]. The monomer, *e.g.*, 10,12–pentacosadiynoic acid, has one π–electron at C≡C bond in the middle of the molecule and one polar COOH group at the end of the molecule. By using laser beam at wavelength of $\lambda_1 = 650\,\text{nm}$, π-electron at C≡C bond is selectively excited whereas the polar COOH group is as it is. Note that traditional photosynthesis is possible by wavelength-dependent reaction, though the photosynthesis is one photon process. The one photon process is a linear polarization process, and the second orientation order parameter S_2 (see Chap. 5) is a main governor of selection of wavelength-dependent reaction. By contrast, SHG is two photon process. The first and third orientation order parameters S_1 and S_3 are main contributors for the selective reaction. SHG is an optical method, and leads to a new photosynthesis way.

9.3.2 *New research field*

Many new research fields are emerging along with the development of novel materials and new advanced technologies, in particular, in the field of organic and molecular electronics. As mentioned in Sec. 9.2, the history of tribo-electrification is very long, and there was an idea to generate electric power using tribo-electrification in the field of engineering related to electrostatics, but the unstable power generation was a bottleneck for the application. However, this situation is changing due to the development of advanced materials and related high technologies. We can say as that paradigm change is happening. As a result, we can see many on-going challenging researches related to the generation of electrification and its application, *e.g.*, tribo-electronic nano-generator (TENG), are developing [18]. Similarly, in the field of conducting polymers, mechanical flexibility is now getting controllable. Accordingly, many new actuators have been proposed, where large deformation of the materials in actuators is well controlled [43–46].

In the field of organic mono- and double- layers, and biological materials, understanding of two-dimensional domain patterns and shapes is progressing on the basis of shape equations of biomolecules, *e.g.*, vesicle and red-blood cell, DNA, monolayers [47]. For example, biconcave shape of red-blood cell, quantized shape of monolayer domains, and other interesting shapes in nature have been successfully investigated by means of theoretical models, on the basis of experimental observations. Now it is possible to control patterns and shapes of two-dimensional domain patterns, biological molecules, *etc.*, both theoretically and experimentally. A preparation method of monolayer [48], bilayer [49], and multilayer has been established, and physico-chemical properties of monolayer [50, 51], bilayer [52], and multilayer [53] and others have been clarified step by step, along with the development of evaluation method.

MDC has been established as one electric method that can clarify the property of monolayers from viewpoint of dielectric physics [51, 54]. SHG is an optical method but it can evaluate electrical response from organic monolayer and multilayers [55, 56]. This situation opens a new research field bridging electronics and organic molecules on molecular level. Organic systems such as monolayer, bilayer and multilayers are basically dielectric materials, into which electrostatic energy can be stored. As that, electrostatic energy will contribute to the deformation of shapes of soft materials such as monolayer, bilayer and biological molecules. That is, the Maxwell–stress contributes to the deformation of the shapes of these systems.

It should be noted here that SHG technique that can explore microscopic origins of electrostatic charge will be effectively utilized to explore the formation and deformation in these systems.

As mentioned above, the SHG measurement coupled with other advanced technologies will lead to open new research field.

References

1. Technical Report of the Inst. of Electrical Engineers of Japan, No. 907, Section 5.6.2, p. 84. (in Japanese)
2. Shimizu, N., Kosaki, M. and Horii, K. (1977). *J. Appl. Phys.*, 48, 2191–2195, 1977; *IEEJ A*, 97, 6–13 (in Japanese), https://doi.org/10.1541/ieejfms1972.97.6.
3. Bamji, S. S. (1999). *IEEE Electrical Insulation Magazine*, 15, no. 3, 9–14, https://doi.org/10.1109/57.768531.
4. Hartman, W. A. and Armstrong, H. L. (1967). *J. Appl. Phys.*, 38, 2393–2395, https://doi.org/10.1063/1.1709898.
5. Darveniza, M. (1959). *Nature*, no. 4663, 743, https://doi.org/10.1038/183743a0.
6. Chiu, C. (1976). *J. Phys. D*, 9, 851–856, https://doi.org/10.1088/0022-3727/9/5/018.
7. Cooper, R. and Elliotto, C. T. (1996). *Pre-electrical light emission from alkali halide crystals*, in Advances in Electronics and Electron Physics, 22(A), pp. 995–1001, https://doi.org/10.1016/S0065-2539(08)61269-X.
8. Pioneer, GD–F1, Nikkan Kogyo Shimbun, 1st October, 1997. Page 9; Nikkan Jidosha Shimbun, 12th November, 1997. Page 5.
9. Sony, XEL–1, The Asahi Shimbun, morning edition, 22nd November 2007, Page 11.
10. Tang, C. W. and VanSlyke, S. A. (1987). Organic electroluminescent diodes, *Appl. Phys. Lett.*, 51, 913–915, https://doi.org/10.1063/1.98799.
11. Taguchi, D., Nakamoto, R., Manaka, T. and Iwamoto, M. (2014). *Jpn. J. Appl. Phys.*, 53, 04EK02, https://doi.org/10.7567/JJAP.53.04EK02.
12. Taguchi, D., Weis, M., Manaka, T. and Iwamoto, M. (2009). *Appl. Phys. Lett.*, 95, 263310, https://doi.org/10.1063/1.3277155.
13. Taguchi, D., Inoue, S., Zhang, L., Li, J., Weis, M., Manaka, T. and Iwamoto, M. (2010). *J. Phys. Chem. Lett.*, 1, 803–807, https://doi.org/10.1021/jz1000215.
14. Taguchi, D., Zhang, L., Li, J., Weis, M., Manaka, T. and Iwamoto, M. (2010). *J. Phys. Chem. C*, 114, 15136–15140, https://doi.org/10.1021/jp104712m.
15. Taguchi, D., Manaka, T. and Iwamoto, M. (2014). *15th International conference on organized molecular films (ICOMF15–LB15)*, S6:ODO2, International Convention Center, Jeju, Korea, 9 July.
16. Taguchi, D., Manaka, T. and Iwamoto, M. (2014). *The 11th international conference on nano-molecular electronics (ICNME2014)*, SIV–O–2, Kobe International Conference Center, Kobe, Japan, 18 December.

17. Boyle, R. (1675). Mechanical Origin or production of electricity (R. Davis Bookseller, Oxford) pp. 1–38; Buys, E. (1771). Nieuw en Volkomen Woordenboek van Konsten en Weeten schappen: Bevattende alle de Takken der Nuttige Kennis (S F Baalde Boekverkooper, Derde Delle Amsterdam) p. 529 "Electriciteit of Brandsteenkracht"; Faraday, M. (1844). Experimental Researches in Electricity, vol. 2, Eighteen series §25 (Richard and John Edward Taylor, London) pp. 106–125; Maxwell, J. C. (1954). A Treatise on Electricity & Magnetism, 3rd Edition, vol. 1 (Dover, New York) pp. 32–34.
18. Zhu, G., Lin, Z.-H., Jing, Q., Bai, P., Pan, C., Yang, Y., Zhou, Y. and Wang, Z. L. (2013). *Nano Letters*, 13, 847–853, https://doi.org/10.1021/nl4001053.
19. Karagozler, M. E., Poupyrev, I., Fedder, G. K. and Suzuki, Y. (2013). *UIST'13 Proceedings of the 26th annual ACM symposium on user interface software and technology*, pp. 23–30, St. Andrews, Scotland, United Kingdom, Oct. 8–11, https://doi.org/10.1145/2501988.2502054.
20. Yamaguchi, Y., Shimizu, K., Matsuzaki, A., Sano, D., Sato, T., Tanaka, Y. and Ishii, H. (2019). *IEICE Trans. Electron.*, E102–C, 168–171, https://doi.org/10.1587/transele.2018OMS0008.
21. Ishii, H., Sugiyama, K., Ito, E. and Seki, K. (1999). Energy level alignment and interfacial electronic structures at organic/metal and organic/organic interfaces, *Adv. Mater.*, 11, 605–625. (Review), https://doi.org/10.1002/(SICI)1521-4095(199906)11:8%3c605::AID-ADMA605%3e3.0.CO;2-Q
22. Tanaka, Y., Ikegami, K., Maruyama, T., Kinjo, H. and Ishii, H. (2018). *Appl. Phys. Express*, 11, 081601, https://doi.org/10.7567/APEX.11.081601.
23. Baytekin, H. T., Patashinski, A. Z., Braicki, M., Baytekin, B., Soh, S. and Grzybowski, B. A. (2011). *Science*, vol. 333, issue 6040, 308–312, https://doi.org/10.1126/science.1201512.
24. Burgo, T. A. L., Ducati, T. R. D., Francisco, K. R., Clinckspoor, K. J., Galembeck, F. and Galembeck, S. E. (2012). *Langmuir*, 28, 7407–7416, https://doi.org/10.1021/la301228j.
25. Chen, X., Taguchi, D., Manaka, T., Iwamoto, M. and Wang, Z. L. (2015). Scientific Reports, 5, 13019, https://doi.org/10.1038/srep13019.
26. Chen, X., Taguchi, D., Manaka, T. and Iwamoto, M. (2016). *Chem. Phys. Lett.*, 646, 64–68, https://doi.org/10.1016/j.cplett.2016.01.002.
27. Taguchi, D., Manaka, T. and Iwamoto, M. (2019). *Appl. Phys. Lett.*, 114, 233301, https://doi.org/10.1063/1.5094171.
28. Itoh, E. and Iwamoto, M. (1997). *J. Appl. Phys.*, 81, 1790–1797, https://doi.org/10.1063/1.364035.
29. Fröhlich, H. (1958). *Theory of Dielectrics*, 2nd ed. (Oxford) pp. 26–30. (Japanese translation by Nagamiya, T. and Nakai, Y. (1960). *Yudentai ron* (Yoshioka Syoten, Kyoto), https://doi.org/10.11501/1378529 Chapter II.6. pp. 28–33.
30. Kubo, R. (1947) *Gomu Dansei* (*Rubber's Elasticity*) (Kawade Shobo, Tokyo) (in Japanese), https://doi.org/10.11501/1139958.
31. Taguchi, D., Manaka, T. and Iwamoto, M. (2021). Visualizing positive and negative charges of triboelectricity generated on polyimide film, to be published in IEICE Transactions on Electronics, June 2021, special section on "Recent Progress on Organic Molecular Electronics", vol. E104-C, no. 6.

32. Taguchi, D., Manaka, T. and Iwamoto, M. (2020). Probing electronic charge and dipolar charge in triboelectric generator by using optical second-harmonic generation, *IEICE Transactions on Electronics*, vol. J103–C, no. 9, 395–402 (in Japanese).
33. Soda, N. (1954) *Masatsu To Junkatsu* (*Tribology and Lubrication*) (Iwanami Shoten, Tokyo) (in Japanese); Soda, N. (1971) *Masatsu No Hanashi* (*Story of tribology*) (Iwanami Shoten, Tokyo) (in Japanese), https://doi.org/10.11501/2467124.
34. Matsukawa, H. (2012) *Masatsu No Butsuri* (*Physics of tribology*) (Iwanami Shoten, Tokyo) Chap. 4, pp. 67–95 (in Japanese); Ando, Y. (2009) *Microtribology Nyuumon* (*Introduction to microtribology*) (Sangyo Tosho, Tokyo) (in Japanese).
35. Sakurai, T. (1967) *X-sen Kessyo Kaiseki, (X-ray analysis of crystal structure)* (Shokabo, Tokyo) (in Japanese), https://doi.org/10.11501/1382441.
36. Kajimoto, N., Manaka, T. and Iwamoto, M. (2006). Decay process of a large surface potential of Alq3 films by heating, *J. Appl. Phys.*, 100, 053707, https://doi.org/10.1063/1.2338137.
37. Yoshizaki, K., Manaka, T. and Iwamoto, M. (2005). Large surface potential of Alq_3 film and its decay, *J. Appl. Phys.*, 97, 023703, https://doi.org/10.1063/1.1835543.
38. Köhler, W., Robello, D. R., Dao, P. T., Willand, C. S. and Williams, D. J. (1990). *J. Chem. Phys.*, 93, 9157, https://doi.org/10.1063/1.459205.
39. Another evident example is the Retro Mass Spectral Synthesis (RMSS); Kametani, Tetsuji (1981). Total synthesis of natural products by retro mass spectral synthesis, Yakugaku Zasshi, 101(1), pp. 1–19. (in Japanese), https://doi.org/10.1248/yakushi1947.101.1_1. In RMSS, a complex molecule is decomposed to small fragments and analysed by mass spectroscopy. By using the fragments as a starting molecule for synthesis, total synthesis route of original natural product has been found.
40. Manaka, T., Kon, H., Ohshima, Y., Zou, G. and Iwamoto, M. (2006). *Chem. Lett.*, 35, 1028–1029, https://doi.org/10.1246/cl.2006.1028.
41. Manaka, T. and Iwamoto, M. (2016). *J. Nanosci. Nanotech.*, 16, 3394–3397, https://doi.org/10.1166/jnn.2016.12276.
42. Kohn, H., Ohshima, Y., Manaka, T. and Iwamoto, M. (2008). *Macromol. Symp.*, 268, 77–80, https://doi.org/10.1002/masy.200850816.
43. Kaneto, K., Kudo, K., Ohmori, Y., Onoda, M. and Iwamoto, M. (1998). *IEICE Trans. Electron.*, E81–C, 1009–1019. Sec. 5. (Review)
44. Onoda, M. and Tada, K. (2004). *IEICE Trans. Electron.*, E87–C, 128–135; Onoda, M., Okamoto, T. and Tada, K. (2000). *IEEJ Transactions on Fundamentals and Materials*, 120, 129–134. (in Japanese), https://doi.org/10.1541/ieejfms1990.120.2_129.
45. Tada, K. and Kaku, M. (2019). Contribution of deformation for electrical resistance of conductive thread artificial muscle, *IEICE Trans. Electron.* J102–C, 271–272. (in Japanese)
46. For example, Nishimura, Y., Wagner, R., Manaka, T. and Iwamoto, M. (2006). *Thin Solid Films*, 499, 40–43, https://doi.org/10.1016/j.tsf.2005.07.051.

47. Iwamoto, M., Yamamoto, T. and OuYang, Z.-C. *Electrical and Geometrical Properties of Organic Monolayers* (World Scientific, Singapore) (in preparation).
48. Blodgett, K. B. and Langmuir, I. (1937). *Phys. Rev.*, 51, 964–982, https://doi.org/10.1103/PhysRev.51.964; Terada, T., Yamamoto, R. and Watanabe, T. (1934). Sc. Pap. I. P. C. R., 23, pp. 173–184 and Pl.XXV–XXX.
49. Kunitake, T. and Okahata, Y. (1977). *J. Am. Chem. Phys.*, 99, 3860–3861, https://doi.org/10.1021/ja00453a066.
50. Langmuir, I. (1917). The constitution and fundamental properties of solids and liquids. II. Liquids, *J. Am. Chem. Soc.*, 39, 1848–1906, https://doi.org/10.1021/ja02254a006; Japanese translation by Fukuda, K. (1975). *Kagaku No Genten No.7 Kaimen Kagaku* (*Originality of chemistry series, No. 7 Interface science*) Sec. 2, The Chemical Society of Japan ed. (Tokyo Daigaku Shuppan Kai, Tokyo) (in Japanese).
51. Iwamoto, M. and Wu, C.-X. (2001) *The Physical Properties of Organic Monolayers* (World Scientific, Singapore), https://doi.org/10.1142/4575.
52. Ou-Yang, Z.-C., Liu, J.-X. and Xie, Y. Z. (1999) *Geometric Methods in the Elastic Theory of Membranes in Liquid Crystal Phases* (World Scientific, Singapore), https://doi.org/10.1142/3579.
53. Boys, C. V. (1920) *Soap Bubbles* (SPCK, London); Japanese translation by Yada, Y. (1959) *C. V. Boys Shabondama No Kagaku* (Maki Shoten, Tokyo) (in Japanese).
54. Ou-yang, W., Weis, M., Chen, X., Manaka, T. and Iwamoto, M. (2009). *J. Chem. Phys.*, 131, 104702, https://doi.org/10.1063/1.3212945.
55. Tamura, R., Manaka, T., Iwamoto, M., Itoh, T. and Watanabe, J. (2005). *Chem. Phys. Lett.*, 407, 337–341, https://doi.org/10.1016/j.cplett.2005.03.083.
56. Matsuoka, Y., Taguchi, D., Manaka, T. and Iwamoto, M. (2014). *Thin Solid Films*, 554, 8–12, https://doi.org/10.1016/j.tsf.2013.04.099.

Appendix A

Electric Field Formed by a Point Charge and a Dipole

The electric field is formed in solids according to the Gauss law. The electric field formed in film with relative dielectric constant ε_s by a point charge q is given by

$$\vec{E}(\vec{r}) = \frac{q}{4\pi\varepsilon_s\varepsilon_0}\frac{\vec{r}}{r^3}, \tag{A.1}$$

where $\vec{r}$ is a positional vector with the origin at the position of point charge. ε_0 is dielectric constant of vacuum. That is, the electric field is diverging from a point charge homogeneously toward all of the directions.

On the other hand, the electric field formed by a dipole $\vec{\mu}$ is given by

$$\vec{E}(\vec{r}) = \frac{1}{4\pi\varepsilon_s\varepsilon_0}\left(-\frac{\vec{\mu}}{r^3} + \frac{3\vec{r}(\vec{\mu}\cdot\vec{r})}{r^5}\right). \tag{A.2}$$

The electric field is dependent on the orientational direction of dipole $\vec{\mu}$.

Appendix B

Dielectric Relaxation and Dielectric Relaxation Time $\tau = \varepsilon/\kappa$

In material with dielectric constant ε and conductivity κ, excess charge distribution decays with relaxation time $\tau = \varepsilon/\kappa$. The conduction current $\vec{i}$ and electric flux $\vec{D}$ are defined as,

$$\vec{D} = \varepsilon \vec{E} \tag{B.1}$$

$$\vec{i} = \kappa \vec{E} \tag{B.2}$$

Using Eq. (B.1) in combination with Eq. (B.2),

$$\frac{\vec{D}}{\varepsilon} = \frac{\vec{i}}{\kappa} (= \vec{E}) \tag{B.3}$$

is obtained, assuming that the electric field $\vec{E}$ is coming out from the electric flux $\vec{D}$. By taking divergence of Eq. (B.3),

$$\frac{\nabla \cdot \vec{D}}{\varepsilon} = \frac{\nabla \cdot \vec{i}}{\kappa} \tag{B.4}$$

with Gauss law $\nabla \cdot \vec{D} = \rho$ and current continuity law,

$$\nabla \cdot \vec{i} = -\frac{d\rho}{dt}$$

is obtained and the following relation is derived.

$$-\frac{d\rho}{\rho} = \frac{dt}{\tau} \quad \text{with } \tau = \frac{\varepsilon}{\kappa}. \tag{B.5}$$

In the case of constant dielectric relaxation time τ, charge distribution satisfying Eq. (B.5) is

$$\rho = \rho_0 e^{-\frac{t}{\tau}}. \tag{B.6}$$

It is instructive here to note that ρ_0 is constant and a function of space, and ρ is decaying function with time τ. This is dielectric relaxation process in electromagnetism. The distribution of ρ in space is relatively the same and its magnitude decays with a relaxation time τ. This is totally different from diffusion process where the gradient of charge density ρ in space is acting as a driving force to change charge distribution, resulting in a development of charge distribution width. In the case of non-constant τ and is function of temperature, see Eq. (3.2) in Chap. 3.

Appendix C

Mathematical Derivation of Integral $\frac{T}{\beta}\int_{T_0}^{T}\frac{e_n}{T}dT = 1$

$P(E', T)$ is a function that gives a peak at $E' = E^*$, where

$$\left.\frac{\partial P}{\partial E'}\right|_{E'=E^*} = 0.$$

Therefore, the following relation is obtained.

$$\frac{e_n}{KT}\exp\left(-\frac{1}{\beta}\int_{T_0}^{T} e_n dT\right) + e_n\left(-\frac{1}{\beta}\int_{T_0}^{T}\frac{e_n}{KT}dT\right) \times \exp\left(-\frac{1}{\beta}\int_{T_0}^{T} e_n dT\right) = 0. \tag{C.1}$$

From Eq. (C.1), the following relation is derived.

$$\frac{T}{\beta}\int_{T_0}^{T}\frac{e_n}{T}dT = 1. \tag{C.2}$$

Equation (C.2) describes the relation Eq. (4.44).

The relation Eq. (C.2) is also derived by carrying out the integration. Starting from left-hand side of Eq. (C.2) with $x = H/KT$ and $H = E' - E_c$, Eq. (C.2) is rewritten as follows:

$$I = \frac{T}{\beta}\int_{T_0}^{T}\frac{e_n}{T}dT = -\frac{\nu T}{\beta}\left(\frac{e^x}{x} - \frac{e^{x_0}}{x_0} - \int_{x_0}^{x} x^{-2}e^x dx\right). \tag{C.3}$$

For $T > T_0$, e^{x_0}/x_0 is negligibly small in comparison with other terms, and Eq. (C.3) is approximately given as

$$\begin{aligned} I &= -\frac{\nu T}{\beta}\left(\frac{e^x}{x} - \int_{x_0}^{x} x^{-2}e^x dx\right) \\ &= \frac{T}{\beta}\left(\frac{KT}{E_c - E^*}e_n + \int_{T_0}^{T}\frac{K}{E_c - E^*}e_n dT\right). \end{aligned} \tag{C.4}$$

The second term in the parenthesis is calculated as

$$\int_{T_0}^{T}\frac{K}{E_c - E^*}e_n dT \cong \frac{K^2T^2}{(E_c - E^*)^2}e_n. \tag{C.5}$$

For $E_c - E^* \gg KT$, Eq. (C.5) becomes nearly zero. At the TSC peak temperature $T = T_m$, the following relationship is satisfied (Eq. (3.5) with $H = E_c - E^*$)).

$$\frac{KT_m^2}{\beta(E_c - E^*)} = e_n^{-1}. \tag{C.6}$$

Using the relations Eq. (C.4), Eq. (C.5), and Eq. (C.6), we get the relation Eq. (C.2) as

$$I = \frac{KT^2}{\beta(E_c - E^*)}e_n = 1. \tag{C.7}$$

Appendix D

Mathematical Derivation of Coefficient $D = 1.2\frac{\beta(E_c - E^*)}{T}$

$P(E', T)$ can be expressed using δ-function as follows:

$$D\delta(E' - E^*) \approx e_n \exp\left(-\frac{1}{\beta}\int_{T_0}^{T} e_n dT\right). \tag{D.1}$$

The function in the right hand-side term,

$$\exp\left(-\frac{1}{\beta}\int_{T_0}^{T} e_n dT\right)$$

is in good agreement with Fermi-Dirac distribution function. This function monotonously decreases by e^{-1} at at $E' = E^*$ and becomes nearly 0 and 1 in the region of energies $E' > E^* + 2KT$ and $E' < E^* - 2KT$, respectively. Accordingly, the following relation is derived.

$$\exp\left(-\frac{1}{\beta}\int_{T_0}^{T} e_n dT\right) = \frac{1}{1 + 1.7\exp\left(2\frac{E'-E^*}{KT}\right)}. \tag{D.2}$$

Using Eq. (D.1) with Eq. (D.2),

$$\begin{aligned} D &= \int_{E_v}^{E_c} \frac{e_n}{1 + 1.7\exp\left(2\frac{E'-E^*}{KT}\right)} dE' \\ &= \int_{E_v}^{E_c} \frac{e_n}{1 + 1.7\exp(2\frac{E_c-E^*}{KT})\exp\left(2\frac{E'-E_c}{KT}\right)} dE' \\ &= KT\nu \int_{\exp}^{1} \left(\frac{E_v - E_c}{KT}\right) \frac{dx}{1 + bx^2}. \end{aligned} \tag{D.3}$$

Here $E' = E^*$ locates in the region of energy between E_c and E_v, and $b = 1.7\exp(2(E_c - E^*)/KT)$ and $x = \exp((E' - E_c)/KT)$. As the integration term is given as $\int 1/(x^2 + c) \cdot dx = 1/\sqrt{c} \cdot \mathrm{Tan}^{-1}(x/\sqrt{c})$,

$$D = \frac{KT\nu}{\sqrt{1.7}} \exp\left(\frac{E^* - E_c}{KT}\right) \cdot \left(\mathrm{Tan}^{-1}\left(\sqrt{1.7}\exp\frac{E_c - E^*}{KT}\right)\right.$$
$$\left. -\mathrm{Tan}^{-1}\left(\sqrt{1.7}\exp\frac{E_v - E^*}{KT}\right)\right). \tag{D.4}$$

For $(E_c - E^*)/KT \gg 1$ and $(E_v - E^*)/KT \ll -1$, and relations $\mathrm{Tan}^{-1}(\sqrt{1.7}\exp(E_c - E^*)/KT) \sim \pi/2$ and $\mathrm{Tan}^{-1}(\sqrt{1.7}\exp(E_v - E^*)/KT) \sim 0$ are derived. Finally,

$$D = 1.2KT\nu \exp\left(\frac{E^* - E_c}{KT}\right). \tag{D.5}$$

Substituting the relation Eq. (C.6) into Eq. (D.5), D is obtained as,

$$D = 1.2\frac{\beta(E_c - E^*)}{T}. \tag{D.6}$$

Appendix E

Determination of Director Tilt Angle

The director tilt angle ϑ is determined in the condensed phase using SHG measurement, by assuming that small domains with C_s–symmetry (director tilt angle is ϑ) are distributed randomly, and have a C_∞–symmetry in average over the area of laser spot. Due to the uniaxial symmetry of the rod-like molecule, β_{333} is dominant among β_{ijk} $(i, j, k = 1, 2, 3$ in molecular coordinate). The tensor components of molecular nonlinear polarizability in Eq. (5.10) are derived as

$$\sigma_{14} = \beta_{123} + \beta_{132} = 0$$
$$\sigma_{25} = \beta_{231} + \beta_{213} = 0$$
$$\sigma_{33} = \beta_{333}$$
$$\sigma_{32} = \beta_{322} = 0$$
$$\sigma_{31} = \beta_{311} = 0$$
$$\sigma_{24} = \beta_{223} + \beta_{232} = 0$$
$$\sigma_{15} = \beta_{131} + \beta_{113} = 0$$
$$\lambda_{14} = \lambda_{25} = \lambda_{36} = \lambda_{15} = \lambda_{24} = \lambda_{23} = 0.$$

Using these nonlinear molecular polarizability components, nonlinear susceptibility $\chi^{(2)}$ is expressed in terms of tilt angle ϑ as

$$s_{14} = 2\chi_{123} = 0$$
$$s_{15} = \frac{2}{5} N_s(S_1 - S_3)\beta_{333} = 2\chi_{113}$$

$$s_{31} = \frac{1}{5}N_s(S_1 - S_3)\beta_{333} = \frac{N_s}{2}\cos\vartheta\sin^2\vartheta\beta_{333} = \chi_{311}$$

$$s_{33} = \frac{1}{5}N_s(3S_1 + 2S_3)\beta_{333} = N_s\cos^3\vartheta\beta_{333} = \chi_{333}$$

$$a_{14} = a_{15} = a_{36} = 0.$$

By using the forth and fifth relation, the director tilt angle ϑ is calculated as

$$\vartheta = \tan^{-1}\sqrt{\frac{2\chi_{311}}{\chi_{333}}}.$$

Utilizing the SHG, it is possible to obtain χ_{311} and χ_{333} by curve fitting the incident light polarization dependence of the intensity of the p- and s-polarized SH signals, and ϑ is calculated. With the director tilt angle ϑ determined from SHG measurements, it is possible to calculate the order parameter S_2 from the collected reflectivity data. To accomplish this, a computer algorithm was utilized. The algorithm calculates the deviation, σ_p, between calculated and collected reflectivity ratio for each collected reflectivity data point as follows:

$$\sigma_p = \frac{|r(S_2)|^2}{|r(0)|^2} - \frac{d_p}{x}, \quad p = 1, 2, 3, \ldots, n \tag{E.1}$$

where r is the reflectivity calculated from Eq. (5.16), d_p is a data point in the collected reflectivity data set d, n represents the total number of data points in d, and x is the average collected reflectivity value when the monolayer is in the gas/LE phase, under the assumption $S_2 = 0$. The purpose for taking the ratios is to negate unknown constants, *e.g.*, the optical loss and the sensitivity of the light detector. All possible values of S_2 ($-0.5 < S_2 < 1$) to the fifth decimal place are inserted into Eq. (E.1). The value of S_2 producing the smallest deviation is the value of the order parameter for the given collected reflectivity point.

Appendix F

Technical Measures to Construct a Stably Operating SHG Measurement System

It is important to keep stable operation of constructed measuring-system, and to make trouble shooting, if necessary, to eliminate troubles in experiment. After setting up the SHG measurement system, SHG signal generated from standard sample should be carefully tested. In this Appendix, SHG measurement system equipped with Q-switched Nd:YAG laser and optical parametric oscillator (OPO) as a probe laser source is mainly described. The OPO scans wavelength in the region from 1300 nm to 800 nm (SHG wavelength 650 nm to 400 nm), and from 637 nm to 488 nm (SHG wavelength 319 nm to 244 nm). Quartz powder is used as a standard sample to check the SHG measurement system. Over these wavelength regions, the second order nonlinear susceptibility of quartz powder is constant. In other words, quartz powder radiates SHG at the same intensity over these wavelength regions, in the presence of laser incident beam. Therefore, SHG spectrum of the quartz powder is available as a reference to test the SHG measurement system. Figure F.1 shows SHG spectrum of quartz sample in visible and ultra-violet SHG wavelength regions. Here, second order susceptibility of quartz powder is constant in both wavelength regions, whereas optical components of measurement system have different optical spectrum. Accordingly, the measured SHG intensity of quartz powder changes with SH wavelength. For the spectra shown here, transmission spectrum of laser

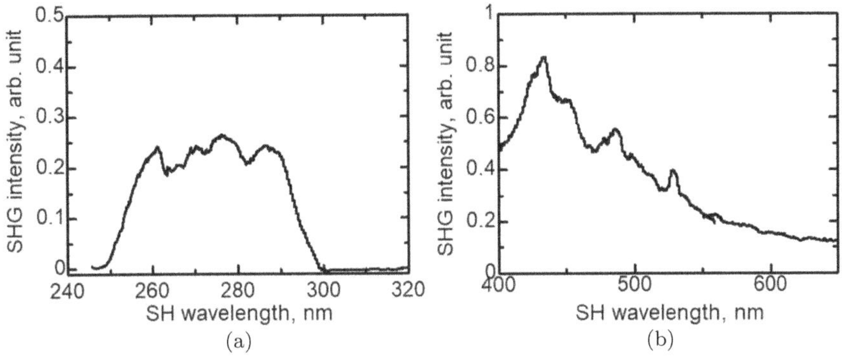

Figure F.1. Consideration of measurement system spectrum by using quartz powder. (a) Ultra-violet SHG wavelength region. (b) Visible SHG wavelength region.

cut filters and wavelength dependence of laser intensity, respectively, governs the spectrum in ultra-violet region and in visible SH wavelength region.

The quartz powder is also useful to make a quick check on SHG measurement system at a selected SHG wavelength. For this purpose, the powder size is smaller than the diameter of laser beam spot on a sample. Here, the quartz powder is made from a block of 1 mm size quartz. That is, by breaking a block of quartz into small pieces using a pestle and mortar for 15 min, quartz powder is made. After dispersing this power into ethanol solution, the solution is suck by a pipette and it is dropped on a glass microscope slide. After the evaporation of ethanol from the drop, quartz powder layer is formed on a glass slide. In SHG measurement, probe laser beam with selected wavelength λ in the region 1300–800 nm is incident onto this layer from the backside of the glass slide, using the optical arrangement of reflection style. A laser beam with a wavelength region 637–488 nm is incident on the quartz powder side, and the reflected SHG is measured with a detector. Commercially available quartz powder as chemical agent and mineral pigment is also SHG active, and it is useful to test the SHG signal. On the other hand, in order to quantitatively measure $\chi^{(2)}$ and $\chi^{(3)}$, a 1 mm thick Y-cut quartz single crystal plate is useful. For the quantitative measurement by using a quartz plate, Maker fringe method can be employed in the manner as described in the book (Ref. [42] in Chap. 6).

F.1 Check point 1: $I_{SHG} - I$ plot

The SHG intensity I_{SHG} is proportional to the square of laser intensity I as

$$\begin{aligned} I_{SHG} &\propto |\vec{P}_{NL}|^2 = |N\beta E_\omega E_\omega|^2 \propto |E_\omega|^4 \propto I^2, \\ I_{SHG} &\propto |\vec{P}_{NL}|^2 = |N\gamma E_0 E_\omega E_\omega|^2 \propto |E_\omega|^4 \propto I^2. \end{aligned} \tag{F.1}$$

Accordingly, SHG intensity I_{SHG} plotted as a function of normalized laser intensity gives a slope $n = 2$ in log–log plot (I_{SHG}–I plot). The n–value of the slope is a check point for the measurement system. Figures F.2(a) and F.2(b) show examples of SHG intensity I_{SHG} from quartz powder and quartz single crystal plate, respectively. The laser intensity I was changed by using neutral density (ND) filter with transmittance T. The lateral axis is represented by normalized laser intensity, where the intensity without using ND filter is chosen unity. ND filter was installed in front of the quartz powder. The laser intensity is $I = I_0$ before passing through a ND filter, and decreases to $I = I_0 T$ after passing through the ND filter. SHG intensity from quartz powder and quartz single crystal plate shows a slope of 2. Note that it is important to use transmittance T of ND filter at laser wavelength, not at the SHG wavelength.

Figure F.3 shows I_{SHG}–I plots of EFISHG. The sample is indium–tin–oxide (ITO)/C60/Al, where C60 film (thickness, 300 nm) is sandwiched between ITO and Al electrodes. External voltage of 5 V is applied to ITO electrode with reference to Al electrode grounded. On keeping the external voltage to be constant, the laser intensity was changed by using ND filters. The I_{SHG}–I plots show a slope of 2. That is, the observed signal intensity increases in proportion to I^2, supporting that the signal is due to SHG. Figure F.4 shows another example of I_{SHG}–I plots for EFISHG. The sample is indium–zinc–oxide (IZO)/polyethylene terephthalate (PET)/Ag device, where the PET film is 260 nm thick. The IZO and Ag electrodes are connected by a conducting wire without applying external

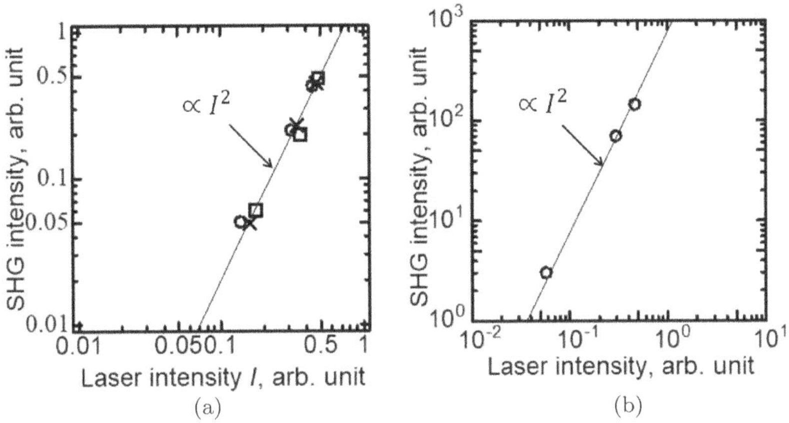

Figure F.2. (a) SHG of quartz powder used to confirm the detection of SHG by the measurement system. Cross, circle, and square indicate SHG intensity at wavelength 490 (laser wavelength 980 nm), 560 (1120), 650 (1300) nm, respectively. (b) SHG of Y-cut single-crystal quartz measured at 532 nm (laser wavelength 1064 nm). Properly arranged SHG measurement system exhibits square dependence of SHG intensity on laser intensity. The log–log plot of SHG intensity and laser intensity follows a slope with 2.

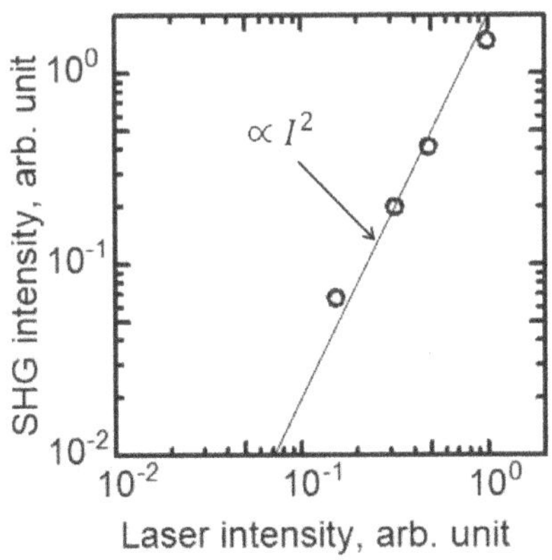

Figure F.3. EFISHG of C60 used to confirm the detection of SHG by the measurement system. SHG wavelength and laser wavelength are 500 and 1000 nm, respectively. C60 thickness is 300 nm. The intensity is proportional to the square of laser intensity.

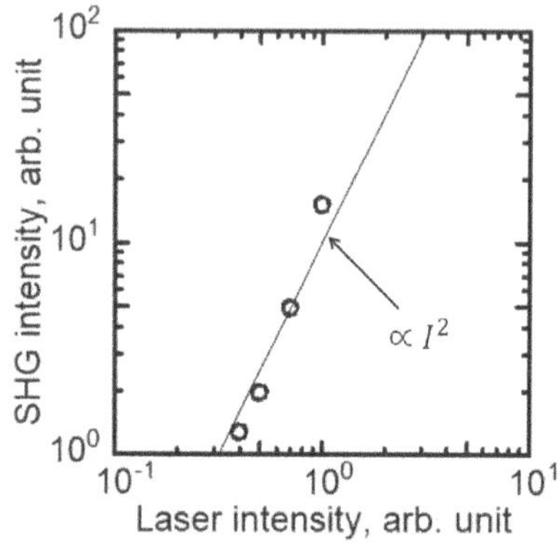

Figure F.4. EFISHG of polyethylene terephthalate (PET) used to confirm the detection of SHG by the measurement system. SHG wavelength and laser wavelength are 285 and 570 nm, respectively. Laser is incident from IZO electrode side of IZO/PET/Ag device with connecting IZO and Ag by a conducting wire. External voltage is zero, but non-zero internal field and dipolar SHG is contributing to the square dependence of measured intensity. This intensity is available to check the measurement system. PET thickness is 260 nm.

voltage, but the presence of non-zero internal electric field in device causes EFISHG. As this EFISHG is superposed on the SHG, I_{SHG}–I plots are available to check the source of SHG signals. The plots for the IZO/PET/Ag device show a slope of 2, and it is concluded that there is no source other than SHG.

On the other hand, if the I_{SHG}–I plots deviates from a slope of 2, we should consider the possibilities of system problem for SHG measurement. Among the possibilities, three of them are described briefly in the followings, to be as a guidance for trouble shooting.

(1) Room light and sun light leakage to detector is a serious problem. This leakage is tested under the condition without radiation of laser beam. These leakage lights will be continuously detected as a non-zero signal output by an optical detector. That is, continuous non-zero light will be detected by the measurement system with laser equipment totally switched off, if this leakage problem is serious. Accordingly, constant light intensity will be superposed on an output of light detector, and will deform the I_{SHG}–I relation. Figure F.5 shows an example, where the slope decreases from 2, along with the

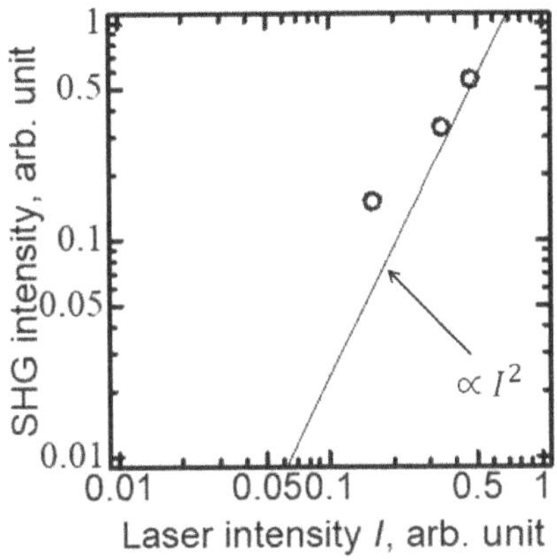

Figure F.5. I_{SHG}–I plot deviates from the slope of 2, if there is a light leakage coming into a detector other than SHG.

laser intensity decreasing toward zero. To remove such continuous lights, it would be better to set up the SHG system in an optical dark room or in an optical dark box. If there is a light source inside of the dark room, such as power-on LED indicator of power meter and other equipment, we need to seal the indicator, display panel, and other light emitting sources, by using a black tape, *e.g.*, electrical vinyl-tape. In summary, if signal output level is non-zero after turning off the laser system, it is concluded that there is a leaking light. Find out the leakage source and remove it until the detector output indicates zero. Note that signal offset level in electrical system gives similar impact on the I_{SHG}–I relation. In this case, offset level should be adjusted to zero.

(2) Another light leakage source is laser radiation from laser chassis. After completely blocking laser beam output at laser beam exit-port of OPO, there will be a leakage from the laser chassis, if it exists. After sealing continuous leak light in the manner as described in case (1), a pulsed light will be detected due to the leakage from laser chassis. The I_{SHG}–I relation shows deviation from a slope 2 in a way similar to case (1). This kind of leakage is obviously displayed on an oscilloscope that monitors output of a detector such as photomultiplier tube (PMT). The pulsed signal on an oscilloscope

is synchronizing with timing of laser pulse emission. With the laser beam path blocked by using a 1 mm thick aluminium plate, this pulsed signal is detected, indicating that the detected laser pulse is not on a laser beam path coming from an exit-port of laser equipment, but a part of the laser is emitting through other paths. Provided that a laser and an OPO are in independent chassis, an OPO will be installed in front of the laser. There is a possibility that a connecting space between the laser and OPO is a source of light leakage. A laser and an OPO may have a hole that is used to maintain air condition inside the equipment and outside laboratory ambient, and laser can radiate through this hole. A chassis of laser and OPO may be equipped with a lid that allows one to open the chassis for changing optical layouts inside the equipment. The lid is fastened by a screw. The screw hole can be a source of laser radiation. It is important to seal all these leak light sources. With a block of main laser path, and if there is an output from detector, it is concluded that there is a leakage. Find out a leakage source until detector output indicates zero.

(3) A laser beam on main beam path is detected by a detector when a system design fails or optical components fail during long-term operation. The I_{SHG}–I plots deviate from a straight line in the manner as plotted in Fig. F.5. In this case, a pulse signal is displayed on an oscilloscope in the way similar to the case (2). By contrast to the leak from a laser chassis, this leakage is totally stopped by blocking the main beam path. The pulse is synchronizing with laser emission timing, and disappears when the laser beam path is blocked. In this case, we need to revise the optical implementation so that only SHG from sample is detected by a detector. For example, PMT with no sensitivity at the laser wavelength will improve this situation.

F.2 Check point 2: $I_{SHG} - E_0$ plot

EFISHG intensity is in proportion to the square of electrostatic field E_0 in device as

$$I_{SHG} \propto |\vec{P}_{NL}|^2 \propto |N\gamma E_0 E_\omega E_\omega|^2 \propto |E_0|^2. \tag{F.2}$$

Accordingly, E_0^2 dependence of the SHG intensity provides a check point for EFISHG measurement. A plot of SHG intensity I_{SHG} as

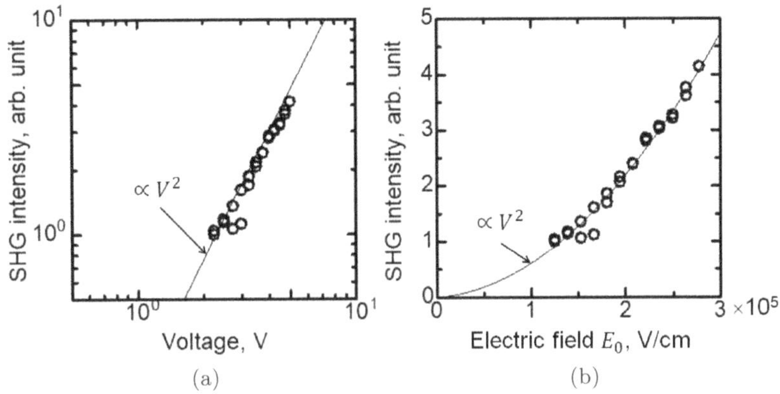

Figure F.6. SHG measurement as a function of voltage applied to a single layer device. Sample is IZO/C60/Al. External voltage is applied to IZO, and Al is grounded as a reference electrode. Layer thickness of C60 is 180 nm. (a) SHG intensity at wavelength of 500 nm is plotted as a function of external voltage. The log–log plot shows the slope of 2, indicating EFISHG process is responsible for the measured SHG. (b) A calibration curve of SHG intensity to electrostatic field formed in the C60 layer.

a function of electric field E_0 in log–log plot gives a slope of 2. Figure F.6 shows an example of I_{SHG}–E_0 plot. The sample is IZO/C60/Al device, where C60 thickness d is 180 nm. External voltage V is applied to IZO in reference to Al grounded. IZO and Al are electrodes, and electric fields in them are zero in steady state. The external electric field is formed in the C60 layer only. In this situation, the electric field in C60 is given as an average field $E_0 = V/d$.

By plotting the same data in linear–linear plot, we can estimate a minimum detectable electrostatic field E_{0m} (or charge density) of the measurement system. Figure F.7(a) gives an example of I_{SHG}–E_0 plot in linear–linear plot. The measured intensity is clipped by a noise level, in the voltage region between -2 V and 2 V. The minimum electric field to be probed by using this SHG measurement system is 1.1×10^5 V/cm. On the other hand, the measurement system with improved SHG collecting efficiency shows I_{SHG}–E_0 plot as illustrated in Fig. F.7(b). External voltage is detectable to the level of 0.1 V, corresponding to electrostatic field 5.6×10^3 V/cm. This result was obtained by using SHG microscope using radially polarized laser beam as a probe light. The laser source for radially polarized laser

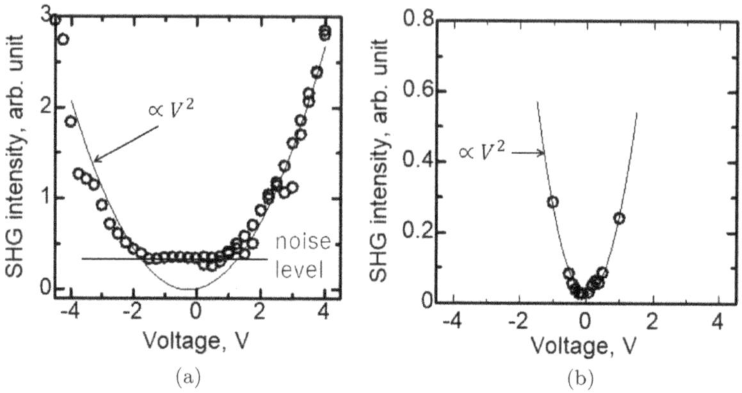

Figure F.7. EFISHG intensity dependence on external voltage, showing the square dependence of measured intensity on voltage. Sample is IZO/C60/Al. External voltage is applied to IZO, and Al is grounded as a reference electrode. C60 thickness is 180 nm. (a) SHG measurement system where noise-level limits the detection of SHG intensity in voltage region $-2\,\mathrm{V} < V < +2\,\mathrm{V}$. (b) SHG measurement system with improved SHG collection efficiency.

beam is Q–switched Nd:YAG laser for this SHG microscope. By using a system equipped with short-pulse laser such as femto-second laser, detectable range could be extended to the order of 10 V/cm.

There are various ways to form an electric field E_0 in materials and devices. Among them is a way to apply external voltage across the film sandwiched between two plane electrodes, as shown in Fig. F.8. In order to get a calibration curve of SHG intensity with respect to electric field, we need to form well-defined electrostatic field in materials and devices. Upon external voltage application to a sandwiched device, there is a possibility that charges are injected into the device from electrodes. The injected charges deform the electric field in the material due to the formation of space charge electric field. We need to avoid the deformation of the electric field to get a calibration curve.

To form well-defined electrostatic field, we need to carry out SHG measurement for devices being free from such space charge field effect. Figure F.8(a) shows a single layer sandwiched between identical two plane electrodes. Electrical capacitors, ferroelectric memory elements, and optoelectrical modulators are examples of single-layer

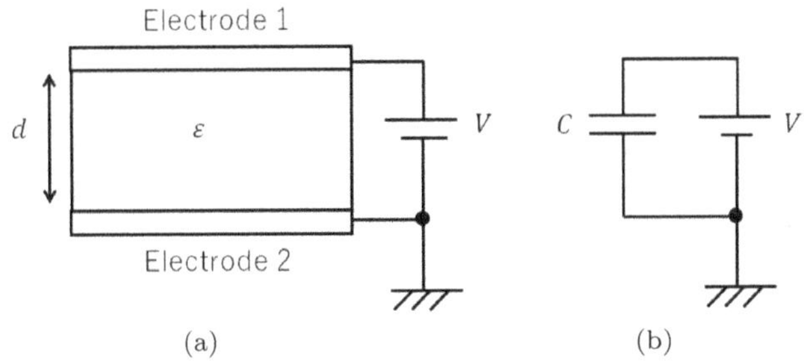

Figure F.8. A single layer device without charge injection. (a) Single layer device arrangement. (b) Circuit model of the single layer device.

devices. The layer of these devices is free from charge injection, and electrostatic field in the layer is well-defined. For a single layer sandwiched between parallel electrodes, voltage V is applied to the electrode 1 in reference to electrode 2 grounded. Assuming that the layer thickness is d, the electric field E_0 formed in the layer is given by

$$E_0 = \frac{V}{d}, \tag{F.3}$$

or

$$E_0 = +\frac{Q}{\varepsilon}, \tag{F.4}$$

with dielectric constant of the film ε. By using an electric circuit model, sandwiched cell has a capacitance C $(= \varepsilon/d)$ as shown in Fig. F.8(b). The direction of electric field is defined positive when the electric field is pointing from electrode 1 to electrode 2. The first expression (Eq. (F.3)) comes from an idea that potential difference between two electrodes is a potential work that is done during a travel of a charge from electrode 1 to electrode 2. The external voltage source keeps the potential difference between the two electrodes constant, and this expression is convenient to calculate the electric field using the applied voltage. The second form (Eq. (F.4)) comes from the Gauss law in electromagnetic theory, where the charge Q on electrode 1 and $-Q$ on electrode 2 reflect displacement of Q in a space between electrode 1 and 2. The charge displacement Q divided by dielectric constant is the electric field.

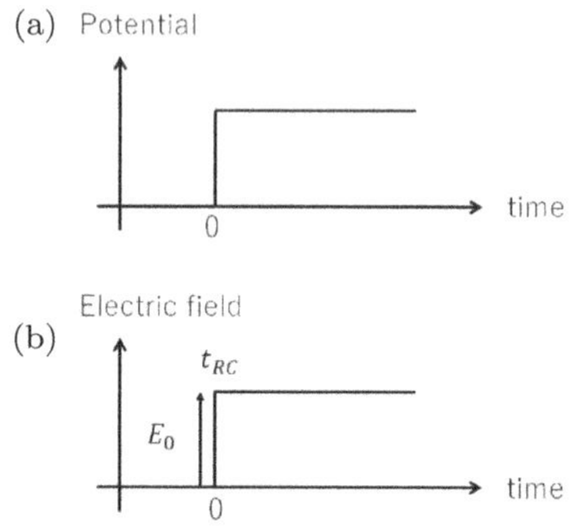

Figure F.9. Electrostatic field in the single layer device without charge injection, after d.c. voltage application at time = 0. (a) Potential across the layer. (b) Electrostatic field in the layer.

Figure F.9 displays electrostatic field in the single-layer after d.c. voltage V is applied to electrode 1, in reference to grounded electrode 2. The electric potential difference across the layer changes from zero to V with circuit time constant, and keeps constant afterwards. In the similar way, electrostatic field in the layer changes from zero to E_0 of Eq. (F.3) with circuit time constant, and is stable afterwards. By measuring SHG intensity in the presence of this stable electrostatic field, calibration curve of SHG intensity to electric field is obtained as shown in Fig. F.6. It is noteworthy that the electrostatic field in the layer is expressed by Eq. (F.3) even when ohmic conduction current flows through the layer. Accordingly, a single-layer device is available to form well-defined electric field of Eq. (F.3) by the fixed potential difference between two electrodes. Low voltage application particularly provides a reliable way that can avoid electric field deformation due to charge injection.

When the applied voltage is high, charge injection occurs, which gives effect on the measurement of calibration curve. Figure F.10 illustrates a special case where charge injection to the single layer device is provoked. Under high voltage application, charge injection occurs in the region near the injecting electrode. In Fig. F.10(a), positive charges inject from the electrode 1 into the layer. The injection depth is δ and smaller than the layer thickness, $\delta \ll d$. Figure F.11 shows the average electrostatic field and electric field

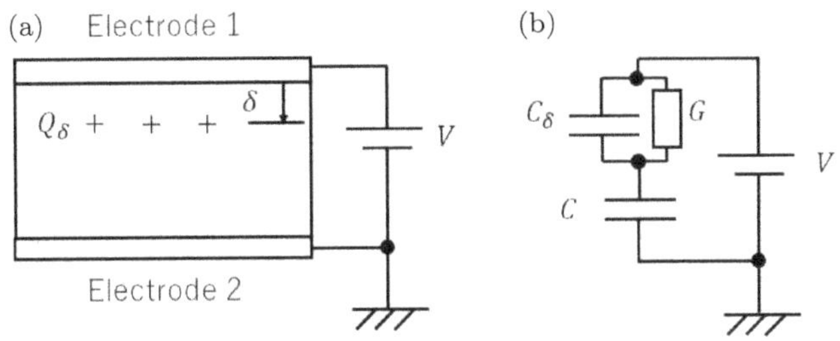

Figure F.10. A single layer device with charge injection. (a) Single layer device arrangement where charge injection occurs from electrode 1 to the region nearby the electrode 1. $\delta(\ll d)$ is injection depth. (b) Circuit model of the single layer device with charge injection.

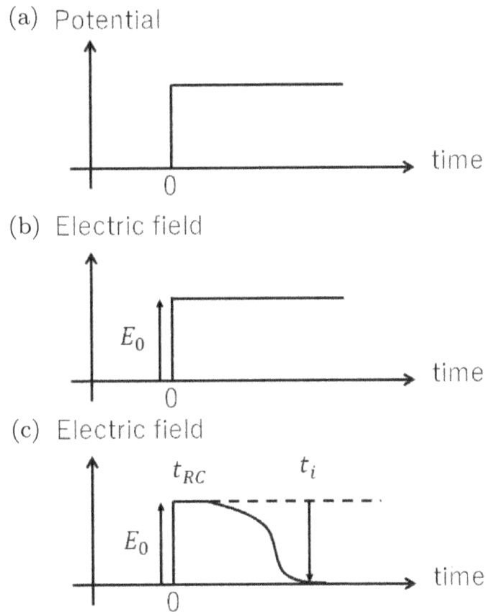

Figure F.11. Electrostatic field in the single layer device with charge injection, after d.c. voltage application at time = 0. (a) Potential across the film. (b) Average electric field in the layer. (c) Electric field in the region from the surface of electrode 1 to charge injection depth δ.

nearby electrode 1 surface after applying d.c. voltage at time 0. Average field is constant and is given by Eq. (F.3) after the potential difference between electrode 1 and electrode 2 saturates at external voltage V (see Fig. F.11(b)). On the other hand, electric field distribution in the layer is deformed due to injected charges. The injected charge forms a space charge field. Assuming that the injected charge density Q_δ is $Q_\delta = CV$, electrostatic field in the region from electrode 1 surface to the injection depth δ is zero (see Fig. F.11(c) after the injection time t_i). Electrostatic field in the region from the front of injection depth δ to electrode 2 surface is E_0 given by Eq. (F.3). In SHG measurement, probe laser incident into the layer where electromagnetic interaction causes SHG radiation. If the distribution of vibrating electric field E_ω of laser beam is constant in the layer, SHG intensity is in proportion to average electric field. With charge injection, electrostatic field in region with thickness δ is deformed from E_0 to zero, but average field does not change, and it is E_0. The average electric field in the layer is well-defined by Eq. (F.3), and we can carry out SHG measurement for making calibration curve.

This situation is no longer reliable for the device where vibrating electric field E_ω of probe laser changes in space due to multiple reflection, optical absorption, and other phenomena. SHG intensity changes depending on the spatial distribution of electrostatic field in the layer. In such a case, electrostatic field given by Eq. (F.3) is not valid to discuss about electrostatic field resulting in measured SHG. For example, attention is needed for a device where an electrode is made of optical light absorbing material such as metal. Example of spatial distribution of vibrating electric field E_ω is displayed in Fig. F.12 for organic semiconductor layer sandwiched between air and metal Al electrode. The simulation assumes that region (i) air with refractive index $n_1 = 1.0$, (ii) organic layer $n_2 = 1.5$ with thickness 100 nm, (iii) metal electrode $n_3 = 0.6 + j3.3$ with thickness 200 nm. For incident laser beam, vibrating electric field E_ω is 1, incidence angle is 45 degree, and p-polarized laser beam is used. Simulation shows that vibrating electric field E_ω at the air/organic interface is 2.3 times higher than the field at organic/metal interface. SHG intensity is proportional to $|E_\omega|^4$. The contribution of electrostatic field at organic/metal interface to SHG intensity is less than 5% compared with the contribution of electric field at air/organic interface. SHG intensity is not in proportion to average electric field, but SHG intensity is proportional to electric field near electrode 1 surface. In this situation, electric field calculated by using Eq. (F.3) is invalid, and we

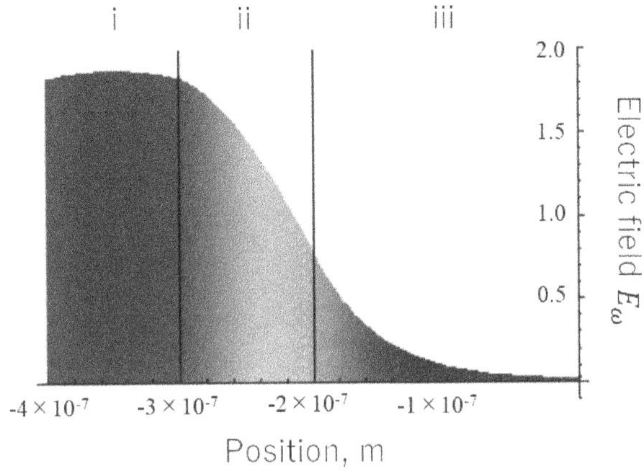

Figure F.12. Computer simulation of vibrating electric field E_ω in the single layer device. The vertical axis is amplitude of vibrating electric field of laser light. The amplitude of incident laser beam is 1. The lateral axis is position in the film thickness direction.

do not know quantitatively the electric field that caused SHG. Note that local field at nanostructured surface and other optical processes contributes to enhance SHG (Refs. [43, 44] in Chap. 6). In this case, calibration method described above is not appropriate, and a novel way of evaluating local field is required.

Figure F.13 shows double-layer device. Organic light-emitting diodes, organic solar cells, and field effect transistors are examples of a double-layer device. The double layer comprising of layer 1 (dielectric constant: ε_1, thickness: d_1) and layer 2 (dielectric constant: ε_2, thickness: d_2) is sandwiched between two identical electrode 1 and electrode 2. Electric field in layer 1 E_1 and layer 2 E_2 is

$$\begin{aligned}
&E_1 = \frac{V}{d^*} \text{ with effective layer thickness} \\
&d^* = d_1 + d_2^* \text{ and } d_2^* = \frac{\varepsilon_1}{\varepsilon_2} d_2, \\
&E_2 = \frac{V}{d^{**}} \text{ with effective layer thickness} \\
&d^{**} = d_1^* + d_2 \text{ and } d_1^* = \frac{\varepsilon_2}{\varepsilon_1} d_1.
\end{aligned} \tag{F.5}$$

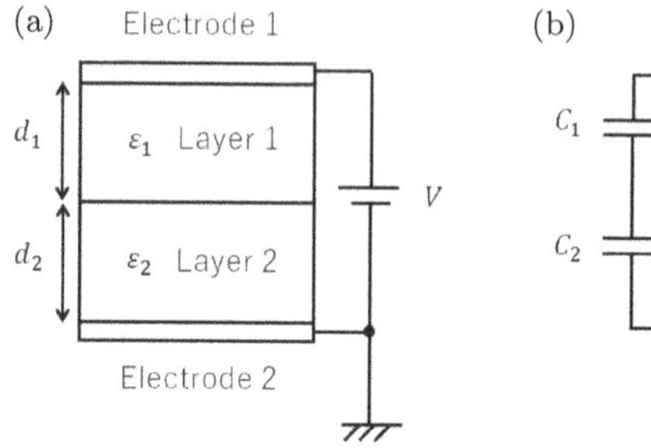

Figure F.13. A double layer device without charge injection. (a) Double layer device arrangement. (b) Circuit model of the double layer device.

$$E_1 = +\frac{Q}{\varepsilon_1}$$
$$E_2 = +\frac{Q}{\varepsilon_2}. \tag{F.6}$$

The first expression (Eq. (F.5)) comes from an idea that potential difference between two electrodes is a potential work that is done during a travel of a charge from electrode 1 to electrode 2. Assuming that the layer 1 and layer 2 have a dielectric constant ε_1, the potential work is eV during a travel of charge e on a route from electrode 1 to electrode 2 whose distance is effective thickness d^*. Assuming that the layer 1 and layer 2 have a dielectric constant ε_2, the potential work is eV during a travel on a route whose distance is effective thickness d^{**}. The second form (Eq. (F.6)) comes from Gauss law. The charge Q on electrode 1 and $-Q$ on electrode 2, reflecting a displacement of Q in the layer 1 and in the layer 2. The displacement divided by dielectric constant is electric field. Average electric field $\bar{E}$ in double-layer is

$$\bar{E} = \frac{E_1 d_1 + E_2 d_2}{d_1 + d_2} = \frac{V}{d}, \quad \text{with } d = d_1 + d_2. \tag{F.7}$$

By using electric circuit model in Fig. F.13(b), these fields are expressed by using electric circuit element C_1 and C_2 which are capacitance of layer 1 and capacitance of layer 2, respectively.

The electric field in layer 1 and layer 2 is,

$$E_1 = \frac{1}{d_1}\frac{C_2}{C_1 + C_2}V,$$
$$E_2 = \frac{1}{d_2}\frac{C_1}{C_1 + C_2}V. \tag{F.8}$$

Under d.c. voltage application to organic double-layer diode, no current flows in reverse direction (Ref. [45] in Chap. 6). In this case, electric field is given by Eq. (F.5) to Eq. (F.8).

As described in Chap. 4, charges accumulate at two-layer interface, and deform electrostatic field in the layer. Consequently, in a double-layer device, electric field in the layer is deformed from E_1 and E_2 expressed by Eq. (F.5) to Eq. (F.8). This situation is different from that of a single-layer device where electric field is well-defined. Figure F.14(a) shows an example of double-layer device where conduction current flows through the layer 1. After applying external voltage V to the device, electrode charges Q and $-Q$ are induced on electrode 1 surface and electrode 2 surface, respectively, at circuit time-constant. Electrical conduction current flows through the layer 1, and positive charge Q_s accumulates at the double-layer interface. The electrical conduction of the layer 1 is modelled by using conductance circuit element G_1 as shown in Fig. F.14(b). After charge

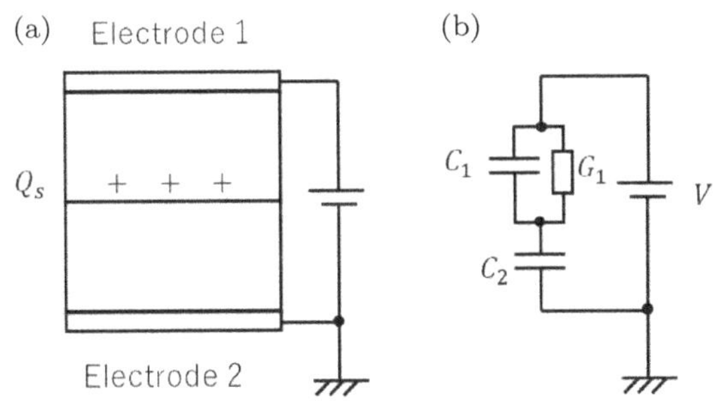

Figure F.14. A double layer device with charge injection. (a) Double layer device arrangement where charge injection occurs from electrode 1 to the double layer interface. (b) Circuit model of the double layer device with charge injection.

accumulation Q_s at the double-layer interface, electric field is

$$E_1' = +\frac{Q}{\varepsilon_1} - \frac{d^* - d_1}{d^*}\frac{Q_s}{\varepsilon_1}\left(= +\frac{Q}{\varepsilon_1} - \frac{d_2}{d^{**}}\frac{Q_s}{\varepsilon_1}\right)$$
$$E_2' = +\frac{Q}{\varepsilon_2} + \frac{d_1}{d^*}\frac{Q_s}{\varepsilon_2}\left(= +\frac{Q}{\varepsilon_2} + \frac{d^{**} - d_2}{d^{**}}\frac{Q_s}{\varepsilon_2}\right) \tag{F.9}$$

and

$$E_1' = \frac{V}{d^*} - \frac{V_s}{d_1}$$
$$E_2' = \frac{V}{d^{**}} + \frac{V_s}{d_2} \tag{F.10}$$

where V_s is a potential built by interfacial charge Q_s in reference to electrode 1 and electrode 2.

By using electric circuit elements in Fig. F.14(b), the field is expressed as

$$E_1' = \frac{1}{d_1}\frac{C_2}{C_1 + C_2}V - \frac{1}{d_1}\frac{Q_s}{C_1 + C_2},$$
$$E_2' = \frac{1}{d_2}\frac{C_1}{C_1 + C_2}V + \frac{1}{d_2}\frac{Q_s}{C_1 + C_2}. \tag{F.11}$$

Figure F.15 illustrates electric field in the layer 1 after applying positive external voltage V to the electrode 1 in reference to grounded electrode 2. After applying d.c. voltage at time 0, potential difference between electrode 1 and electrode 2 is established to V at circuit time-constant t_{RC}. Electrostatic field in layer 1 and layer 2 is formed as E_1 and E_2 expressed by Eq. (F.5) to Eq. (F.8). Afterwards, electrical current flows through the layer 1 and conveys charges, causing charge accumulation at the double layer interface. The time-constant t_{MW} for this charge accumulation process is given by relaxation time of Maxwell–Wagner effect described in Chap. 4. At the same time, accumulated charge Q_s forms space charge field in the layers. The electric fields in the layers change to E_1' and E_2' expressed by Eq. (F.9) to Eq. (F.11). From standpoint of experiment, the electric field E_1' and E_2' in Eq. (F.9) to Eq. (F.11) is unknown in prior to experiment.

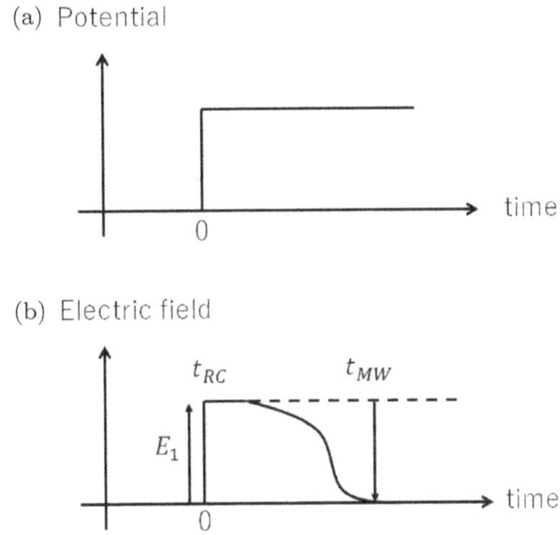

Figure F.15. Electrostatic field in the double layer device with charge injection, after d.c. voltage application at time = 0. (a) Potential across the double layer. (b) Electric field in the layer 1 of double-layer.

To make a calibration curve, we need to carry out experiment in a condition of no charge accumulation $Q_s = 0$. In this condition, electric field is well-defined and expressed as Eq. (F.5) to Eq. (F.8) in double layer materials and devices. The SHG intensity is used to plot calibration curves of these electric field E_1 and E_2, not E_1' and E_2'. The SHG intensity I_{SHG} changes in proportion to the square of electric field E_1 and E_2 formed in the layer. To probe electric field E_1 and E_2 without charge accumulation, there are several ways in SHG measurement. Among them, the time-resolved SHG measurement by using AC square voltage application is one way. In the time-resolved measurement, timing of AC square voltage and pulse laser are synchronized. The SHG intensity is recorded with sweeping the timing τ of laser incidence in reference to AC square voltage. By plotting SHG intensity as a function of timing τ, a transient of SHG intensity, that reflects a transient of electric field in the layer shown in Fig. F.15(b), is obtained. The transient of SHG allows one to identify the circuit time-constant t_{RC} and t_{MW} (Ref. [46] in Chap. 6). By setting the timing τ in time range $t_{RC} \ll \tau \ll t_{MW}$, SHG intensity is proportional to the square of E_1 and E_2, being free of deformation by charge accumulation. For example, organic field-effect transistors, organic solar cells, organic light-emitting diodes, and others, the electric field given by Eq. (F.5) to Eq. (F.8) is successfully probed by using the time-resolved SHG measurement.

F.3 Check point 3: Technical considerations based on sample properties

There are several nonlinear optical processes other than SHG process, which lead to emission of overlapping signals in the optical measurement. This is a serious problem for stable operation of SHG measurement system. In order to catch only SHG signals, we need to eliminate other optical signals. For example, the signal of two-photon photoluminescence (TPL) overlaps with SHG signal. This photoluminescence is emitted from luminescent materials when the energy of two photons transfers to an electron in molecule and the electron occupies excited state, then the excited electron returns to a ground state with spontaneous photon emission. The SHG is stimulated emission, while two-photon photoluminescence is spontaneous emission. Accordingly, we can distinguish SHG from TPL in many ways.

One way is to pay attention to time-dependence of SHG and TPL. Time duration of SHG is mostly the same as that of probe laser pulse. That is, Q–switched Nd–YAG laser with 4 ns pulse duration causes SHG radiation with the time duration of 4 ns. This is immediately displayed on an oscilloscope that monitors output of photomultiplier tube (PMT) used as a detector of SHG. TPL is spontaneous emission. Its emission gradually decays with lifetime of excited state. The corresponding TPL signal appears on an oscilloscope with sustained signal longer than pulse duration of laser beam, if the lifetime is longer than laser pulse duration. By using a box-car integrator to pick up SHG signal only in laser pulse duration, TPL signal is reduced.

Another way is to focus on difference in wavelength of SHG and TPL. SHG is stimulated light emission process, and occurs in the wavelength region corresponding to optical absorption spectrum of materials. TPL spectrum overlaps with photoluminescence spectrum for spontaneous emission. Spontaneous emission band is shifted to longer wavelength region due to Stokes shift, in comparison with original absorption spectrum. One can avoid TPL coming into a detector, by setting the SHG wavelength to shorter wavelength region away from the photoluminescence band (Ref. [47] in Chap. 6). In TPL, the photoluminescence photon energy is distributed in broad wavelength range. By contrast to TPL, SHG is monochromatic and its wavelength is just the double of probe laser. Using an optical filter in front of a detector with very narrow transmission band is effective to catch SHG only.

Optical damaging of sample impacts on SHG signals, and a noise is superposed. This damage generates random spike-like signals displayed on an oscilloscope. The spikes initiate at the timing of laser pulse, and continue randomly afterwards. The damage develops and does not recover. As a result, successive two measurements in identical condition give different experimental results. Figure F.16 is an example obtained by using an SHG microscope system. The result shows the measurement of IZO/pentacene/Al device, where the pentacene thickness is 160 nm. External voltage of -5 V is applied to IZO electrode with reference to grounded Al electrode. SHG wavelength is

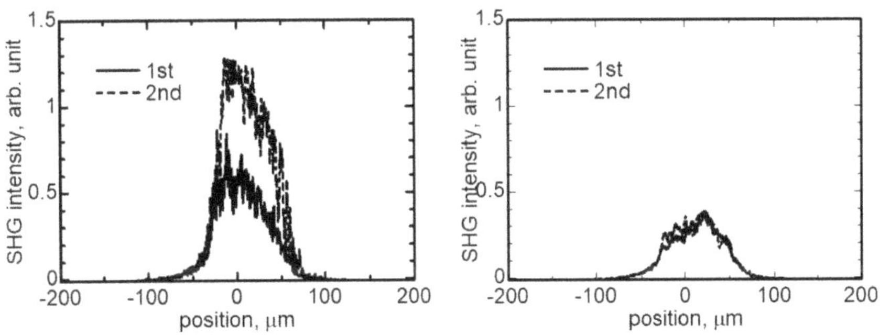

(a) (b)

Figure F.16. Consideration of laser damage in SHG microscope. Two successive measurements in identical condition are repeated. (a) The 2$^{\rm nd}$ scan shows enhancement of SHG intensity, which is caused by laser damage of Al electrode. (b) The 2$^{\rm nd}$ scan is same as the 1$^{\rm st}$ scan. After reducing laser intensity, laser damage is avoided.

560 nm (laser wavelength 1120 nm). Position of microscope objective is scanned through a focus point twice. The microscope objective lens focuses laser beam on a sample. The high laser intensity at focusing point induces the damage to Al electrode. After reducing the laser intensity, this damage is avoided.

Electrical breakdown of air is induced when a focusing light intensity is strong. The breakdown generates spike-like signals on an oscilloscope. We can directly check the occurrence of breakdown by looking that the focusing point emits sparkling light with accompanying sounds, in the way same as in the case of static electrical shock. After reducing the laser intensity, electrical breakdown is avoided.

Electroluminescence (EL) overlaps with SHG signals for SHG measurement of organic light-emitting diode (OLED) (Ref. [2] in Chap. 6). This EL is evident since the signal is observed without laser irradiation. For electroluminescence, holes and electrons are injected from electrodes, and recombine in OLED. By using AC square voltage whose duration is less than transit time of holes (or electrons) crossing the layer of diode, EL radiation is not enhanced, and we can probe SHG in prior to recombination process. Selecting proper SHG wavelength in the same way as in the case of TPL is also an effective way to probe electric field during operation of OLED.

These processes add a signal to SHG, but as described above, we can eliminate additional signals one-by-one for stable SHG measurement.

Appendix G

Internal Electric Field $\boldsymbol{E_i}$ Formed by Dissimilar Electrodes

Internal electric field E_i is formed in a device where single layer or multilayer is sandwiched between two dissimilar electrodes, *e.g.*, see Refs. [30, 31, 32, 107, 108, 109] of Chap. 7. For example, a C60 layer is sandwiched between Al electrode and IZO electrode where Al and IZO work functions are $\phi_1 = 3.4\,\text{eV}$ and $\phi_2 = 4.9\,\text{eV}$, respectively. Assuming that the C60 layer functions as a dielectric layer, energy band diagram of this single layer device is illustrated in Fig. G.1. The electrode 1 and electrode 2 correspond to Al and IZO electrode, respectively. After connecting electrode 1 and electrode 2, internal electric field E_i is formed in the dielectric layer in short-circuit condition $V = 0$. We define positive internal field $E_i > 0$ is pointing from low work function electrode, *e.g.*, Al, to high work function electrode, *e.g.*, IZO. The internal electric field E_i is given by

$$E_i = \frac{\phi_2 - \phi_1}{eL}$$

where e is elementary charge, L is distance between two dissimilar electrodes. Figure G.1(b) shows energy diagram in prior to contacting materials in the device. Figure G.1(c) illustrates energy diagram of the device after contacting electrode and the dielectric layer, and connecting electrode 1 and electrode 2 in short-circuit condition. By connecting electrode 1 and electrode 2, Fermi energy of the two electrodes aligns. Consequently, non-zero internal electric field E_i is formed in the layer.

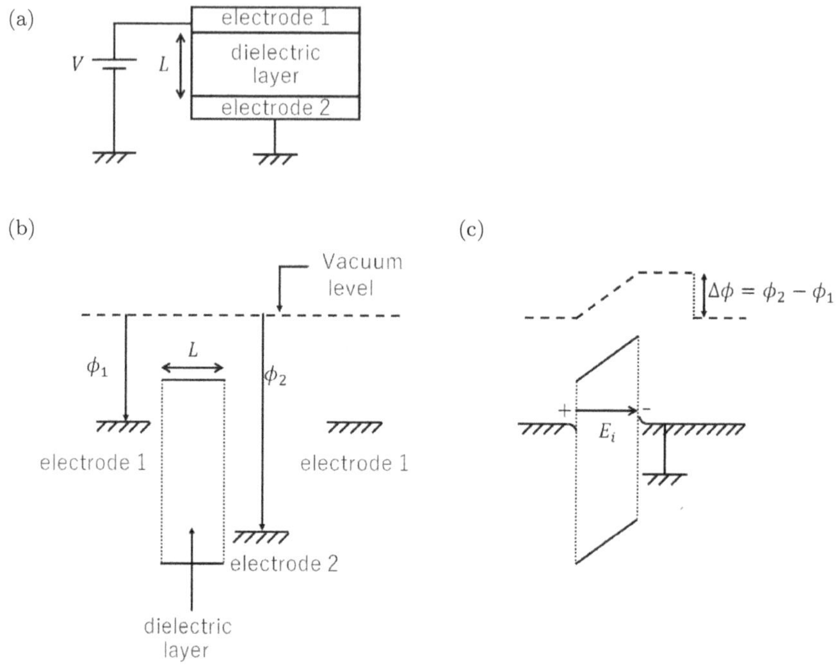

Figure G.1. Internal electric field formed in a device with dissimilar two electrodes where the two electrodes' work functions are different $\phi_1 \neq \phi_2$. (a) Device structure with a dielectric layer, separating the two electrodes with distance L. Energy diagram of the device (b) in prior to contacting materials in the device, and (c) after contacting and connecting two dissimilar electrodes by using conducting wire ($V = 0$), resulting in formation of non-zero internal electric field E_i across the dielectric layer.

With external voltage V applied to electrode 1 in reference to electrode 2 grounded, total electric field E in the layer is

$$E = \bar{E} + E_i$$

where $\bar{E}$ is external electric field induced by the voltage source V (see Eq. (7.16)). Consequently, in the presence of internal field E_i, EFISHG intensity $I_{2\omega}$ is

$$I_{2\omega} \propto |E|^2 = |\bar{E} + E_i|^2.$$

Total electric field is non-zero ($E = E_i \neq 0$) under zero external voltage $V = 0$ where external electric field is zero ($\bar{E} = 0$). In EFISHG measurement, non-zero SHG intensity $I_{2\omega}$ is detected under short-circuit condition, if there is non-zero internal field $E_i \neq 0$. This SHG is different from SHG due to dipolar alignment, *i.e.*, $\chi^{(2)} \neq 0$ (see Chap. 6). The minimum of SHG intensity at $\bar{E} = \bar{E}^*$ in $I_{2\omega}$–$\bar{E}$ plot gives internal electric field as $E_i = -\bar{E}^* = -V/L$. SHG measurement is available to probe this internal electric field (see Ref. [107] of Chap. 7).

Index

www.ingramcontent.com/pod-product-compliance
Lightning Source LLC
LaVergne TN
LVHW011656200926
848484LV00004B/956